LONDON MATHEMATICAL SOCIETY LECTURE N

Managing Editor: Professor M. Reid, Mathematics Institute,
University of Warwick, Coventry CV4 7AL, United Kingdom

The titles below are available from booksellers, or from Cambridge
http://www.cambridge.org/mathematics

315 Structured ring spectra, A. BAKER & B. RICHTER (eds)
316 Linear logic in computer science, T. EHRHARD, P. RUET, J.-Y. GIRARD & P. SCOTT (eds)
317 Advances in elliptic curve cryptography, I.F. BLAKE, G. SEROUSSI & N.P. SMART (eds)
318 Perturbation of the boundary in boundary-value problems of partial differential equations, D. HENRY
319 Double affine Hecke algebras, I. CHEREDNIK
320 L-functions and Galois representations, D. BURNS, K. BUZZARD & J. NEKOVÁŘ (eds)
321 Surveys in modern mathematics, V. PRASOLOV & Y. ILYASHENKO (eds)
322 Recent perspectives in random matrix theory and number theory, F. MEZZADRI & N.C. SNAITH (eds)
323 Poisson geometry, deformation quantisation and group representations, S. GUTT *et al* (eds)
324 Singularities and computer algebra, C. LOSSEN & G. PFISTER (eds)
325 Lectures on the Ricci flow, P. TOPPING
326 Modular representations of finite groups of Lie type, J.E. HUMPHREYS
327 Surveys in combinatorics 2005, B.S. WEBB (ed)
328 Fundamentals of hyperbolic manifolds, R. CANARY, D. EPSTEIN & A. MARDEN (eds)
329 Spaces of Kleinian groups, Y. MINSKY, M. SAKUMA & C. SERIES (eds)
330 Noncommutative localization in algebra and topology, A. RANICKI (ed)
331 Foundations of computational mathematics, Santander 2005, L.M PARDO, A. PINKUS, E. SÜLI & M.J. TODD (eds)
332 Handbook of tilting theory, L. ANGELERI HÜGEL, D. HAPPEL & H. KRAUSE (eds)
333 Synthetic differential geometry (2nd Edition), A. KOCK
334 The Navier–Stokes equations, N. RILEY & P. DRAZIN
335 Lectures on the combinatorics of free probability, A. NICA & R. SPEICHER
336 Integral closure of ideals, rings, and modules, I. SWANSON & C. HUNEKE
337 Methods in Banach space theory, J.M.F. CASTILLO & W.B. JOHNSON (eds)
338 Surveys in geometry and number theory, N. YOUNG (ed)
339 Groups St Andrews 2005 I, C.M. CAMPBELL, M.R. QUICK, E.F. ROBERTSON & G.C. SMITH (eds)
340 Groups St Andrews 2005 II, C.M. CAMPBELL, M.R. QUICK, E.F. ROBERTSON & G.C. SMITH (eds)
341 Ranks of elliptic curves and random matrix theory, J.B. CONREY, D.W. FARMER, F. MEZZADRI & N.C. SNAITH (eds)
342 Elliptic cohomology, H.R. MILLER & D.C. RAVENEL (eds)
343 Algebraic cycles and motives I, J. NAGEL & C. PETERS (eds)
344 Algebraic cycles and motives II, J. NAGEL & C. PETERS (eds)
345 Algebraic and analytic geometry, A. NEEMAN
346 Surveys in combinatorics 2007, A. HILTON & J. TALBOT (eds)
347 Surveys in contemporary mathematics, N. YOUNG & Y. CHOI (eds)
348 Transcendental dynamics and complex analysis, P.J. RIPPON & G.M. STALLARD (eds)
349 Model theory with applications to algebra and analysis I, Z. CHATZIDAKIS, D. MACPHERSON, A. PILLAY & A. WILKIE (eds)
350 Model theory with applications to algebra and analysis II, Z. CHATZIDAKIS, D. MACPHERSON, A. PILLAY & A. WILKIE (eds)
351 Finite von Neumann algebras and masas, A.M. SINCLAIR & R.R. SMITH
352 Number theory and polynomials, J. MCKEE & C. SMYTH (eds)
353 Trends in stochastic analysis, J. BLATH, P. MÖRTERS & M. SCHEUTZOW (eds)
354 Groups and analysis, K. TENT (ed)
355 Non-equilibrium statistical mechanics and turbulence, J. CARDY, G. FALKOVICH & K. GAWEDZKI
356 Elliptic curves and big Galois representations, D. DELBOURGO
357 Algebraic theory of differential equations, M.A.H. MACCALLUM & A.V. MIKHAILOV (eds)
358 Geometric and cohomological methods in group theory, M.R. BRIDSON, P.H. KROPHOLLER & I.J. LEARY (eds)
359 Moduli spaces and vector bundles, L. BRAMBILA-PAZ, S.B. BRADLOW, O. GARCÍA-PRADA & S. RAMANAN (eds)
360 Zariski geometries, B. ZILBER
361 Words: Notes on verbal width in groups, D. SEGAL
362 Differential tensor algebras and their module categories, R. BAUTISTA, L. SALMERÓN & R. ZUAZUA
363 Foundations of computational mathematics, Hong Kong 2008, F. CUCKER, A. PINKUS & M.J. TODD (eds)
364 Partial differential equations and fluid mechanics, J.C. ROBINSON & J.L. RODRIGO (eds)
365 Surveys in combinatorics 2009, S. HUCZYNSKA, J.D. MITCHELL & C.M. RONEY-DOUGAL (eds)
366 Highly oscillatory problems, B. ENGQUIST, A. FOKAS, E. HAIRER & A. ISERLES (eds)
367 Random matrices: High dimensional phenomena, G. BLOWER
368 Geometry of Riemann surfaces, F.P. GARDINER, G. GONZÁLEZ-DIEZ & C. KOUROUNIOTIS (eds)
369 Epidemics and rumours in complex networks, M. DRAIEF & L. MASSOULIÉ
370 Theory of p-adic distributions, S. ALBEVERIO, A.YU. KHRENNIKOV & V.M. SHELKOVICH
371 Conformal fractals, F. PRZYTYCKI & M. URBAŃSKI
372 Moonshine: The first quarter century and beyond, J. LEPOWSKY, J. MCKAY & M.P. TUITE (eds)
373 Smoothness, regularity and complete intersection, J. MAJADAS & A. G. RODICIO
374 Geometric analysis of hyperbolic differential equations: An introduction, S. ALINHAC

375 Triangulated categories, T. HOLM, P. JØRGENSEN & R. ROUQUIER (eds)
376 Permutation patterns, S. LINTON, N. RUŠKUC & V. VATTER (eds)
377 An introduction to Galois cohomology and its applications, G. BERHUY
378 Probability and mathematical genetics, N. H. BINGHAM & C. M. GOLDIE (eds)
379 Finite and algorithmic model theory, J. ESPARZA, C. MICHAUX & C. STEINHORN (eds)
380 Real and complex singularities, M. MANOEL, M.C. ROMERO FUSTER & C.T.C WALL (eds)
381 Symmetries and integrability of difference equations, D. LEVI, P. OLVER, Z. THOMOVA & P. WINTERNITZ (eds)
382 Forcing with random variables and proof complexity, J. KRAJÍČEK
383 Motivic integration and its interactions with model theory and non-Archimedean geometry I, R. CLUCKERS, J. NICAISE & J. SEBAG (eds)
384 Motivic integration and its interactions with model theory and non-Archimedean geometry II, R. CLUCKERS, J. NICAISE & J. SEBAG (eds)
385 Entropy of hidden Markov processes and connections to dynamical systems, B. MARCUS, K. PETERSEN & T. WEISSMAN (eds)
386 Independence-friendly logic, A.L. MANN, G. SANDU & M. SEVENSTER
387 Groups St Andrews 2009 in Bath I, C.M. CAMPBELL *et al* (eds)
388 Groups St Andrews 2009 in Bath II, C.M. CAMPBELL *et al* (eds)
389 Random fields on the sphere, D. MARINUCCI & G. PECCATI
390 Localization in periodic potentials, D.E. PELINOVSKY
391 Fusion systems in algebra and topology, M. ASCHBACHER, R. KESSAR & B. OLIVER
392 Surveys in combinatorics 2011, R. CHAPMAN (ed)
393 Non-abelian fundamental groups and Iwasawa theory, J. COATES *et al* (eds)
394 Variational problems in differential geometry, R. BIELAWSKI, K. HOUSTON & M. SPEIGHT (eds)
395 How groups grow, A. MANN
396 Arithmetic differential operators over the p-adic integers, C.C. RALPH & S.R. SIMANCA
397 Hyperbolic geometry and applications in quantum chaos and cosmology, J. BOLTE & F. STEINER (eds)
398 Mathematical models in contact mechanics, M. SOFONEA & A. MATEI
399 Circuit double cover of graphs, C.-Q. ZHANG
400 Dense sphere packings: a blueprint for formal proofs, T. HALES
401 A double Hall algebra approach to affine quantum Schur–Weyl theory, B. DENG, J. DU & Q. FU
402 Mathematical aspects of fluid mechanics, J.C. ROBINSON, J.L. RODRIGO & W. SADOWSKI (eds)
403 Foundations of computational mathematics, Budapest 2011, F. CUCKER, T. KRICK, A. PINKUS & A. SZANTO (eds)
404 Operator methods for boundary value problems, S. HASSI, H.S.V. DE SNOO & F.H. SZAFRANIEC (eds)
405 Torsors, étale homotopy and applications to rational points, A.N. SKOROBOGATOV (ed)
406 Appalachian set theory, J. CUMMINGS & E. SCHIMMERLING (eds)
407 The maximal subgroups of the low-dimensional finite classical groups, J.N. BRAY, D.F. HOLT & C.M. RONEY-DOUGAL
408 Complexity science: the Warwick master's course, R. BALL, V. KOLOKOLTSOV & R.S. MACKAY (eds)
409 Surveys in combinatorics 2013, S.R. BLACKBURN, S. GERKE & M. WILDON (eds)
410 Representation theory and harmonic analysis of wreath products of finite groups, T. CECCHERINI-SILBERSTEIN, F. SCARABOTTI & F. TOLLI
411 Moduli spaces, L. BRAMBILA-PAZ, O. GARCÍA-PRADA, P. NEWSTEAD & R.P. THOMAS (eds)
412 Automorphisms and equivalence relations in topological dynamics, D.B. ELLIS & R. ELLIS
413 Optimal transportation, Y. OLLIVIER, H. PAJOT & C. VILLANI (eds)
414 Automorphic forms and Galois representations I, F. DIAMOND, P.L. KASSAEI & M. KIM (eds)
415 Automorphic forms and Galois representations II, F. DIAMOND, P.L. KASSAEI & M. KIM (eds)
416 Reversibility in dynamics and group theory, A.G. O'FARRELL & I. SHORT
417 Recent advances in algebraic geometry, C.D. HACON, M. MUSTAŢĂ & M. POPA (eds)
418 The Bloch–Kato conjecture for the Riemann zeta function, J. COATES, A. RAGHURAM, A. SAIKIA & R. SUJATHA (eds)
419 The Cauchy problem for non-Lipschitz semi-linear parabolic partial differential equations, J.C. MEYER & D.J. NEEDHAM
420 Arithmetic and geometry, L. DIEULEFAIT *et al* (eds)
421 O-minimality and Diophantine geometry, G.O. JONES & A.J. WILKIE (eds)
422 Groups St Andrews 2013, C.M. CAMPBELL *et al* (eds)
423 Inequalities for graph eigenvalues, Z. STANIĆ
424 Surveys in combinatorics 2015, A. CZUMAJ *et al* (eds)
425 Geometry, topology and dynamics in negative curvature, C.S. ARAVINDA, F.T. FARRELL & J.-F. LAFONT (eds)
426 Lectures on the theory of water waves, T. BRIDGES, M. GROVES & D. NICHOLLS (eds)
427 Recent advances in Hodge theory, M. KERR & G. PEARLSTEIN (eds)
428 Geometry in a Fréchet context, C. T. J. DODSON, G. GALANIS & E. VASSILIOU
429 Sheaves and functions modulo p, L. TAELMAN
430 Recent progress in the theory of the Euler and Navier-Stokes equations, J.C. ROBINSON, J.L. RODRIGO, W. SADOWSKI & A. VIDAL-LÓPEZ (eds)
431 Harmonic and subharmonic function theory on the real hyperbolic ball, M. STOLL
432 Topics in graph automorphisms and reconstruction (2nd Edition), J. LAURI & R. SCAPELLATO
433 Regular and irregular holonomic D-modules, M. KASHIWARA & P. SCHAPIRA
434 Analytic semigroups and semilinear initial boundary value problems (2nd Edition), K. TAIRA
435 Graded rings and graded Grothendieck groups, R. HAZRAT
436 Groups, graphs and random walks, T. CECCHERINI-SILBERSTEIN, M. SALVATORI & E. SAVA-HUSS (eds)
437 Dynamics and analytic number theory, D. BADZIAHIN, A. GORODNIK & N. PEYERIMHOFF (eds)
438 Random Walks and Heat Kernels on Graphs, MARTIN T. BARLOW

London Mathematical Society Lecture Note Series: 438

Random Walks and Heat Kernels on Graphs

MARTIN T. BARLOW
University of British Columbia, Canada

CAMBRIDGE
UNIVERSITY PRESS

University Printing House, Cambridge CB2 8BS, United Kingdom

One Liberty Plaza, 20th Floor, New York, NY 10006, USA

477 Williamstown Road, Port Melbourne, VIC 3207, Australia

4843/24, 2nd Floor, Ansari Road, Daryaganj, Delhi – 110002, India

79 Anson Road, #06–04/06, Singapore 079906

Cambridge University Press is part of the University of Cambridge.

It furthers the University's mission by disseminating knowledge in the pursuit of education, learning, and research at the highest international levels of excellence.

www.cambridge.org
Information on this title: www.cambridge.org/9781107674424
DOI: 10.1017/9781107415690

First published 2017

Printed in the United Kingdom by Clays, St Ives plc

A catalogue record for this publication is available from the British Library.

Library of Congress Cataloging-in-Publication Data
Names: Barlow, M. T.
Title: Random walks and heat kernels on graphs / Martin T. Barlow, University of British Columbia, Canada.
Description: Cambridge : Cambridge University Press, [2017] |
Series: London Mathematical Society lecture note series ; 438 |
Includes bibliographical references and index.
Identifiers: LCCN 2016051295 | ISBN 9781107674424
Subjects: LCSH: Random walks (Mathematics) | Graph theory. | Markov processes. | Heat equation.
Classification: LCC QA274.73 .B3735 2017 | DDC 511/.5–dc23
LC record available at https://lccn.loc.gov/2016051295

ISBN 978-1-107-67442-4 Paperback

To my mother, Yvonne Barlow, and in memory of my father, Andrew Barlow.

Contents

Preface

The topic of random walks on graphs is a vast one, and has close connections with many other areas of probability, as well as analysis, geometry, and algebra. In the probabilistic direction, a random walk on a graph is just a reversible or symmetric Markov chain, and many results on random walks on graphs also hold for more general Markov chains. However, pursuing this generalisation too far leads to a loss of concrete interest, and in this text the context will be restricted to random walks on graphs where each vertex has a finite number of neighbours.

Even with these restrictions, there are many topics which this book does not cover – in particular the very active field of relaxation times for finite graphs. This book is mainly concerned with infinite graphs, and in particular those which have polynomial volume growth. The main topic is the relation between geometric properties of the graph and asymptotic properties of the random walk. A particular emphasis is on properties which are stable under minor perturbations of the graph – for example the addition of a number of diagonal edges to the Euclidean lattice $\mathbb{Z}^2$. The precise definition of 'minor perturbation' is given by the concept of a rough isometry, or quasi-isometry. One example of a property which is stable under these perturbations is transience; this stability is proved in Chapter 2 using electrical networks. A considerably harder theorem, which is one of the main results of this book, is that the property of satisfying Gaussian heat kernel bounds is also stable under rough isometries.

The second main theme of this book is deriving bounds on the transition density of the random walk, or the heat kernel, from geometric information on the graph. Once one has these bounds, many properties of the random walk can then be obtained in a straightforward fashion.

Chapter 1 gives the basic definition of graphs and random walks, as well as that of rough isometry. Chapter 2 explores the close connections between random walks and electrical networks. The key contribution of network theory is to connect the hitting properties of the random walk with the effective resistance between sets. The effective resistance can be bounded using variational principles, and this introduces tools which have no purely probabilistic counterpart.

Chapter 3 introduces some geometric and analytic inequalities which will play a key role in the remainder of the book – two kinds of isoperimetric inequality, Nash inequalities, and Poincaré inequalities. It is shown how the two analytic inequalities (i.e. Nash and Poincaré) can be derived from the geometric information given by the isoperimetric inequalities.

Chapter 4 studies the transition density of the discrete time random walk, or discrete time heat kernel. Two initial upper bounds on this are proved – an 'on-diagonal' upper bound from a Nash inequality, and a long range bound, the Carne–Varopoulos bound, which holds in very great generality. Both these proofs are quite short. To obtain full Gaussian or sub-Gaussian bounds much more work is needed. The second half of Chapter 4 makes the initial steps, and introduces conditions (bounds on exit times from balls, and a 'near-diagonal lower bound') which if satisfied will lead to full Gaussian bounds.

Chapter 5 introduces the continuous time random walk, which is in many respects an easier and more regular object to study than the discrete time random walk. In this chapter it is defined from the discrete time walk and an independent Poisson process – another construction, from the transition densities, is given in the Appendix. It is proved that Gaussian or sub-Gaussian heat kernel bounds hold for the discrete time random walk if and only if they hold for the continuous time walk.

Chapter 6 proves sub-Gaussian or Gaussian bounds for two classes of graphs. The first are 'strongly recurrent' graphs; given the work done in Chapter 4 the results for these come quite easily from volume and resistance estimates. The second class are graphs which satisfy a Poincaré inequality – these include $\mathbb{Z}^d$ and graphs which are roughly isometric to $\mathbb{Z}^d$. The proof of Gaussian bounds for these graphs is based on the tools developed by Nash to study divergence form PDE; these methods also work well in the graph context.

Chapter 7 gives a brief introduction to potential theory. Using the method of balayage, the heat kernel bounds of Chapter 6 lead to a Harnack inequality, and this in turn is used to prove the strong Liouville property (i.e. all positive harmonic functions are constant) for graphs which are roughly isometric to $\mathbb{Z}^d$.

This book is based on a course given at the first Pacific Institute for the Mathematical Sciences (PIMS) summer school in Probability, which was held at UBC in 2004. It is written for first year graduate students, though it is probably too long for a one semester course at most institutions. A much shorter version of this course was also given in Kyoto in 2005, and I wish to thank both PIMS and the Research Institute for the Mathematical Sciences (RIMS) for their invitations to give these courses.

I circulated a version of these notes several years ago, and I would like to thank those who have read them and given feedback, and in particular, Alain Sznitman and Gerard Ben-Arous, for encouraging me to complete them. I also wish to thank Alain for making available to me some lecture notes of his own on this material.

I have learnt much from a number of previous works in this area – the classical text of Spitzer [Spi], as well as the more recent works [AF, LL, LP, Wo].

I am grateful to those who have sent me corrections, and in particular to Yoshihiro Abe, Alma Sarai Hernandez Torres, Guillermo Martinez Dibene, Jun Misumi, and Zichun Ye.

1

Introduction

1.1 Graphs and Weighted Graphs

We start with some basic definitions. To be quite formal initially, a graph is a pair $\Gamma = (\mathbb{V}, E)$. Here $\mathbb{V}$ is a finite or countably infinite set, and E is a subset of $\mathscr{P}_2(\mathbb{V}) = \{\text{subsets } A \text{ of } \mathbb{V} \text{ with two elements}\}$. Given a set A we write $|A|$ for the number of elements of A. The elements of $\mathbb{V}$ are called *vertices*, and the elements of E *edges*. In practice, we think of the elements of $\mathbb{V}$ as points (often embedded in some space such as $\mathbb{R}^d$), and the edges as lines between the vertices.

Now for some general definitions.

(1) We write $x \sim y$ to mean $\{x, y\} \in E$, and say that y is a *neighbour* of x. Since all edges have two distinct elements, we have not allowed edges between a point x and itself. Also, we do not allow multiple edges between points, though we could; anything which can be done with multiple edges can be done better with weights, which will be introduced shortly.

(2) A path γ in Γ is a sequence $x_0, x_1, \dots, x_n$ with $x_{i-1} \sim x_i$ for $1 \le i \le n$. The length of a path γ is the number of edges in γ. A path $\gamma = (x_0, \dots, x_n)$ is a *cycle* or *loop* if $x_0 = x_n$. A path is *self-avoiding* if the vertices in the path are distinct.

(3) We define $d(x, y)$ to be the length n of the shortest path $x = x_0, x_1, \dots, x_n = y$. If there is no such path then we set $d(x, y) = \infty$. (This is the *graph* or *chemical metric* on Γ.) We also write for $x \in \mathbb{V}$, $A \subset \mathbb{V}$,

$$d(x, A) = \min\{d(x, y) : y \in A\}.$$

(4) Γ is *connected* if $d(x, y) < \infty$ for all x, y.

(5) Γ is *locally finite* if $N(x) = \{y : y \sim x\}$ is finite for each $x \in \mathbb{V}$, that is, every vertex has a finite number of neighbours. Γ has *bounded geometry* if $\sup_x |N(x)| < \infty$.

(6) Define balls in Γ by

$$B(x,r) = \{y : d(x,y) \le r\}, \quad x \in \mathbb{V}, \quad r \in [0,\infty).$$

Note that while the metric $d(x,\cdot)$ is integer valued, we allow r here to be real. Of course $B(x,r) = B(x,\lfloor r \rfloor)$.

(7) We define the *exterior boundary of* A by

$$\partial A = \{y \in A^c : \text{ there exists } x \in A \text{ with } x \sim y\}.$$

Set also

$$\begin{aligned} \partial_i A &= \partial(A^c) = \{y \in A : \text{ there exists } x \in A^c \text{ with } x \sim y\}, \\ \overline{A} &= A \cup \partial A, \\ A^o &= A - \partial_i A. \end{aligned}$$

We use the notation $A_n \uparrow\uparrow \mathbb{V}$ to mean that A_n is an increasing sequence of finite sets with $\cup_n A_n = \mathbb{V}$.

Notation We write c_i, C_i for (strictly) positive constants, which will remain fixed in each argument. In some proofs we will write c, c' etc. for positive constants, which may change from line to line. Given functions $f, g : A \to \mathbb{R}$ we write $f \asymp g$ to mean that there exists $c \ge 1$ such that

$$c^{-1} f(x) \le g(x) \le c f(x) \quad \text{for all } x \in A.$$

We write $\mathbf{1}$ for the function f which is identically 1, and $\mathbf{1}_A$ for the indicator function (or characteristic function) of the set A. We write $\mathbf{1}_x$ for $\mathbf{1}_{\{x\}}$.

From now on we will always assume:

$$\Gamma \text{ is locally finite,} \tag{H1}$$

$$\Gamma \text{ is connected;} \tag{H2}$$

and in addition from time to time we will assume:

$$\Gamma \text{ has bounded geometry.} \tag{H3}$$

We will treat *weighted graphs*.

Definition 1.1 We assume there exist weights (also called *conductances*) μ_{xy}, $x, y \in \mathbb{V}$ satisfying:

(i) $\mu_{xy} = \mu_{yx}$,
(ii) $\mu_{xy} \geq 0$ for all $x, y \in \mathbb{V}$,
(iii) if $x \neq y$ then $\mu_{xy} > 0$ if and only if $x \sim y$.

We call (Γ, μ) a *weighted graph*. Note that condition (iii) allows us to have $\mu_{xx} > 0$. Since $E = \big\{\{x, y\} : x \neq y, \mu_{xy} > 0\big\}$, we can recover the set of edges from μ.

The *natural weights* on Γ are given by

$$\mu_{xy} = \begin{cases} 1 & \text{if } x \sim y, \\ 0 & \text{otherwise.} \end{cases}$$

Whenever we discuss a graph without explicit mention of weights, we will assume we are using the natural weights.

Let $\mu_x = \mu(x) = \sum_y \mu_{xy}$, and extend μ to a measure on $\mathbb{V}$ by setting

$$\mu(A) = \sum_{x \in A} \mu_x. \tag{1.1}$$

Since Γ is locally finite we have:

(i) $|B(x, r)| < \infty$ for any x and r,
(ii) $\mu(A) < \infty$ for any finite set A.

We will sometimes need additional conditions on the weights.

Definition 1.2 We say (Γ, μ) has *bounded weights* if there exists $C_1 < \infty$ such that $C_1^{-1} \leq \mu_e \leq C_1$ for all $e \in E$, and $\mu_{xx} \leq C_1$ for all $x \in \mathbb{V}$. We say (Γ, μ) has *controlled weights* if there exists $C_2 < \infty$ such that

$$\frac{\mu_{xy}}{\mu_x} \geq \frac{1}{C_2} \quad \text{whenever } x \sim y.$$

We introduce the conditions:

$$(\Gamma, \mu) \text{ has bounded weights,} \tag{H4}$$

$$(\Gamma, \mu) \text{ has controlled weights.} \tag{H5}$$

Controlled weights is called 'the p_0 condition' in [GT1, GT2]; note that it holds if (H3) and (H4) hold. While a little less obvious than (H4), it turns out that controlled weights is the 'right' condition in many circumstances. As an example of its use, we show how it gives bounds on both the cardinality and the measure of a ball.

Lemma 1.3 *Suppose (Γ, μ) satisfies (H5) with constant C_2. Then (H3) holds and, for $n \geq 0$, $|B(x,n)| \leq 2C_2^n$, $\mu(B(x,n)) \leq 2C_2^n \mu_x$.*

Proof Since $\mu_{xy} \geq \mu_x/C_2$ if $x \sim y$, we have $|N(x)| \leq C_2$ for all x and thus (H3) holds. Further, we have $C_2 \geq 2$ unless $|\mathbb{V}| \leq 2$. Then writing $S(x,n) = \{y : d(x,y) = n\}$, we have $|S(x,n)| \leq C_2|S(x,n-1)|$, and so by induction $|B(x,n)| \leq (C_2^{n+1} - 1)/(C_2 - 1) \leq 2C_2^n$. Also

$$\mu(N(x)) = \sum_{y \sim x} \mu_y \leq C_2 \sum_{y \sim x} \mu_{xy} = C_2 \mu_x.$$

Hence again by induction $\mu(B(x,n)) \leq 2C_2^n \mu_x$. □

Examples 1.4 (1) The Euclidean lattice $\mathbb{Z}^d$. Here $\mathbb{V} = \mathbb{Z}^d$, and $x \sim y$ if and only if $|x - y| = 1$. Strictly speaking we should denote this graph by $(\mathbb{Z}^d, E_d)$, where E_d is the set of edges defined above.

(2) The d-ary tree. ('Binary' if $d = 2$.) This is the unique infinite graph with $|N(x)| \equiv d + 1$, and with no cycles.

(3) We will also consider the 'rooted binary tree' $\mathbb{B}$. Let $\mathbb{B}_0 = \{o\}$, and for $n \geq 1$ let $\mathbb{B}_n = \{0,1\}^n$. Then the vertex set is given by $\mathbb{B} = \cup_{n=0}^{\infty} \mathbb{B}_n$. For $x = (x_1, \dots, x_n) \in \mathbb{B}_n$ with $n \geq 2$ set $a(x) = (x_1, \dots, x_{n-1})$ – we call $a(x)$ the *ancestor of* x. (We set $a(z) = o$ for $z \in \mathbb{B}_1$.) Then the edge set of the rooted binary tree is given by

$$E(\mathbb{B}) = \big\{\{x, a(x)\} : x \in \mathbb{B} - \mathbb{B}_0\big\}.$$

(4) The *canopy tree* is a subgraph of $\mathbb{B}$ defined as follows. Let

$$\mathbb{U} = \{o\} \cup \{x = (x_1, \dots, x_n) \in \mathbb{B} : x_1 = \dots = x_n = 0\}.$$

Define $f : \mathbb{B} \to \mathbb{B}$ by taking $f(x)$ to be the point in $\mathbb{U}$ closest to x, and set $\mathbb{B}_{\mathrm{Can}} = \{x \in \mathbb{B} : d(x, f(x)) \leq d(o, f(x))\}$. Then the canopy tree is the subgraph of $(\mathbb{B}, E(\mathbb{B}))$ generated by $\mathbb{B}_{\mathrm{Can}}$. The canopy tree has exponential volume growth, but only one self-avoiding path from o to infinity.

(5) Let $\mathcal{G}$ be a finitely generated group, and let $\Lambda = \{g_1, \dots, g_n\}$ be a set of generators, not necessarily minimal. Write $\Lambda^* = \{g_1, \dots, g_n, g_1^{-1}, \dots, g_n^{-1}\}$. Let $\mathbb{V} = \mathcal{G}$ and let $\{g, h\} \in E$ if and only if $g^{-1}h \in \Lambda^*$. Then $\Gamma = (\mathbb{V}, E)$ is the *Cayley graph* of the group $\mathcal{G}$ with generators Λ.

$\mathbb{Z}^d$ is the Cayley graph of the group $\mathbb{Z} \oplus \dots \oplus \mathbb{Z}$, with generators g_k, $1 \leq k \leq d$; here g_k has 1 in the kth place and is zero elsewhere. Since different sets of generators will give rise to different Cayley graphs, in general a group $\mathcal{G}$ will have many different Cayley graphs. The d-ary tree is the Cayley graph of the group generated by $\Lambda = \{x_1, x_2, \dots, x_{d+1}\}$ with the relations $x_1^2 = \cdots = x_{d+1}^2 = 1$. Since the ternary tree is also the Cayley graph of the free group with two generators, two distinct (non-isomorphic) groups can have the same Cayley graph.

(6) Let $\alpha > 0$. Consider the graph $\mathbb{Z}_+ = \{0, 1, \dots\}$, but with weights $\mu^{(\alpha)}_{n,n+1} = \alpha^n$. This satisfies (H1)–(H3) and (H5), but (H4) fails unless $\alpha = 1$. If $\alpha < 1$ then $(\mathbb{Z}_+, \mu^{(\alpha)})$ is an infinite graph, but $\mu^{(\alpha)}(\mathbb{Z}_+) < \infty$.

(7) Given a weighted graph $(\mathbb{V}, E, \mu)$, the *lazy* weighted graph $(\mathbb{V}, E, \mu^{(L)})$ is defined by setting $\mu^{(L)}_{xy} = \mu_{xy}$ if $y \neq x$, and

$$\mu^{(L)}_{xx} = \sum_{y \neq x} \mu_{xy}.$$

Definition 1.5 We introduce the following operations on graphs or weighted graphs. Let (Γ, μ) be a weighted graph.

(1) *Subgraphs.* Let $H \subset \mathbb{V}$. Then the *subgraph induced by H* is the (weighted) graph with vertex set H, edge set

$$E_H = \big\{\{x, y\} \in E : x, y \in H\big\},$$

and weights (if appropriate) given by

$$\mu^H_{xy} = \mu_{xy}, \quad x, y \in H.$$

We define $\mu^H_x = \sum_{y \in H} \mu^H_{xy}$; note that $\mu^H_x \leq \mu_x$. Clearly (H, E_H) need not be connected, even if Γ is.

(2) *Collapse of a subset to a point.* Let $A \subset \mathbb{V}$. The graph obtained by collapsing A to a point a (where $a \notin \mathbb{V}$) is the graph $\Gamma' = (\mathbb{V}', E')$ given by

$$\begin{aligned} \mathbb{V}' &= (\mathbb{V} - A) \cup \{a\}, \\ E' &= \big\{\{x, y\} \in E : x, y \in \mathbb{V} - A\big\} \cup \big\{\{x, a\} : x \in \partial A\big\}. \end{aligned}$$

We define weights on $(\mathbb{V}', E')$ by setting $\mu'_{xy} = \mu_{xy}$ if $x, y \in \mathbb{V} - A$, and

$$\mu'_{xa} = \sum_{b \in A} \mu_{xb}, \quad x \in \mathbb{V} - A.$$

Note that $\mu'_{xa} \leq \mu_x < \infty$. In general this graph need not be locally finite, but will be if $|\partial A| < \infty$. It is easy to check that if Γ is connected then so is Γ'.

(3) *Join of two graphs.* Let $\Gamma_i = (\mathbb{V}_i, E_i), i = 1, 2$ be two graphs. (We regard $\mathbb{V}_1, \mathbb{V}_2$ as being distinct sets.) If $x_i \in \mathbb{V}_i$, the *join of* Γ_1 *and* Γ_2 *at* x_1, x_2 is the graph $(\mathbb{V}, E)$ given by

$$\mathbb{V} = \mathbb{V}_1 \cup \mathbb{V}_2, \qquad E = E_1 \cup E_2 \cup \big\{\{x_1, x_2\}\big\}.$$

We define weights on $(\mathbb{V}, E)$ by keeping the weights on Γ_1 and Γ_2, and giving the new edge $\{x_1, x_2\}$ weight 1.

(4) *Graph products.* Let Γ_i be graphs. There are two natural products, called the *Cartesian product* (denoted $\Gamma_1 \,\Box\, \Gamma_2$) and the *direct* or *tensor product* $\Gamma_1 \times \Gamma_2$. In both cases we take the vertex set of the product to be $\mathbb{V}_1 \times \mathbb{V}_2 = \{(x_1, x_2) : x_1 \in \mathbb{V}_1, x_2 \in \mathbb{V}_2\}$. For the Cartesian product we define

$$\begin{aligned} E_{\Box} &= \big\{\{(x, y_1), (x, y_2)\} : x \in \mathbb{V}_1, \{y_1, y_2\} \in E_2\big\} \\ &\quad \cup \big\{\{(x_1, y), (x_2, y)\} : \{x_1, x_2\} \in E_1, y \in \mathbb{V}_2\big\}, \end{aligned}$$

and for the tensor product

$$E_{\times} = \big\{\{(x_1, y_1), (x_2, y_2)\} : \{x_1, x_2\} \in E_1, \{y_1, y_2\} \in E_2\big\}.$$

Note that $\mathbb{Z}^2 = \mathbb{Z} \,\Box\, \mathbb{Z}$. If $\mu^{(i)}$ are weights on Γ_i then we define weights on the product by

$$\begin{aligned} \mu^{\Box}_{(x,y_1),(x,y_2)} &= \mu^{(2)}_{y_1,y_2}, \quad \mu^{\Box}_{(x_1,y),(x_2,y)} = \mu^{(1)}_{x_1,x_2}, \\ \mu^{\times}_{(x_1,y_1),(x_2,y_2)} &= \mu^{(1)}_{x_1,x_2}\mu^{(2)}_{y_1,y_2}. \end{aligned}$$

1.2 Random Walks on a Weighted Graph

Set

$$\mathscr{P}(x, y) = \frac{\mu_{xy}}{\mu_x}. \tag{1.2}$$

Let X be the discrete time Markov chain $X = (X_n, n \geq 0, \mathbb{P}^x, x \in \mathbb{V})$ with transition matrix $(\mathscr{P}(x, y))$, defined on a space $(\Omega, \mathscr{F})$ – see Appendix A.2. Here $\mathbb{P}^x$ is the law of the chain with $X_0 = x$, and the transition probabilities are given by

$$\mathbb{P}^z(X_{n+1} = y | X_n = x) = \mathscr{P}(x, y).$$

We call X the *simple random walk* (SRW) on (Γ, μ) and $(\mathscr{P}(x, y))$ the *transition matrix* of X. At each time step X moves from a vertex x to a neighbour y with probability proportional to μ_{xy}. Since we can have $\mu_{xx} > 0$ we do in general allow X to remain at a vertex x with positive probability. If ν

is a probability measure on $\mathbb{V}$ we write $\mathbb{P}^\nu$ for the law of X started with distribution ν.

Note that $(\mathscr{P}(x,y))$ is μ-*symmetric*, that is:

$$\mu_x \mathscr{P}(x,y) = \mu_y \mathscr{P}(y,x) = \mu_{xy}. \tag{1.3}$$

Set

$$\mathscr{P}_n(x,y) = \mathbb{P}^x(X_n = y).$$

The condition (1.3) implies that X is reversible.

Lemma 1.6 *Let $x_0, x_1, \dots, x_n \in \mathbb{V}$. Then*

$$\begin{aligned}\mu_{x_0}\mathbb{P}^{x_0}&(X_0 = x_0, X_1 = x_1, \dots, X_n = x_n)\\ &= \mu_{x_n}\mathbb{P}^{x_n}(X_0 = x_n, X_1 = x_{n-1}, \dots, X_n = x_0).\end{aligned}$$

Proof Using the μ-symmetry of $\mathscr{P}$ it is easy to check that both sides equal $\prod_{i=1}^n \mu_{x_{i-1},x_i}$. □

Definition 1.7 Let $A \subset \mathbb{V}$. Define the hitting times

$$\begin{aligned}T_A &= \min\{n \ge 0 : X_n \in A\},\\ T_A^+ &= \min\{n \ge 1 : X_n \in A\}.\end{aligned}$$

Here we adopt the convention that $\min \varnothing = +\infty$, so that $T_A = \infty$ if and only if X never hits A. Note also that if $X_0 \notin A$ then $T_A = T_A^+$. Write $T_x = T_{\{x\}}$ and

$$\tau_A = T_{A^c} = \min\{n \ge 0 : X_n \notin A\} \tag{1.4}$$

for the time of the first exit from A. If $A, B \subset \mathbb{V}$ we will write $\{T_A < T_B\}$ for the event that could more precisely be denoted $\{T_A < T_B, T_A < \infty\}$. If A, B are disjoint then the sample space Ω can be written as the disjoint union

$$\begin{aligned}\Omega = &\{T_A < T_B, T_A < \infty\} \cup \{T_B < T_A, T_B < \infty\}\\ &\cup \{T_A = T_B = \infty\} \cup \{T_A = T_B < \infty\}.\end{aligned}$$

The following theorem is proved in most first courses on Markov chains – see for example [Dur, KSK, Nor], or Appendix A.2.

Theorem 1.8 *Let Γ be connected, locally finite, infinite.*
(T) The following five conditions are equivalent:

(a) There exists $x \in \mathbb{V}$ *such that* $\mathbb{P}^x(T_x^+ < \infty) < 1$.
(b) For all $x \in \mathbb{V}$, $\mathbb{P}^x(T_x^+ < \infty) < 1$.
(c) For all $x \in \mathbb{V}$,

$$\sum_{n=0}^{\infty} \mathscr{P}_n(x, x) < \infty.$$

(d) For all $x, y \in \mathbb{V}$ *with* $x \neq y$ *either* $\mathbb{P}^x(T_y < \infty) < 1$ *or* $\mathbb{P}^y(T_x < \infty) < 1$.
(e) For all $x, y \in \mathbb{V}$,

$$\mathbb{P}^x(X \textit{ hits } y \textit{ only finitely often}) = \mathbb{P}^x\Big(\sum_{n=0}^{\infty} \mathbf{1}_{(X_n=y)} < \infty\Big) = 1.$$

(R) The following five conditions are equivalent:

(a) There exists $x \in \mathbb{V}$ *such that* $\mathbb{P}^x(T_x^+ < \infty) = 1$.
(b) For all $x \in \mathbb{V}$, $\mathbb{P}^x(T_x^+ < \infty) = 1$.
(c) For all $x \in \mathbb{V}$,

$$\sum_{n=0}^{\infty} \mathscr{P}_n(x, x) = \infty.$$

(d) For all $x, y \in \mathbb{V}$, $\mathbb{P}^x(T_y < \infty) = 1$.
(e) For all $x, y \in \mathbb{V}$,

$$\mathbb{P}^x(X \textit{ hits } y \textit{ infinitely often}) = \mathbb{P}^x\Big(\sum_{n=0}^{\infty} \mathbf{1}_{(X_n=y)} = \infty\Big) = 1.$$

Definition 1.9 If (T)(a)–(e) of Theorem 1.8 hold we say Γ (or X) is *transient*; otherwise (R)(a)–(e) of Theorem 1.8 hold and we say Γ or X is *recurrent*. The *type problem* for a graph Γ is to determine whether it is transient or recurrent. A special case is given by the Cayley graphs of groups, and is sometimes called *Kesten's problem*: which groups have recurrent Cayley graphs? In this case, recall that the Cayley graph depends on both the group $\mathscr{G}$ and the set of generators Λ.

Let us start with the Euclidean lattices. Polya [Pol] proved the following in 1921.

Theorem 1.10 $\mathbb{Z}^d$ *is recurrent if* $d \leq 2$ *and transient if* $d \geq 3$.

Polya used the fact that X_n is a sum of independent random variables, and that therefore the Fourier transform of $\mathscr{P}_n(0, x)$ is a product. Inverting the Fourier transform, he deduced that $\mathscr{P}_{2n}(0, 0) \sim c_d n^{-d/2}$, and hence proved his theorem.

Most modern textbooks give a combinatorial proof. For $d = 1$ we have

$$\mathscr{P}_{2n}(0,0) = 2^{-2n}\binom{2n}{n} \sim \frac{1}{(\pi n)^{1/2}};$$

this is easy using Stirling's formula. For $d = 2$ we have

$$\mathscr{P}_{2n}((0,0),(0,0)) = 4^{-2n}\sum_{k=0}^{n}\binom{2n}{k}^2 \sim \frac{1}{\pi n},$$

after a bit more work. (Or we can use the trick of writing $Z_n = (\frac{1}{2}(X_n + Y_n), \frac{1}{2}(X_n - Y_n))$, where X and Y are independent SRW on $\mathbb{Z}$.) In general we obtain $c_d n^{-d/2}$, so the series converges when $d \geq 3$. This expression is to be expected: by the local limit theorem (a variant of the central limit theorem), we have that the distribution of X_n is approximately the multivariate Gaussian $N(0, n^{1/2}d^{-1/2}I_d)$. (Of course the fact that, after rescaling, the SRW X on $\mathbb{Z}^d$ with $d \geq 3$ converges weakly to a transient process does not prove that X is transient, since transience/recurrence is not preserved by weak convergence.)

The advantage of the combinatorial argument is that it is elementary. On the other hand, the details are a bit messy for $d \geq 3$. More significantly, the technique is not robust. Consider the following three examples:

(1) The SRW on the hexagonal lattice in $\mathbb{R}^2$.
(2) The SRW on a graph derived from a Penrose tiling.
(3) The graph (Γ, μ) where $\Gamma = \mathbb{Z}^d$, and the weights μ satisfy $\mu_{xy} \in [c^{-1}, c]$.

Example (1) could be handled by the combinatorial method, but the details would be awkward, since one now has to count six kinds of steps. Also it is plainly a nuisance to have to give a new argument for each new lattice. Polya's method does work for this example, but both methods look hopeless for (2) or (3), since they rely on having an exact expression for $\mathscr{P}_n(x,x)$ or its transform. (For the convergence of SRW on a Penrose tiling to Brownian motion see [T3, BT].)

We will be interested in how the geometry of Γ is related to the long run behaviour of X. As far as possible we want techniques which are *stable* under various perturbations of the graph.

Definition 1.11 Let P be some property of a weighted graph (Γ, μ), or the SRW X on it. P is *stable under bounded perturbation of weights (weight stable)* if whenever (Γ, μ) satisfies P, and μ' are weights on Γ such that

$$c^{-1}\mu_{xy} \leq \mu'_{xy} \leq c\mu_{xy}, \quad x, y \in \mathbb{V},$$

then (Γ, μ') satisfies P. (We say the weights μ and μ' are *equivalent*.)

Definition 1.12 Let (X_i, d_i), $i = 1, 2$ be metric spaces. A map $\varphi : X_1 \to X_2$ is a *rough isometry* if there exist constants C_1, C_2 such that

$$C_1^{-1}(d_1(x, y) - C_2) \le d_2(\varphi(x), \varphi(y)) \le C_1(d_1(x, y) + C_2), \tag{1.5}$$

$$\bigcup_{x \in X_1} B_{d_2}(\varphi(x), C_2) = X_2. \tag{1.6}$$

If there exists a rough isometry between two spaces they are said to be *roughly isometric*. (One can check that this is an equivalence relation.)

This concept was introduced for groups by Gromov [Grom1, Grom2] under the name *quasi-isometry*, and in the context of manifolds by Kanai [Kan1, Kan2]. A rough isometry between two spaces implies that the two spaces have the same large scale structure. However, to get any useful consequences of two spaces being roughly isometric one also needs some kind of local regularity. This is usually done by considering rough isometries within a family of spaces satisfying some fixed local regularity condition. (For example, Kanai assumed the manifolds had bounded geometry: that is, that the Ricci curvature was bounded below by a constant.) Sometimes one has to be careful not to forget such 'hidden' side conditions.

Example $\mathbb{Z}^d$, $\mathbb{R}^d$, and $[0, 1] \times \mathbb{R}^d$ are all roughly isometric.

The following will be proved in Chapter 2 – see Proposition 2.59.

Proposition 1.13 *Let $\mathcal{G}$ be a finitely generated infinite group, and Λ, Λ' be two sets of generators. Let Γ, Γ' be the associated Cayley graphs. Then Γ and Γ' are roughly isometric.*

For rough isometries of weighted graphs, the natural additional regularity condition is to require that both graphs have controlled weights. Using Lemma 1.3 and the condition (1.7) below this allows one to relate the measures of balls in the two graphs.

Definition 1.14 Let (Γ_i, μ_i), $i = 1, 2$ be weighted graphs satisfying (H5). A map $\varphi : \mathbb{V}_1 \to \mathbb{V}_2$ is a *rough isometry* (between (Γ_1, μ_1) and (Γ_2, μ_2)) if:

(1) φ is a rough isometry between the metric spaces $(\mathbb{V}_1, d_{\Gamma_1})$ and $(\mathbb{V}_2, d_{\Gamma_2})$ (with constants C_1 and C_2);
(2) there exists $C_3 < \infty$ such that for all $x \in \mathbb{V}_1$

$$C_3^{-1}\mu_1(x) \le \mu_2(\varphi(x)) \le C_3\mu_1(x). \tag{1.7}$$

Two weighted graphs are *roughly isometric* if there is a rough isometry between them; one can check that this is also an equivalence relation. We define *stability under rough isometries* of a property P of Γ or X in the obvious way.

Example Consider the graphs $(\mathbb{Z}_+, \mu^{(\alpha)})$ defined in Examples 1.4, and let $\varphi : \mathbb{Z}_+ \to \mathbb{Z}_+$ be the identity map. Then φ is (of course) a rough isometry between the graphs $\mathbb{Z}_+$ and $\mathbb{Z}_+$, but if $\alpha \neq \beta$ it is not a rough isometry between the weighted graphs $(\mathbb{Z}_+, \mu^{(\alpha)})$ and $(\mathbb{Z}_+, \mu^{(\beta)})$. We will see in Chapter 2 that $(\mathbb{Z}_+, \mu^{(\alpha)})$ is recurrent if and only if $\alpha \leq 1$, and that the type of a graph is stable under rough isometries (of weighted graphs).

Question Is there any 'interesting' (global) property of the random walk on a weighted graph (Γ, μ) which is weight stable but not stable under rough isometries? (Properties such as Γ being bipartite or a tree are not global.)

1.3 Transition Densities and the Laplacian

Rather than working with the transition probabilities $\mathscr{P}(x, y)$, it is often more convenient to consider the *transition density* of X with respect to the measure μ; we will also call this the (discrete time) *heat kernel* of the graph Γ. We set

$$p_n(x, y) = \frac{\mathscr{P}_n(x, y)}{\mu_y} = \frac{\mathbb{P}^x(X_n = y)}{\mu_y}.$$

We write $p(x, y) = p_1(x, y)$, and note that

$$p_0(x, y) = \frac{\mathbf{1}_x(y)}{\mu_x}, \qquad p(x, y) = p_1(x, y) = \frac{\mu_{xy}}{\mu_x \mu_y}.$$

It is easy to verify:

Lemma 1.15 *The transition densities of X satisfy*

$$p_{n+m}(x, y) = \sum_z p_n(x, z) p_m(z, y) \mu_z, \tag{1.8}$$

$$p_n(x, y) = p_n(y, x),$$

$$\sum_y p_n(x, y) \mu_y = \sum_x \mu_x p_n(x, y) = 1.$$

Define the function spaces

$$C(\mathbb{V}) = \mathbb{R}^{\mathbb{V}} = \{f : \mathbb{V} \to \mathbb{R}\},$$
$$C_0(\mathbb{V}) = \{f : \mathbb{V} \to \mathbb{R} \text{ such that } f \text{ has finite support}\}.$$

Let $C_+(\mathbb{V}) = \{f : \mathbb{V} \to \mathbb{R}_+\}$, and write $C_{0,+} = C_0 \cap C_+$ etc. For $f \in C(\mathbb{V})$ we write

$$\text{supp}(f) = \{x : f(x) \neq 0\}.$$

For $p \in [1, \infty)$ and $f \in C(\mathbb{V})$ let

$$||f||_p^p = \sum_{x \in \mathbb{V}} |f(x)|^p \mu_x,$$

and

$$L^p(\mathbb{V}, \mu) = \{f \in C(\mathbb{V}) : ||f||_p < \infty\}.$$

We set $||f||_\infty = \sup_x |f(x)|$ and $L^\infty(\mathbb{V}, \mu) = \{f : ||f||_\infty < \infty\}$. If $p \in [1, \infty)$ and $A_n \uparrow\uparrow \mathbb{V}$ then $f\mathbf{1}_{A_n} \to f$ in L^p for any $f \in L^p$, so that C_0 is dense in L^p.

We write $\langle f, g \rangle$ for the inner product on $L^2(\mathbb{V}, \mu)$:

$$\langle f, g \rangle = \sum_{x \in \mathbb{V}} f(x) g(x) \mu_x.$$

We define the operators

$$P_n f(x) = \sum_{y \in \mathbb{V}} \mathscr{P}_n(x, y) f(y) = \sum_{y \in \mathbb{V}} p_n(x, y) f(y) \mu_y = \mathbb{E}^x f(X_n), \tag{1.9}$$

and write $P = P_1$. Since we are assuming (H1), we have that $P_n f$ is defined for any $f \in C(\mathbb{V})$. Using Lemma 1.15 we obtain

$$P_n = P^n.$$

The (probabilistic) Laplacian Δ is defined on $C(\mathbb{V})$ by

$$\begin{aligned} \Delta f(x) &= \frac{1}{\mu_x} \sum_{y \in \mathbb{V}} \mu_{xy}(f(y) - f(x)) \\ &= \sum_y p(x, y)(f(y) - f(x))\mu_y = \mathbb{E}^x f(X_1) - f(x). \end{aligned}$$

In operator terms we have

$$\Delta = P - I.$$

Notation We write $\|A\|_{p \to p'}$ to denote the norm of A as an operator from L^p to $L^{p'}$, that is,

$$\|A\|_{p \to p'} = \sup\{\|Af\|_{p'} : \|f\|_p \le 1\}.$$

Let q' be the conjugate index of p'. Since, by the duality of $L^{p'}$ and $L^{q'}$,

$$||f||_{p'} = \sup\{||fg||_1 : ||g||_{q'} \le 1\},$$

we also have

$$\|A\|_{p\to p'} = \sup\{\|gAf\|_1 : \|f\|_p \le 1,\ \|g\|_{q'} \le 1\}. \tag{1.10}$$

Proposition 1.16 *(a)* $P1 = 1$.
(b) For $f \in C(\mathbb{V})$, $|Pf| \le P|f|$.
(c) For each $p \in [1,\infty]$, $||P||_{p\to p} \le 1$, *and* $||\Delta||_{p\to p} \le 2$.

Proof (a) and (b) are easy. For (c), let first $p \in [1,\infty)$ and q be the conjugate index. Let $f \in L^p(\mathbb{V},\mu)$. Then, using Hölder's inequality,

$$\begin{aligned}||Pf||_p^p &= \sum_x |Pf(x)|^p \mu_x = \sum_x \Big|\sum_y p(x,y)f(y)\mu_y\Big|^p \mu_x \\ &\le \sum_x \Big(\big(\sum_y p(x,y)|f(y)|^p\mu_y\big)^{1/p}\big(\sum_y p(x,y)1^q\mu_y\big)^{1/q}\Big)^p \mu_x \\ &= \sum_x\sum_y p(x,y)|f(y)|^p\mu_x\mu_y = \sum_y |f(y)|^p\mu_y = ||f||_p^p.\end{aligned}$$

If $p = \infty$ the result is easy:

$$|Pf(x)| = |\sum_y p(x,y)f(y)\mu_y| \le ||f||_\infty \sum_y p(x,y)\mu_y = ||f||_\infty.$$

As $\Delta = P - I$ the second assertion is immediate. □

Remark Note that this proof did not actually use the symmetry of $p(x,y)$, but just the fact that $\sum_x \mu_x p(x,y) = 1$, that is, that μ is an invariant measure for P.

Remark 1.17 The *combinatorial Laplacian* is defined by Δ_{Com} by

$$\Delta_{\text{Com}} f(x) = \sum_{y\in\mathbb{V}} \mu_{xy}(f(y) - f(x)). \tag{1.11}$$

In matrix terms $\Delta_{\text{Com}} = A - D$, where $A_{xy} = \mu_{xy}$, and D is the diagonal matrix with $D_{xx} = \mu_x$. If Γ has its natural weights, then A is the adjacency matrix of Γ. It is easy to check that A and hence Δ_{Com} is self-adjoint with respect to counting measure on $\mathbb{V}$.

If $M = \sup_x \mu_x < \infty$ then one has

$$\begin{aligned}||Af||_p^p &= \sum_x \mu_x |\sum_y A_{xy} f(y)|^p \\ &\le \sum_x \mu_x\Big[\big(\sum_y A_{xy}|f(y)|^p\big)^{1/p}\big(\sum_y A_{xy}1^q\big)^{1/q}\Big]^p\end{aligned}$$

$$\le \sum_x \mu_x \sum_y A_{xy}|f(y)|^p M^{p/q} \le M^{1+p/q} \sum_y |f(y)|^p \mu_y$$
$$= M^p ||f||_p^p,$$

which implies that $||A||_{p\to p} \le M$. Since $||D||_{p\to p} \le M$ this gives

$$||\Delta_{\mathrm{Com}}||_{p\to p} \le 2M.$$

One can then define the 'delayed RW' on (Γ, μ) by considering the random walk associated with the operator

$$P = I - \frac{1}{M}\Delta_{\mathrm{Com}}.$$

This is the SRW on the weighted graph $(\mathbb{V}, E, \mu')$, where $\mu'_{xy} = M^{-1}\mu_{xy}$ if $x \ne y$ and $\mu'_{xx} = 1 - \sum_{y\ne x} \mu'_{xy}$.

If $M = \infty$ then the operator Δ_{Com} is not naturally associated with any discrete time random walk. However it can be associated in a natural way with a continuous time random walk – see Remark 5.7.

Proposition 1.18 *P and Δ are self-adjoint on L^2.*

Proof This is clear for P since

$$\langle Pf, g\rangle = \sum_x \Big(\sum_y p(x, y) f(y)\mu_y\Big) g(x)\mu_x$$
$$= \sum_x \sum_y p(x, y) f(y) g(x)\mu_y\mu_x$$
$$= \sum_y \sum_x p(y, x) g(x)\mu_x f(y)\mu_y = \langle f, Pg\rangle.$$

Hence $\Delta = P - I$ is also self-adjoint. □

Remark 1.19 See Example 1.35 in Section 1.6 for a graph Γ and functions u, v such that both Δu and Δv have finite support, but

$$\sum_x v(x)\Delta u(x)\mu_x \ne \sum_x u(x)\Delta v(x)\mu_x.$$

(Of course the functions u, v are not in L^2.)

On probability spaces one has $L^{p_2} \subset L^{p_1}$ if $p_2 > p_1$. On discrete spaces such as Γ the inclusions are in the opposite direction, provided that μ_x is bounded below.

Lemma 1.20 *Suppose that $\mu_x \geq a > 0$ for each $x \in \mathbb{V}$. Then if $p_2 > p_1$ and $f \in C(\mathbb{V})$,*

$$||f||_{p_2} \leq a^{1/p_2 - 1/p_1} ||f||_{p_1}.$$

Proof By replacing f by $f/||f||_{p_1}$ we can assume that $||f||_{p_1} = 1$. If $x \in \mathbb{V}$ then $|f(x)|^{p_1} \leq 1/\mu_x \leq 1/a$. Hence

$$\begin{aligned} ||f||_{p_2}^{p_2} &= \sum_x |f(x)|^{p_1} |f(x)|^{p_2 - p_1} \mu_x \leq \sum_x |f(x)|^{p_1} a^{-(p_2 - p_1)/p_1} \mu_x \\ &\leq a^{1 - p_2/p_1}. \end{aligned}$$

□

Write $p_n^x(\cdot) = p_n(x, \cdot)$. We can rewrite the equations in Lemma 1.15 as:

$$p_{n+m}(x, y) = \langle p_n^x, p_m^y \rangle, \tag{1.12}$$

$$P p_n^x(y) = p_{n+1}^x(y), \tag{1.13}$$

$$\Delta p_n^x(y) = (P - I) p_n^x(y) = p_{n+1}^x(y) - p_n^x(y). \tag{1.14}$$

Equation (1.14) is the discrete time heat equation. Note that (1.12) implies that $||p_n^x||_2^2 \leq \mu_x^{-1}$ for all $x \in \mathbb{V}, n \geq 0$.

1.4 Dirichlet or Energy Form

For $f, g \in C(\mathbb{V})$ define the quadratic form

$$\mathcal{E}(f, g) = \frac{1}{2} \sum_{x \in \mathbb{V}} \sum_{y \in \mathbb{V}} \mu_{xy} (f(x) - f(y))(g(x) - g(y)), \tag{1.15}$$

whenever this sum converges absolutely. This is the discrete analogue of $\int \nabla f \cdot \nabla g$. Let

$$H^2(\mathbb{V}, \mu) = H^2(\mathbb{V}) = H^2 = \{f \in C(\mathbb{V}) : \mathcal{E}(f, f) < \infty\}.$$

Choose a 'base point' $o \in \mathbb{V}$ and define

$$||f||_{H^2}^2 = \mathcal{E}(f, f) + f(o)^2.$$

The sum in (1.15) is well defined for $f, g \in H^2(\mathbb{V})$. If $A \subset \mathbb{V}$ we will sometimes use the notation

$$\mathcal{E}_A(f, g) = \frac{1}{2} \sum_{x \in A} \sum_{y \in A} (f(x) - f(y))(g(x) - g(y)) \mu_{xy}. \tag{1.16}$$

Proposition 1.21 *(a) If $x \sim y$ then*

$$|f(x) - f(y)| \leq \mu_{xy}^{-1/2} \mathcal{E}(f, f)^{1/2}.$$

(b) $\mathcal{E}(f,f) = 0$ *if and only if* f *is constant.*

(c) *If* $f \in L^2$,

$$\mathcal{E}(f,f) \leq 2||f||_2^2. \tag{1.17}$$

In particular $L^2 \subset H^2$.

(d) *If* (f_n) *is Cauchy in* H^2 *then* (f_n) *converges pointwise and in* H^2 *to a function* $f \in H^2$.

(e) *Convergence in* H^2 *implies pointwise convergence.*

(f) $H^2(\mathbb{V})$ *is a Hilbert space with inner product*

$$\langle f, g\rangle_H = \mathcal{E}(f,g) + f(o)g(o).$$

(g) *Let* (f_n) *be a sequence of functions with* $\sup_n ||f_n||_{H^2} < \infty$. *Then there exists a subsequence* f_{n_k} *and a function* $f \in H^2$ *such that* $f_{n_k} \to f$ *pointwise, and*

$$||f||_{H^2} \leq \liminf_{k\to\infty} ||f_{n_k}||_{H^2}. \tag{1.18}$$

(h) *The conclusion of (g) holds if* $A_n \uparrow\uparrow \mathbb{V}$ *with* $o \in A_1$, *and* $f_n : A_n \to \mathbb{R}$ *with* $\sup_n(|f_n(o)|^2 + \mathcal{E}_{A_n}(f,f)) < \infty$.

Proof (a) This is immediate from the definition of $\mathcal{E}$.
(b) As Γ is connected, this is immediate from (a).
(c) We have

$$\begin{aligned}\mathcal{E}(f,f) &= \tfrac{1}{2}\sum_x\sum_y (f(x)-f(y))^2\mu_{xy} \leq \sum_x\sum_y (f(x)^2 + f(y)^2)\mu_{xy}\\ &= 2||f||_2^2.\end{aligned}$$

(d) Let (f_n) be Cauchy in H^2. Then $f_n(o)$ is Cauchy in $\mathbb{R}$, and so converges. Let $A = \{x \in \mathbb{V} : f_n(x) \text{ converges}\}$. If $x \in A$ and $y \sim x$ then, applying (a) to $f_n - f_m$,

$$|f_n(y) - f_m(y)| \leq \mu_{xy}^{-1/2}\mathcal{E}(f_n - f_m, f_n - f_m)^{1/2} + |f_n(x) - f_m(x)|.$$

Letting $m, n \to \infty$ it follows that $y \in A$, and so, as Γ is connected, $A = \mathbb{V}$ and there exists a function f such that f_n converges to f pointwise. By Fatou's lemma we have

$$||f_n - f||_{H^2}^2 \leq \liminf_{m\to\infty} ||f_n - f_m||_{H^2}^2,$$

and hence $||f_n - f||_{H^2} \to 0$.
(e) This is immediate from (d).
(f) It is clear that H^2 is an inner product space; (d) then implies that H^2 is complete, so is a Hilbert space.

(g) and (h) We have that $(|f_n(o)|, n \geq 1)$ is bounded. Hence, using (a) it follows that $(|f_n(x)|, n \geq 1)$ is bounded for each $x \in \mathbb{V}$. A standard diagonalisation argument then implies that there exists a subsequence (n_k) such that (f_{n_k}) converges pointwise. Using Fatou's lemma then gives (1.18). □

Let $H_0^2(\mathbb{V})$ be the closure of $C_0(\mathbb{V})$ in $H^2(\mathbb{V})$. As $C_0(\mathbb{V}) \subset L^2(\mathbb{V}) \subset H^2(\mathbb{V})$, $H_0^2(\mathbb{V})$ is also the closure of $L^2(\mathbb{V})$ in $H^2(\mathbb{V})$. One always has $L^2 \subset H_0^2 \subset H^2$. If $\mu(\mathbb{V})$ is infinite then H^2 is strictly larger than L^2, since the constant function 1 is in H^2. All three spaces can be distinct – see Example 2.67, which shows that this holds for the join of $\mathbb{Z}_+$ and $\mathbb{Z}^3$. We will see later (Theorem 2.36) that $H^2 = H_0^2$ if and only if Γ is recurrent.

Exercise 1.22 (a) Show that $H_0^2(\mathbb{Z}_+) = H^2(\mathbb{Z}_+)$.
(b) Show that if $f \in H_0^2$ then $f_+, f_- \in H_0^2$.
(c) Show that if $f \in H_{0,+}^2$ and $f_n \to f$ in H^2 then $f_n \wedge f \to f$ in H^2.

Proposition 1.16 shows that Δ is defined on L^2. However, in fact it maps the larger space H^2 to L^2.

Proposition 1.23 *Let $f \in H^2$. Then $\Delta f \in L^2$, and*

$$||\Delta f||_2^2 \leq 2\mathscr{E}(f, f) \leq 2||f||_{H^2}^2.$$

Proof We have

$$\begin{aligned}
||\Delta f||_2^2 &= \sum_x |\Delta f(x)|^2 \mu_x \\
&= \sum_x \mu_x^{-1} \Big(\sum_y (f(y) - f(x)) \mu_{xy} \Big)^2 \\
&\leq \sum_x \mu_x^{-1} \Big(\sum_y (f(y) - f(x))^2 \mu_{xy} \Big) \Big(\sum_y \mu_{xy} \Big) \\
&= \sum_x \sum_y (f(y) - f(x))^2 \mu_{xy} = 2\mathscr{E}(f, f).
\end{aligned}$$
□

We will frequently use the following.

Theorem 1.24 (Discrete Gauss–Green Theorem) *Let $f, g \in C(\mathbb{V})$ satisfy*

$$\sum_{x \in \mathbb{V}} \sum_{y \in \mathbb{V}} |g(x)||f(x) - f(y)| \mu_{xy} < \infty. \tag{1.19}$$

Then

$$\mathcal{E}(f,g) = -\sum_x (\Delta f(x)) g(x)\mu_x, \tag{1.20}$$

and the sum in (1.20) converges absolutely. In particular (1.20) holds if $f \in H^2$ *and* $g \in L^2$*, or if at least one of* f, g *is in* $C_0(\mathbb{V})$.

Proof The condition (1.19) implies that both sides of (1.20) converge absolutely. We have

$$\mathcal{E}(f,g) = \tfrac{1}{2}\sum_x\sum_y (f(y) - f(x))g(y)\mu_{xy} - \tfrac{1}{2}\sum_x\sum_y (f(y) - f(x))g(x)\mu_{xy},$$

and both sums on the right side converge absolutely. Interchanging the indices x and y in the second sum, and using the fact that $\mu_{xy} = \mu_{yx}$, gives

$$\mathcal{E}(f,g) = -\sum_x\sum_y (f(y) - f(x))g(x)\mu_{xy} = -\sum_x g(x)\Delta f(x)\mu_x. \qquad \square$$

Remark (1) Let $\mathbb{V} = \mathbb{N}$, $f(n) = \sum_{k=1}^n (-1)^k k^{-1}$, and $g(n) = 1$ for all n. Then $f, g \in H^2$, but $|\Delta f(n)| \sim n^{-1}$. Thus while the sum in $\mathcal{E}(f,g)$ converges absolutely, the sum on the right side of (1.20) does not. Since $|f(n+1) - f(n)| \sim n^{-1}$, (1.19) fails. In fact $f, g \in H^2_0$ – this is easy to verify directly, or one can use Theorem 2.36.

(2) Example 1.35 (Section 1.6) shows that even if both sides of (1.20) converge absolutely equality need not hold.

(3) Applying (1.20) to p^x_n, p^y_m (which are both in C_0) gives

$$\begin{aligned}\mathcal{E}(p^x_n, p^y_m) &= \langle -\Delta p^x_n, p^y_m\rangle = \langle p^x_n - p^x_{n+1}, p^y_m\rangle \\ &= p_{n+m}(x,y) - p_{n+m+1}(x,y).\end{aligned}$$

(4) $\mathcal{E}(f,f)$ can be regarded as the energy of the function f.

(5) The bilinear form $\mathcal{E}$, with domain L^2, is a regular Dirichlet form on $L^2(\mathbb{V},\mu)$ – see [FOT].

Corollary 1.25 *We have*

$$\mathcal{E}(\mathbf{1}_x, \mathbf{1}_y) = \begin{cases} -\mu_{xy}, & y \neq x, \\ \mu_x - \mu_{xx}, & y = x. \end{cases}$$

Proof We have

$$\mathscr{E}(\mathbf{1}_x, \mathbf{1}_y) = \langle \mathbf{1}_x, -\Delta \mathbf{1}_y \rangle = -\mu_x \Delta \mathbf{1}_y(x) = -\sum_z \mu_{xz}(\mathbf{1}_y(x) - \mathbf{1}_y(z)),$$

and a simple calculation completes the proof. □

As another application of (1.20) we give the following:

Lemma 1.26 *Let (Γ, μ) be a weighted graph with $\mu(\mathbb{V}) < \infty$. Then (Γ, μ) is recurrent.*

Proof Let $z \in \mathbb{V}$, and $\varphi(x) = \mathbb{P}^x(T_z = \infty)$. Then $0 \le \varphi \le 1$, $\varphi(z) = 0$, and $\Delta\varphi(x) = 0$ if $x \neq z$ – see Theorem 2.5. As $\mu(\mathbb{V}) < \infty$, the function $\varphi \in L^2(\mathbb{V})$ and so

$$-\mathscr{E}(\varphi, \varphi) = \langle \Delta\varphi, \varphi \rangle = \varphi(z)\Delta\varphi(z)\mu_z = 0.$$

So by Proposition 1.21 φ is constant, and hence is identically zero. □

Lemma 1.27 *(a) Let $f \in H^2(\mathbb{V})$, $a \in \mathbb{R}$, and $g = (f - a)_+$, $h = f \wedge a$. Then $g, h \in H^2(\mathbb{V})$ and $\mathscr{E}(g, g) \le \mathscr{E}(f, f)$, $\mathscr{E}(h, h) \le \mathscr{E}(f, f)$.*
(b) Let $f_1, \dots, f_n \in H^2(\mathbb{V})$ and $g = \min\{f_1, \dots, f_n\}$. Then

$$\mathscr{E}(g, g) \le \sum_{i=1}^n \mathscr{E}(f_i, f_i).$$

(c) Let $f \in H^2(\mathbb{V})$. Then

$$\mathscr{E}(f_+, f_+) + \mathscr{E}(f_-, f_-) \le \mathscr{E}(f, f) \le 2\mathscr{E}(f_+, f_+) + 2\mathscr{E}(f_-, f_-). \quad (1.21)$$

Proof (a) Since $|(b_1 - a)^+ - (b_2 - a)^+| \le |b_1 - b_2|$ for $a, b_1, b_2 \in \mathbb{R}$, we have for $x, y \in \mathbb{V}$

$$(g(x) - g(y))^2 \le (f(x) - f(y))^2,$$

and summing over x, y gives the inequality $\mathscr{E}(g, g) \le \mathscr{E}(f, f)$. Since $f \wedge a = a - (a - f)_+$ we have

$$\begin{aligned}\mathscr{E}(f, f) &= \mathscr{E}(a - f, a - f) \ge \mathscr{E}((a - f)_+, (a - f)_+) \\ &= \mathscr{E}(a - (a - f)_+, a - (a - f)_+) = \mathscr{E}(h, h).\end{aligned}$$

(b) Let $x \sim y$, and suppose that $g(x) \le g(y)$, and $g(x) = f_j(x)$. Then

$$0 \le g(y) - g(x) = g(y) - f_j(x) \le f_j(y) - f_j(x).$$

Hence

$$|g(y) - g(x)|^2 \le \max_i |f_i(y) - f_i(x)|^2 \le \sum_i |f_i(y) - f_i(x)|^2,$$

and again summing over x, y completes the proof of (b).
(c) If $a, b \in \mathbb{R}$ then $(a - b)^2 \ge (a_+ - b_+)^2 + (a_- - b_-)^2$; summing this inequality as in (a) gives the left side of (1.21). The right side is immediate from Cauchy–Schwarz. □

Property (a) of Lemma 1.27 gives what is called in [FOT] the *Markov* property of the quadratic form $\mathscr{E}$. To explain the significance of this, let $\mathbb{V}$ be a countable set, and let $(a_{xy}, x, y \in \mathbb{V})$ satisfy $a_{xy} = a_{yx}$, $a_{xx} = 0$, $\sum_y |a_{xy}| < \infty$ for all x. For $f, g \in C_0(\mathbb{V})$ define the quadratic form $\mathscr{A}(f, g)$ associated with (a_{xy}) in the same way as in (1.15):

$$\mathscr{A}(f, g) = \tfrac{1}{2} \sum_{x \in \mathbb{V}} \sum_{y \in \mathbb{V}} a_{xy}(f(x) - f(y))(g(x) - g(y)). \tag{1.22}$$

(Note that this sum converges absolutely when $f, g \in C_0(\mathbb{V})$.) If $a_{xy} \ge 0$ for all x, y then of course $\mathscr{A}(f, f) \ge 0$ for all $f \in C_0(\mathbb{V})$, but the converse is false.

Lemma 1.28 *Let $\mathscr{A}$ be the quadratic form on $C_0(\mathbb{V})$ associated with (a_{xy}) by (1.22). Then $a_{xy} \ge 0$ for all x, y if and only if $\mathscr{A}$ has the property:*

$$\mathscr{A}(f^+, f^+) \le \mathscr{A}(f, f) \quad \textit{for all } f \in C_0(\mathbb{V}). \tag{1.23}$$

Proof The forward implication is immediate from Lemma 1.27(a). Suppose now that (1.23) holds. Then if $a'_{xy} = a_{xy}$, $y \ne x$, and $a'_{xx} = -\sum_y a_{xy}$,

$$\mathscr{A}(f, g) = -\sum_x \sum_y f(x) a'_{xy} g(y).$$

Let $x, y \in \mathbb{V}$, with $x \ne y$, $\lambda > 0$, and set $f_\lambda = \mathbf{1}_y - \lambda \mathbf{1}_x$. Then

$$\mathscr{A}(f_\lambda, f_\lambda) = -\lambda^2 a'_{xx} + 2\lambda a_{xy} - a'_{yy} = \lambda^2 |a'_{xx}| + 2\lambda a_{xy} + |a'_{yy}|.$$

Since $\lambda > 0$ we have $f_\lambda^+ = f_0$, and so by (1.23)

$$0 \le \mathscr{A}(f_\lambda, f_\lambda) - \mathscr{A}(f_\lambda^+, f_\lambda^+) = \lambda^2 |a'_{xx}| + 2\lambda a_{xy},$$

and thus $a_{xy} \ge 0$. □

We will not make much use of the next result, but include it since, combined with Theorems 2.36 and 2.37, it will show that $H_0^2 = H^2$ if and only if Γ is recurrent.

Proposition 1.29 *We have* $\mathbf{1} \in H_0^2$ *if and only if* $H^2 = H_0^2$.

Proof The backward implication is obvious. Suppose that $\mathbf{1} \in H_0^2$. Then there exist $u_n \in C_0(\mathbb{V})$ with $||1 - u_n||_{H^2} \le n^{-3}$. By Lemma 1.27(a) we can assume that $0 \le u_n \le 1$. Set $v_n = nu_n$; then $\mathscr{E}(v_n, v_n) \le n^{-1}$, but since $u_n \to 1$ pointwise, $v_n(x) \to \infty$ for all $x \in \mathbb{V}$. Now let $f \in H_+^2$, and set $f_n = f \wedge v_n$, so that $f_n \in C_0(\mathbb{V})$, and for each $x \in \mathbb{V}$ we have $f_n(x) = f(x)$ for all sufficiently large n.

Let $\varepsilon > 0$, and choose a finite set A such that if $B = \mathbb{V} - A$ then $\mathscr{E}_B(f, f) < \varepsilon$. Now choose $n > 1/\varepsilon$ such that $f_n(x) = f(x)$ for all $x \in \overline{A}$. Then since $f - f_n = (f - v_n)_+$

$$\begin{aligned}\mathscr{E}(f - f_n, f - f_n) &= \mathscr{E}_B(f - f_n, f - f_n)\\ &\le \mathscr{E}_B(f - v_n, f - v_n)\\ &\le 2\mathscr{E}_B(f, f) + 2\mathscr{E}_B(v_n, v_n) \le 4\varepsilon.\end{aligned}$$

We therefore deduce that $f_n \to f$ in H^2, so that $f \in H_0^2$. □

1.5 Killed Process

Let $A \subset \mathbb{V}$, and recall from (1.4) the definition of τ_A, the time of the first exit from A. Set

$$p_n^A(x, y) = \mu_y^{-1} \mathbb{P}^x (X_n = y, n < \tau_A). \tag{1.24}$$

This is the heat kernel for the process X killed on exiting from A. Define also the operators

$$\begin{aligned}I_A f(x) &= \mathbf{1}_A(x) f(x),\\ P_A f(x) &= \sum_{y \in A} p_1^A(x, y) f(y) \mu_y,\\ \Delta_A &= P_A - I_A,\\ P_n^A f(x) &= \mathbb{E}^x \big(f(X_n); n < \tau_A \big) = \sum_y p_n^A(x, y) f(y) \mu_y.\end{aligned}$$

Note that $P_0^A = I_A$.

Lemma 1.30 *(a)* $p_n^A(x, y) = 0$ *if either* $x \notin A$ *or* $y \notin A$.
(b) If $x, y \in A$ *then*

$$p_{n+1}^A(x, y) = \sum_{z \in \mathbb{V}} p_n^A(x, z) p_1^A(z, y) \mu_z = \sum_{z \in A} p_n^A(x, z) p(z, y) \mu_z, \tag{1.25}$$

$$p_{n+1}^A(x,y) - p_n^A(x,y) = \Delta_y p_n^A(x,y),$$
$$p_n^A(x,y) = p_n^A(y,x). \tag{1.26}$$

Thus $p_n^A(x,\cdot)$ *satisfies the discrete time heat equation on* A *with zero boundary conditions on* ∂A.

(c) We have $P_n^A = (P_A)^n$.

(d) We have

$$P_A f = I_A P I_A f, \quad \Delta_A f = I_A \Delta I_A f.$$

Proof (a) This is clear from the definition. Using this, we have

$$\begin{aligned}\mu_y p_{n+1}^A(x,y) &= \mathbb{P}^x(X_{n+1} = y, \tau_A > n+1)\\ &= \sum_{z\in\mathbb{V}} \mathbb{P}^x(X_{n+1} = y, \tau_A > n+1 | X_n = z, \tau_A > n) p_n^A(x,z)\mu_z\\ &= \sum_{z\in A} \mathbb{P}^x(X_{n+1} = y, \tau_A > n+1 | X_n = z, \tau_A > n) p_n^A(x,z)\mu_z,\end{aligned}$$

giving the first equality in (1.25). The second follows since $p_1^A(x,z) = p(x,z)$ if $x, z \in A$. The other equalities in (b), as well as (c), then follow easily from (1.25). (d) is again clear from the definition. □

1.6 Green's Functions

Let $A \subset \mathbb{V}$. We define the Green's functions (or potential kernel density) by

$$g_A(x,y) = \sum_{n=0}^{\infty} p_n^A(x,y).$$

It is clear from (1.26) that $g_A(x,y)$ is symmetric in x, y. We write g for $g_{\mathbb{V}}$, and as above we may write $g_A^x(\cdot) = g_A(x,\cdot)$. Define the number of hits by X on y (or local time of X at y) by $L_0^y = 0$ and

$$L_n^y = \sum_{r=0}^{n-1} \mathbf{1}_{\{X_r = y\}}, \quad n \ge 1.$$

Then we also have

$$\mathbb{E}^x L_{\tau_A}^y = \sum_{n=0}^{\infty} \mathbb{P}^x(X_n = y, \tau_A > n) = \sum_{n=0}^{\infty} p_n^A(x,y)\mu_y = g_A(x,y)\mu_y. \tag{1.27}$$

If Γ is recurrent then Theorem 1.8 implies that $g(x,y) = \infty$ for all $x, y \in \mathbb{V}$.

Theorem 1.31 *Let $A \subset \mathbb{V}$ and suppose that either Γ is transient or $A \neq \mathbb{V}$. Then*

$$g_A(x, y) = \mathbb{P}^x(T_y < \tau_A) g_A(y, y), \text{ for all } x, y \in \mathbb{V}, \tag{1.28}$$

$$g_A(y, y)\mu_y = \frac{1}{\mathbb{P}^y(\tau_A \le T_y^+)} < \infty \text{ for all } y \in A. \tag{1.29}$$

In particular $g_A(x, y) < \infty$ for all $x, y \in \mathbb{V}$.

Proof Applying the strong Markov property of X at T_y (see Appendix A.2),

$$g_A(x, y)\mu_y = \mathbb{E}^x(L^y_{\tau_A}) = \mathbb{E}^x\left(\mathbf{1}_{(T_y<\tau_A)}\mathbb{E}^y L^y_{\tau_A}\right) = \mathbb{P}^x(T_y < \tau_A) g_A(y, y)\mu_y,$$

proving (1.28).

We have $p = \mathbb{P}^y(T_y^+ < \tau_A) < 1$. This is clear from the definition if Γ is transient, while if Γ is recurrent then there exists $z \in \mathbb{V} - A$, and therefore $\mathbb{P}^y(T_y^+ > \tau_A) \ge \mathbb{P}^y(T_y^+ > T_z) > 0$. By the strong Markov property of X,

$$\mathbb{P}^y(L^y_{\tau_A} = k) = (1 - p)p^{k-1}, \quad k \ge 1.$$

Thus $L^y_{\tau_A}$ has a geometric distribution, and, summing the series in k,

$$\mu_y g_A(y, y) = \mathbb{E}^y(L^y_{\tau_A}) = \sum_{k=1}^{\infty} k p^{k-1}(1-p) = \frac{1}{1-p} = \frac{1}{\mathbb{P}^y(\tau_A \le T_y^+)} < \infty.$$

Note that our hypotheses allow $\mathbb{P}^y(\tau_A = T_y^+ = \infty) > 0$. □

The Green's function satisfies the following equations.

Lemma 1.32 *Let $x, y \in A$. Then*

$$P g_A^x(y) = g_A(x, y) - \mu_x^{-1}\mathbf{1}_x(y), \tag{1.30}$$

$$\Delta g_A^x(y) = -\mu_x^{-1}\mathbf{1}_x(y). \tag{1.31}$$

Proof We have

$$\begin{aligned} P g_A^x(y) &= \sum_{z\in\mathbb{V}} p(y, z)\mu_z g_A(x, z) \\ &= \sum_{z\in\mathbb{V}} \sum_{n=0}^{\infty} p(y, z) p_n^A(x, z)\mu_z \\ &= \sum_{n=0}^{\infty} p_{n+1}^A(x, y) = g_A(x, y) - p_0^A(x, y). \end{aligned}$$

Substituting for $p_0(x, y)$ gives (1.30). Since $\Delta = P - I$, (1.31) is immediate. □

Remark The term μ_x^{-1} in (1.31) may look odd, but arises because we are considering densities with respect to μ.

We also define the potential operator

$$G_A f(x) = \sum_{n=0}^{\infty} P_n^A f(x),$$

and write $G = G_{\mathbb{V}}$. We have

$$G_A f(x) = \sum_y g_A(x, y) f(y) \mu_y = \mathbb{E}^x \Big(\sum_{n=0}^{\tau_A - 1} f(X_n) \Big).$$

Unless Γ is recurrent and $A = \mathbb{V}$ we have $G_A : L^1(\mathbb{V}) \to C(\mathbb{V})$, since by Theorem 1.31

$$|G_A f(x)| \le \sum_y g_A(x, y) |f(y)| \mu_y \le g_A(x, x) ||f||_1. \tag{1.32}$$

Remark For $\mathbb{Z}^d$ (with natural weights) one has if $d \ge 3$

$$g(x, y) \sim \frac{c_d}{|x - y|^{d-2}}.$$

Hence $g(x, \cdot) \in L^2$ if $d \ge 5$, but not for $d = 3, 4$. Also, as we will see in Corollary 3.44, G does not map $L^2(\mathbb{Z}^d)$ to $L^2(\mathbb{Z}^d)$ for any d.

Lemma 1.33 *(a) Let $A \subset \mathbb{V}$ with $A \neq \mathbb{V}$ if Γ is recurrent. Then*

$$\Delta_A G_A f = -I_A f, \text{ for } f \in L^1(\mathbb{V}).$$

In particular $\Delta G_A f(x) = -f(x)$ for $x \in A$.

(b) If Γ is transient then

$$\Delta G f = -f, \text{ for } f \in L^1(\mathbb{V}).$$

Proof (a) We have

$$\Delta_A G_A = (P_A - I_A) \sum_{n=0}^{\infty} (P_A)^n = -I_A;$$

the sums here converge by (1.32). To prove the second assertion, since $G_A f$ has support A, by Lemma 1.30 $-I_A f = \Delta_A G_A f = I_A \Delta G_A f$, so $-f$ and $\Delta G_A f$ agree on A.

(b) This is immediate from (a). □

Theorem 1.34 *Let $A \subset \mathbb{V}$ with $A \neq \mathbb{V}$ if Γ is recurrent.*

(a) If $A_n \uparrow\uparrow A$ then $g_{A_n}(x, y) \uparrow g_A(x, y)$ pointwise.
(b) For any $x \in A$, $g_A(x, \cdot) \in H_0^2(\mathbb{V})$.
(c) If $f \in H_0^2(\mathbb{V})$ with $\operatorname{supp}(f) \subset A$ then

$$\mathscr{E}(f, g_A^x) = f(x). \tag{1.33}$$

In particular $\mathscr{E}(g_A^x, g_A^y) = g_A(x, y)$.

Proof (a) Since $\tau_{A_n} \uparrow \tau_A$ this is clear from (1.27) by monotone convergence.
(b) Suppose first that A is finite, and let $f : \mathbb{V} \to \mathbb{R}$ with $\operatorname{supp}(f) \subset A$. By (1.31) we have $\Delta g_A^x(y) = 0$ for $y \in A$, $y \neq x$, while $f(y) = 0$ if $y \in \mathbb{V} - A$. Therefore by Theorem 1.24

$$\mathscr{E}(f, g_A^x) = \langle f, -\Delta g_A^x \rangle = \langle f, \mu_x^{-1} \mathbf{1}_x \rangle = f(x). \tag{1.34}$$

Now let A be any set satisfying the hypotheses of the theorem. The conditions on Γ and A imply that g_A^x is finite. Let $A_n = B(x, n) \cap A$, so that $A_n \uparrow\uparrow A$. If $m \leq n$ then $\operatorname{supp}(g_{A_m}^x) \subset A_n$, so by (1.34) $\mathscr{E}(g_{A_m}^x, g_{A_n}^x) = g_{A_m}(x, x)$. Therefore

$$\mathscr{E}(g_{A_m}^x - g_{A_n}^x, g_{A_m}^x - g_{A_n}^x) = g_{A_n}(x, x) - g_{A_m}(x, x),$$

which implies that $(g_{A_n}^x)$ is Cauchy in H^2. Since $g_{A_n}^x$ converges pointwise to g_A^x, by Proposition 1.21(d) we also have convergence in H^2, and thus $g_A^x \in H_0^2$.
(c) Let $f \in H_0^2$ with $\operatorname{supp}(f) \subset A$. Choose $f_n \in C_0(\mathbb{V})$ with $f_n \to f$ in H^2. As f_n has finite support, the support of f_n is contained in A_m for m large enough. So by (1.34) $\mathscr{E}(f_n, g_{A_m}^x) = f_n(x)$, and taking the limit, first as $m \to \infty$, and then as $n \to \infty$, gives (1.33). □

Example 1.35 Suppose that $\mathbb{V}$ is transient; by Lemma 1.26 we have $\mu(\mathbb{V}) = \infty$. Fix $x \in \mathbb{V}$ and let $f = 1$. By the above we have $f \in H^2$, $g^x \in H_0^2$, and therefore $\mathscr{E}(f, g^x)$ is defined. (We will see in Theorem 2.37 that $f \notin H_0^2$.) Since $\Delta f = 0$ and $\Delta g^x = -\mu_x^{-1} \mathbf{1}_x$, both Δf and Δg^x have finite support, and therefore both sums $\langle f, \Delta g^x \rangle$ and $\langle \Delta f, g^x \rangle$ are defined and converge absolutely. However, by direct calculation one has

$$\mathscr{E}(f, g^x) = 0, \quad \sum_y (-\Delta f(y)) g^x(y) \mu_y = 0,$$
$$\sum_y f(y)(-\Delta g^x(y)) \mu_y = f(x) = 1.$$

Thus (1.33) may fail if $f \in H^2(\mathbb{V})$.

1.7 Harmonic Functions, Harnack Inequalities, and the Liouville Property

Definition 1.36 Let $A \subset \mathbb{V}$ and f be defined on $\overline{A} = A \cup \partial A$. We say f is

$$\begin{aligned} &\textit{superharmonic} \text{ in } A \text{ if } \Delta f \le 0 \quad (\text{i.e. } Pf \le f),\\ &\textit{subharmonic} \text{ in } A \text{ if } \Delta f \ge 0 \quad (\text{i.e. } Pf \ge f),\\ &\textit{harmonic} \text{ in } A \text{ if } \Delta f = 0 \quad (\text{i.e. } Pf = f). \end{aligned}$$

Clearly f is superharmonic if and only if $-f$ is subharmonic. By Lemma 1.32 $g_A(x, \cdot)$ is superharmonic in A, and harmonic in $A - \{x\}$. If $f \ge 0$ then by Lemma 1.33 $G_A f$ is superharmonic in A.

If h_1, h_2 are superharmonic then so is $h = h_1 \wedge h_2$. For if $h(x) = h_1(x)$ then

$$h(x) = h_1(x) \ge \sum_y p(x, y) h_1(y) \mu_y \ge \sum_y p(x, y) h(y) \mu_y.$$

If g is superharmonic then

$$\mathbb{E}^x(g(X_n)|X_{n-1}, \dots, X_0) = \sum_y p(X_{n-1}, y) g(y) \mu_y = Pg(X_{n-1}) \le g(X_{n-1}),$$

so $g(X_n)$ is a supermartingale. (See Appendix A.1.)

Theorem 1.37 (Maximum principle) *Let $A \subset \mathbb{V}$ be connected, and h be subharmonic in A.*

(a) If there exists $x \in A$ such that

$$h(x) = \max_{z \in \overline{A}} h(z)$$

then h is constant on $\overline{A}$.

(b) If A is finite then h attains its maximum on $\overline{A}$ on ∂A, that is,

$$\max_{\partial A} h = \max_{\overline{A}} h.$$

Proof (a) Let $B = \{y \in \overline{A} : h(y) = h(x)\}$. If $y \in B \cap A$ then $h(z) \le h(y)$ whenever $z \sim y$, so

$$0 \le \mu_y \Delta h(y) = \sum_{z \sim y} (h(z) - h(y)) \mu_{yz} \le 0.$$

Thus if $y \in A \cap B$ and $z \sim y$ then $z \in B$. Hence, as A is connected, $B = \overline{A}$.

(b) Let x be a point in $\overline{A}$ where h attains its maximum. If $x \in \partial A$ we are done; if not then $x \in A$ and by (a) h is constant in $\overline{A}$. □

Example The maximum principle fails in general if A is infinite. Let $\Gamma = \mathbb{Z}^3$, $A = \mathbb{Z}^3 - \{0\}$, and $h(x) = \mathbb{P}^x(T_0 = \infty)$. Then h is harmonic in A, but $\sup_A h = 1$ while $\sup_{\partial A} h = h(0) = 0$.

Remark 1.38 If h is superharmonic then applying the maximum principle to $-h$ gives a corresponding *minimum principle*:

If there exists $x \in A$ such that $h(x) = \min_{y \in \overline{A}} h(y)$ then h is constant on $\overline{A}$.

Definition 1.39 (Γ, μ) has the *Liouville property* if all bounded harmonic functions on Γ are constant. (Γ, μ) has the *strong Liouville property* if all positive harmonic functions on Γ are constant.

For the probabilistic interpretation of the Liouville property in terms of the tail behaviour of the random walk X see the end of this chapter. Example 2.67 shows that the join of $\mathbb{Z}_+$ and $\mathbb{Z}^3$ at their origins satisfies the Liouville property but not the strong Liouville property.

Theorem 1.40 *Let (Γ, μ) be recurrent.*
(a) Any positive superharmonic function on $\mathbb{V}$ is constant.
(b) (Γ, μ) has the strong Liouville property.

Proof (a) There is a quick probabilistic proof using the martingale convergence theorem (see Theorem A.4). If $h \geq 0$ is superharmonic then $M_n = h(X_n)$ is a positive supermartingale. So M_n converges $\mathbb{P}^x$-a.s. to a limit M_∞, and as for any $y \in \mathbb{V}$ the set $\{n : X_n = y\}$ is $\mathbb{P}^x$-a.s. unbounded, we have $\mathbb{P}^x(h(y) = M_\infty) = 1$ for all $y \in \mathbb{V}$. Thus h is constant.

One can also give an analytic proof based on the maximum principle. Let $u \geq 0$ be superharmonic. So $Pu \leq u$, and iterating $P^{n+1}u \leq P^n u \leq \cdots \leq Pu \leq u$. Set $v = u - Pu \geq 0$. Then $P^k v \geq 0$, and

$$\sum_{k=0}^{n} P^k v = u - P^{n+1}u \leq u.$$

Hence

$$\sum_y g(x, y)v(y)\mu_y = Gv(x) = \lim_{n\to\infty} \sum_{k=0}^{n} P^k v(x) \leq u(x) < \infty.$$

Since $g(x, y) = \infty$ for all x, y we must have $v(y) = 0$ for all y, so that $u = Pu$, and therefore u is harmonic.

We have proved that if $u \geq 0$ is superharmonic then u is harmonic. Now let $f \geq 0$ be superharmonic. Suppose f is non-constant, so there exist $x, y \in \mathbb{V}$, $\lambda > 0$ such that $f(x) < \lambda < f(y)$. Then the function $v = f \wedge \lambda$ is superharmonic, so harmonic. Further, v attains its maximum at y, so by the maximum principle (applied on any finite connected set containing x and y) v is constant, a contradiction.

(b) Since a harmonic function is superharmonic, this is immediate from (a). $\square$

Proposition 1.41 ('Foster's criterion' – see [Fos, Mer]) *Let $A \subset \mathbb{V}$ be finite. (Γ, μ) is recurrent if and only if there exists a non-negative function h which is superharmonic in $\mathbb{V} - A$ and which satisfies*

$$|\{x : h(x) < M\}| < \infty \text{ for all } M. \tag{1.35}$$

Proof By collapsing A to a single point, we can assume that $A = \{o\}$.

Suppose a function h with the properties above exists. Then for any initial point $x \in \mathbb{V}$ the process $M_n = h(X_{n \wedge T_o})$ is a non-negative supermartingale, so converges $\mathbb{P}^x$-a.s. to a r.v. M_∞. If X is transient then we can find $x \in \mathbb{V}$ so that $\mathbb{P}^x(T_o < \infty) < 1$, and using (1.35) it follows that $\mathbb{P}^x(M_\infty = \infty) > 0$, which contradicts the convergence theorem for supermartingales. (See Corollary A.5.)

Now assume that (Γ, μ) is recurrent. Let $B_n = B(o, n)$, and let $h_n(x) = \mathbb{P}^x(\tau_{B_n} < T_o)$. Then h_n is superharmonic in $\mathbb{V} - \{o\}$, equals 1 on B_n^c, and satisfies

$$\lim_{n \to \infty} h_n(x) = 0 \text{ for all } x.$$

Since B_k is finite for each k, there exists a sequence $n_k \to \infty$ such that

$$h_{n_k}(x) \leq 2^{-k} \text{ for all } x \in B_k.$$

Let $h = \sum_k h_{n_k}$. Then $h(x) < \infty$ for all x, and h is non-negative and superharmonic. Let $M \in \mathbb{N}$. Then $h_{n_j}(x) = 1$ if $x \in B_{n_M}^c$ and $j \leq M$, so $h(x) \geq M$ for all $x \in B_{n_M}^c$. Thus h satisfies (1.35). $\square$

We can extend the minimum principle to infinite sets if Γ is recurrent.

Theorem 1.42 *Let Γ be recurrent, and $A \subset \mathbb{V}$. If h is bounded and superharmonic in A then*

$$\min_{\partial A} h = \min_{\overline{A}} h.$$

Proof Let $b = \min_{\partial A} h$, and set $g = h \wedge b$ on $\overline{A}$, and $g = b$ on $\overline{A}^c$. Then g is superharmonic in A, while if $y \in A^c$ then $b = g(y) \geq Pg(y)$. So g is superharmonic in the whole of $\mathbb{V}$, and thus g is constant and equal to b. Hence $h \geq b$ on A. □

Proposition 1.43 (Local Harnack inequality) *Let (Γ, μ) satisfy (H5) with constant C_1. Then if $h \geq 0$ is harmonic in $\{x, y\}$ with $x \sim y$,*

$$C_1^{-1} h(y) \leq h(x) \leq C_1 h(y). \tag{1.36}$$

Proof We have

$$h(x) = \sum_{z \sim x} \frac{\mu_{xz}}{\mu_x} h(z) \geq C_1^{-1} h(y).$$

Interchanging x and y gives the second inequality. □

Iterating (1.36) we deduce that if $h \geq 0$ and is harmonic on $B(x, r)$, and $y \in B(x, r)$, then

$$h(y) \leq C_1^{d(x,y)} h(x).$$

Definition 1.44 (Γ, μ) satisfies a *(scale invariant) elliptic Harnack inequality* (EHI) if there exists a constant C_H such that, given $x_0 \in \mathbb{V}$, $R \geq 1$, $h \geq 0$ on $\overline{B}(x_0, 2R)$ and harmonic in $B(x_0, 2R)$ then

$$h(x) \leq C_H h(y) \quad \text{for all } x, y \in B(x_0, R),$$

or equivalently

$$\max_{B(x_0,R)} h \leq C_H \min_{B(x_0,R)} h. \tag{1.37}$$

Remark (1) Note that $h(x) \geq 0$ for $x \in B(x_0, 2R)$ does not imply that $h \geq 0$ on $\overline{B}(x_0, 2R)$. We sometimes have to be careful in this way in the discrete setup.

(2) We have considered the balls $B(x, R) \subset \overline{B}(x, 2R)$ for simplicity. An easy exercise shows that if $K > 1$ and the EHI holds in the form stated above, then (1.37) holds (with a different constant $C_H(K)$) for any non-negative function h defined on $\overline{B}(x_0, KR)$ and harmonic in $B(x_0, KR)$, provided that $R \geq 2/(K - 1)$.

If $f : A \to \mathbb{R}$ write $\mathrm{Osc}(f, A) = \sup_A f - \inf_A f$.

Proposition 1.45 *Suppose* (Γ, μ) *satisfies EHI with constant* C_H. *Then if* $\rho = 1/(2C_H)$ *and h is harmonic in* $B(x_0, 2R)$,

$$\operatorname{Osc}(h, B(x_0, R)) \leq (1 - \rho) \operatorname{Osc}(h, \overline{B}(x_0, 2R)). \tag{1.38}$$

Proof Write $B = B(x_0, R)$, $D = \overline{B}(x_0, 2R)$. Replacing h by $a + bh$ (where $a, b \in \mathbb{R}$) we can arrange that

$$\min_D h = 0, \qquad \max_D h = 1.$$

Suppose that $h(x_0) \geq \frac{1}{2}$. Then by the EHI

$$\tfrac{1}{2} \leq \max_B h \leq C_H \min_B h.$$

Thus $h(x) \geq 1/(2C_H)$ on B, so

$$\operatorname{Osc}(h, B) \leq 1 - 1/(2C_H) = (1 - \rho) \operatorname{Osc}(h, D).$$

If $h(x_0) \leq \frac{1}{2}$ then apply the argument above to $h' = 1 - h$. □

Remark The oscillation inequality (1.38) is slightly weaker than the EHI: [Ba2] shows that some random walks on 'lamplighter' type graphs satisfy (1.38) without satisfying the EHI. A fairly weak lower bound on hitting probabilities combined with (1.38) is enough to give the EHI – see for example [FS].

Theorem 1.46 *Suppose* (Γ, μ) *satisfies the EHI. Then it satisfies the strong Liouville property.*

Proof Let $h \geq 0$ be harmonic in $\mathbb{V}$. Fix $x_0 \in \mathbb{V}$ and let $A_n = B(x_0, 2^n)$, $\lambda_n = \operatorname{Osc}(h, A_n)$. We begin by using the EHI in $A_n \subset A_{n+1}$:

$$\max_{A_n} h \leq C_H \min_{A_n} h \leq C_H h(x_0).$$

As n is arbitrary it follows that h is bounded by $c_1 = C_H h(x_0)$.

Now by Proposition 1.45,

$$\lambda_n \leq (1 - \rho) \operatorname{Osc}(h, \overline{A}_{n+1}) \leq (1 - \rho) \operatorname{Osc}(h, A_{n+2}) = (1 - \rho)\lambda_{n+2}.$$

Iterating, and using the fact that $\lambda_m \leq \operatorname{Osc}(h, \mathbb{V}) \leq c_1$ for any m, we deduce that, for any $n \geq 1$ and $k \geq 1$,

$$\lambda_n \leq (1 - \rho)^k \lambda_{n+2k} \leq (1 - \rho)^k c_1.$$

So $\lambda_n = 0$ for all n, proving that h is constant. □

Remark T. Lyons [Ly2] proved that the Liouville property and the strong Liouville property are not weight stable. The EHI has been recently proved to be weight stable – see [BMu].

Example An example of a graph which does not satisfy the Liouville property is the join of two copies of $\mathbb{Z}^3$ at their origins. Write $\mathbb{Z}^3_{(i)}$, $i = 1, 2$ for the two copies, and 0_i for their origins. Let

$$F = \{X \text{ is ultimately in } \mathbb{Z}^3_{(1)}\} = \bigcup_{n=1}^{\infty} \bigcap_{m=n}^{\infty} \{X_m \in \mathbb{Z}^3_{(1)}\},$$

and let $h(x) = \mathbb{P}^x(F)$. Then h is harmonic, and is non-constant since

$$h(x) \geq \mathbb{P}^x(X \text{ never hits } 0_1) \text{ for } x \in \mathbb{Z}^3_{(1)},$$
$$h(x) \leq \mathbb{P}^x(X \text{ hits } 0_2) \text{ for } x \in \mathbb{Z}^3_{(2)}.$$

1.8 Strong Liouville Property for $\mathbb{R}^d$

In general, proving the EHI requires some hard work. In Chapter 7 we will see how it can be proved from estimates on the transition density of X. A corollary is that the Liouville property and the strong Liouville property hold for graphs which are roughly isometric to $\mathbb{Z}^d$. Here is a proof of the strong Liouville property for $\mathbb{R}^d$, $d \geq 3$, using the same ideas.

We say a function $h \in C^2(\mathbb{R}^d)$ is *harmonic* on $\mathbb{R}^d$ if $\Delta h(x) = 0$ for all x; here Δ is the usual Laplacian on $\mathbb{R}^d$. Let S_r be the sphere centre 0 and radius r, and let σ_r be surface measure on S_r. By Green's third identity, if $v(x) = c_1(|x|^{2-d} - r^{2-d})$ is the fundamental solution of the Laplace equation inside S_r, then

$$h(0) = \int_{S_r} h(y) \frac{\partial v}{\partial n}(y) d\sigma_r(y),$$

and it follows that

$$h(0) = \sigma(S_r)^{-1} \int_{S_r} h d\sigma_r.$$

Let $k_t(x, y) = (2\pi t)^{-d/2} \exp(-|x - y|^2/2t)$ be the usual Gaussian density, and

$$K_t f(x) = \int k_t(x, y) f(y) dy.$$

Then the spherical symmetry of $k_t(x, \cdot)$ implies that $h(x) = K_t h(x)$ for all $x \in \mathbb{R}^d$ and $t > 0$. We have

$$\frac{k_t(x, y)}{k_{2t}(0, y)} = 2^{d/2} \exp\left(\frac{|y|^2 - 2|x - y|^2}{4t}\right)$$
$$= 2^{d/2} \exp\left(\frac{2|x|^2 - |y - 2x|^2}{4t}\right) \le 2^{d/2} e^{|x|^2/2t} \le e 2^{d/2}$$

if $2t \ge |x|^2$.

Now let $h \ge 0$ be harmonic in $\mathbb{R}^d$, and suppose h is non-constant. Replacing h by $h - \inf h$ if necessary, we can assume that $\inf h = 0$. Let $x \in \mathbb{R}^d$, and choose $t \ge \frac{1}{2}|x|^2$; then

$$h(x) = \int k_t(x, y)h(y)dy \le e2^{d/2} \int k_{2t}(0, y)h(y) = e2^{d/2}h(0).$$

So h is bounded, and (using translation invariance) we also have that $h(x) \le e2^{d/2}h(y)$ for all $x, y \in \mathbb{R}^d$. Since h is non-constant there exists $x_0 \in \mathbb{R}^d$ with $h(x_0) > 0$, so we deduce that $h(y) \ge e^{-1}2^{-d/2}h(x_0)$ for all $y \in \mathbb{R}^d$, and hence that $\inf h > 0$, a contradiction.

1.9 Interpretation of the Liouville Property

We will need more of the theory of Markov processes than in the rest of this chapter: see Appendix A.2 for further details. As before we assume that the random walk X is defined on a probability space $(\Omega, \mathcal{F})$, with probability measures $(\mathbb{P}^x, x \in \mathbb{V})$ such that $\mathbb{P}^x(X_0 = x) = 1$. We also assume that we have the *shift operators* $\theta_n : \Omega \to \Omega$ satisfying

$$X_m(\theta_n \omega) = X_{n+m}(\omega), \qquad n, m \ge 0, \; \omega \in \Omega.$$

For a r.v. ξ we set $(\xi \circ \theta_n)(\omega) = \xi(\theta_n \omega)$. We have

$$\mathbf{1}_F \circ \theta_n = \mathbf{1}_{\theta_n^{-1}(F)}. \tag{1.39}$$

We will sometimes write $F \circ \theta_n$ for $\theta_n^{-1}(F)$. We define the σ-fields

$$\mathcal{F}_n = \sigma(X_k, k \le n), \quad \mathcal{G}_n = \sigma(X_k, k \ge n), \quad \mathcal{F} = \mathcal{G}_0 = \sigma(X_k, k \ge 0). \tag{1.40}$$

The *invariant* σ-field $\mathcal{I}$ is defined by:

$$F \in \mathcal{I} \text{ if and only if } F \in \mathcal{F} \text{ and } \theta_n^{-1}(F) = F \text{ for all } n.$$

The *tail* σ-field $\mathscr{T}$ is defined by:

$$\mathscr{T} = \bigcap_{n=0}^{\infty} \mathscr{G}_n. \tag{1.41}$$

Lemma 1.47 $\mathscr{I} \subset \mathscr{T}$.

Proof We have that $F \in \mathscr{G}_n$ if and only if $F = \theta_n^{-1}(F_n)$ for some $F_n \in \mathscr{F}_0$; see Lemma A.25. Let $F \in \mathscr{I}$. Then $F = \theta_n^{-1}(F)$, so $F \in \mathscr{G}_n$ for any n, and hence $F \in \mathscr{T}$. □

A σ-field $\mathscr{G}$ is $\mathbb{P}^x$-*trivial* if $\mathbb{P}^x(G) \in \{0, 1\}$ for every $G \in \mathscr{G}$.

Theorem 1.48 *The following are equivalent:*
(a) Γ satisfies the Liouville property.
(b) $\mathscr{I}$ is $\mathbb{P}^\nu$-trivial for all probability distributions ν on $\mathbb{V}$.
(c) There exists $x \in \mathbb{V}$ such that $\mathscr{I}$ is $\mathbb{P}^x$-trivial.

Proof Let $F \in \mathscr{I}$, and set $h(x) = \mathbb{P}^x(F)$. Then as $F = F \circ \theta_n$, by the Markov property of X (see Theorem A.10),

$$\mathbb{E}^\nu(\mathbf{1}_F|\mathscr{F}_n) = \mathbb{E}^\nu(\mathbf{1}_F \circ \theta_n|\mathscr{F}_n) = \mathbb{E}^{X_n}(\mathbf{1}_F) = h(X_n).$$

Thus, taking expectations with respect to $\mathbb{E}^x$ and setting $n = 1$,

$$h(x) = \mathbb{E}^x h(X_1) = \sum_y \mathscr{P}(x, y)h(y),$$

so that h is harmonic.

Now assume (a). Then h is constant, while by the martingale convergence theorem

$$c = h(X_n) = \mathbb{E}^\nu(\mathbf{1}_F|\mathscr{F}_n) \to \mathbf{1}_F, \quad \mathbb{P}^\nu\text{-a.s.}$$

So $\mathbb{P}^\nu(\mathbf{1}_F = c) = 1$, and thus F is $\mathbb{P}^\nu$-trivial, so that (b) holds.

That (b) implies (c) is immediate.

Suppose (c) holds, and let h be a bounded harmonic function. Let $M_k = h(X_k)$, so M is a bounded martingale, and by the martingale convergence theorem (see Theorem A.7), for each $x \in \mathbb{V}$, $M_k \to M_\infty$ $\mathbb{P}^x$-a.s. and in L^1. Let $n \geq 0$. Now $M_k \circ \theta_n = M_{k+n}$, so $M_\infty = M_\infty \circ \theta_n$, and thus M_∞ is $\mathscr{I}$-measurable. As $\mathscr{I}$ is $\mathbb{P}^x$-trivial, it follows that $\mathbb{P}^x(M_\infty = c) = 1$ for some constant c. Further, as $M_n \to M_\infty$ in L^1,

$$h(X_n) = M_n = \mathbb{E}^x(M_\infty|\mathscr{F}_n) = c$$

for any n, so $h(y) = c$ for any y such that $\mathbb{P}^x(X_n = y) > 0$ for some $n \geq 0$. As Γ is connected we deduce (a). □

Combining Theorems 1.40 and 1.48 we obtain:

Corollary 1.49 *Let Γ be recurrent. Then $\mathcal{I}$ is $\mathbb{P}^\nu$-trivial for all ν.*

While in general the Liouville property can be hard to prove, the special structure of the SRW on $\mathbb{Z}^d$ as a sum of i.i.d. random variables enables us to give two quick proofs of the Liouville property; one direct, and one via the Hewitt–Savage 01 law. We begin with a direct argument from [DM].

Theorem 1.50 *Let $\mathcal{G}$ be a countable Abelian group and Λ be a set of generators. Let ν be a probability measure on Λ, with $\nu_z = \nu(z) > 0$ for all $z \in \Lambda$. If h is bounded and harmonic then h is constant.*

Remark In this context h harmonic means that $h = Ph$, where

$$Ph(x) = \sum_{y \in \Lambda} h(x+y)\nu_y.$$

In order that the associated random walk be given by edge weights, we need the additional conditions that if $x \in \Lambda$ then $-x \in \Lambda$, and that $\nu_x = \nu_{-x}$ for all $x \in \Lambda$. We can then set $\mu_{x,x+y} = \mu_{x+y,x+y+(-y)} = \nu_y$.

Proof Suppose that h is non-constant. Then since Λ generates $\mathcal{G}$ there exist $w \in \Lambda$ and $x' \in \mathcal{G}$ such that $h(x') \neq h(x'+w)$. Set

$$h_w(x) = h(x) - h(x+w);$$

replacing h by $-h$ if necessary we can assume that $M = \sup h_w > 0$. Then

$$\begin{aligned} Ph_w(x) &= \sum_{y \in \Lambda} h_w(x+y)\nu_y \\ &= \sum_{y \in \Lambda} (h(x+y) - h(x+w+y))\nu_y = h(x) - h(x+w) = h_w(x), \end{aligned}$$

so that h_w is also harmonic. Then

$$h_w(x) \leq \nu_w h_w(x+w) + (1-\nu_w)M,$$

which gives that

$$M - h_w(x+w) \leq \nu_w^{-1}(M - h_w(x)).$$

Iterating this equation we obtain

$$M - h_w(x + nw) \le \nu_w^{-n}(M - h_w(x)), \qquad \text{for all } x \text{ and } n \ge 0\,. \qquad (1.42)$$

Let $N \ge 1$. Since $\inf_x(M - h_w(x)) = 0$, there exists $x_0 \in \mathcal{G}$ such that $M - h_w(x_0) \le \frac{1}{2}M\nu_w^N$. So by (1.42), for $0 \le n \le N$,

$$M - h_w(x_0 + nw) \le \nu_w^{-n}(M - h_w(x_0)) \le \tfrac{1}{2}M,$$

which implies that

$$h_w(x_0 + nw) \ge \tfrac{1}{2}M, \quad 0 \le n \le N.$$

Then

$$\begin{aligned} 2||h||_\infty \ge h(x_0) - h(x_0 + Nw) &= \sum_{n=1}^{N} h(x_0 + nw) - h(x_0 + (n-1)w) \\ &= \sum_{n=1}^{N} h_w(x_0 + (n-1)w) \ge \tfrac{1}{2}NM, \end{aligned}$$

a contradiction if N is chosen large enough. □

Remark The example of the free group with two generators shows this result is not true for non-Abelian groups. To see where the proof fails, suppose that $\mathcal{G}$ is non-Abelian, and the random walk acts by multiplication on the right, so that $Ph(x) = \sum_{y\in\Lambda} h(xy)\nu_y$. If we define $h_w(x) = h(x) - h(xw)$ then h_w is not harmonic. If instead we set $h_w(x) = h(x) - h(wx)$ then, while h_w is harmonic, the relation $h(xw^n) - h(xw^{n-1}) = h_w(xw^{n-1})$, used in the last line of the proof, fails in general.

We now introduce the Hewitt–Savage 01 law. A finite permutation $\pi : \mathbb{N} \to \mathbb{N}$ is a 1-1 onto map from $\mathbb{N}$ to $\mathbb{N}$ such that $\{i : \pi(i) \ne i\}$ is finite. It follows that there exists $m = m_\pi$ such that $\pi(j) = j$ for all $j \ge m$. Now let $(\xi_n, n \in \mathbb{N})$ be random variables on a probability space $(\Omega, \mathcal{F}, \mathbb{P})$. For simplicity we assume that ξ_i are discrete, with values in a finite set S, that $\Omega = S^{\mathbb{N}}$, and that ξ_i are the coordinate maps, so that if $\omega = (\omega_1, \omega_2, \dots) \in \Omega$, then $\xi_i(\omega) = \omega_i$. Given a finite permutation π we define the map $M_\pi : \Omega \to \Omega$ by $M_\pi(\omega) = (\omega_{\pi(1)}, \omega_{\pi(2)}, \dots)$; unpacking this notation we have $\xi_i(M_\pi(\omega)) = \xi_{\pi(i)}(\omega)$. If $F \in \mathcal{F}$ then $M_\pi^{-1}(F) = \{\omega : M_\pi(\omega) \in F\}$. We say that F is *permutable* if $F = M_\pi^{-1}(F)$ for every π.

Theorem 1.51 (Hewitt–Savage 01 law – see [Dur], Th. 4.1.1.) *(a) The collection of permutable events is a σ-field, $\mathcal{F}_E$, called the exchangeable σ-field.*
(b) Let $(\xi_n, n \in \mathbb{N})$ be i.i.d. Then $\mathcal{F}_E$ is trivial.

Corollary 1.52 *Let X be the SRW on $\mathbb{Z}^d$. Then $\mathcal{T}$ and $\mathcal{I}$ are trivial, and X satisfies the Liouville property.*

Proof Let S be the set of the $2d$ neighbours of 0 in $\mathbb{Z}^d$, and let $\Xi = (\xi_n, n \in \mathbb{N})$ be i.i.d.r.v. with distribution given by $\mathbb{P}(\xi_n = x) = 1/2d$ for each $x \in S$. Define $X_0 = 0$, and

$$X_n = \sum_{i=1}^{n} \xi_i, \quad n \geq 1.$$

Let $\mathcal{G}_n$ and $\mathcal{T}$ be as in (1.40) and (1.41), and let $G \in \mathcal{T}$. Let π be a finite permutation and let

$$X_n^\pi = \sum_{i=1}^{n} \xi_{\pi(i)}, \quad n \geq 1.$$

Then there exists $m \geq 1$ such that $\pi_i = i$ for all $i \geq m$. Hence if $n \geq m$ then $X_n^\pi = X_n$. Since $G \in \mathcal{G}_m$ it follows that $M_\pi^{-1}(G) = G$.

Therefore G is permutable, and so $G \in \mathcal{F}_E$. Hence $\mathcal{T} \subset \mathcal{F}_E$, and so by Theorem 1.51 $\mathcal{T}$ is trivial. Thus $\mathcal{I}$ is trivial, and so by Theorem 1.48 X satisfies the Liouville property. □

Since the Liouville property is not stable under rough isometries, this result is of no help in proving the Liouville property for graphs which are roughly isometric to $\mathbb{Z}^d$.

Example It is easy to give examples of graphs for which $\mathcal{T} \neq \mathcal{I}$. Let $\Gamma = \mathbb{Z}^3$, with the usual edges except for one additional edge e^* between $(0, 0, 0)$ and $(1, 1, 0)$. Let

$$A_0 = \{x = (x_1, x_2, x_3) \in \mathbb{Z}^d : x_1 + x_2 + x_3 \text{ is even}\},$$

and $A_1 = \mathbb{Z}^d - A_0$. As the SRW on $\mathbb{Z}^3$ is transient, it only traverses e^* a finite number of times. So if

$$F_i = \{X_{2n} \in A_i \text{ for all large } n\},$$

then $\mathbb{P}^x(F_0 \cup F_1) = 1$. The events F_i are in $\mathscr{T}$ and have $\mathbb{P}^x(F_i) > 0$ for all x, so $\mathscr{T}$ is non-trivial. On the other hand $\mathscr{I}$ is trivial by Theorem 7.19.

Remark 1.53 Appendix A.4 gives a '02 law' which gives necessary and sufficient conditions to have $\mathscr{T} = \mathscr{I}$. As a corollary we have that $\mathscr{T} = \mathscr{I}$ for the lazy random walk, that is the simple random walk on the lazy graph defined in Examples 1.4(7).

2

Random Walks and Electrical Resistance

2.1 Basic Concepts

Doyle and Snell [DS] made explicit the connection between random walks and electrical resistance, which had been implicit in many previous works, such as [BD, NW].

Informally an electrical network is a set of wires and nodes. A weighted graph $\Gamma = (\mathbb{V}, E, \mu)$ can be interpreted as an electrical network: the vertices are 'nodes' and the edge $\{x, y\}$ corresponds to a wire with conductivity μ_{xy}. (We only consider 'pure resistor' networks – no impedances or capacitors.)

Definition 2.1 A *flow* on Γ is a map $I : \mathbb{V} \times \mathbb{V} \to \mathbb{R}$ such that:

$$I_{xy} = 0 \text{ if } x \not\sim y,$$
$$I_{xy} = -I_{yx} \text{ for all } x, y.$$

Consider a single wire $\{x, y\}$. We take its resistance to be μ_{xy}^{-1} (which is why we called the weights μ_{xy} conductances). If x, y are at potential $f(x), f(y)$ respectively, then Ohm's law ('$I = V/R$') gives the current as

$$I_{xy} = \mu_{xy}(f(y) - f(x)) \tag{2.1}$$

(my currents flow uphill), and the power dissipation ('$E = V^2/R$') is

$$\mu_{xy}(f(y) - f(x))^2. \tag{2.2}$$

Note that I defined by (2.1) is a flow, and that summing (2.2) over edges gives $\mathscr{E}(f, f)$.

Electrical engineering books regard currents as primary (they correspond to physical processes) and potentials as derived from the currents by (2.2). But in this mathematical treatment it is easier to take potentials as our main object.

Definition 2.2 (a) Let $f \in C(\mathbb{V})$. Define the flow ∇f by

$$(\nabla f)_{xy} = \mu_{xy}(f(y) - f(x)). \tag{2.3}$$

(b) Given a flow I set

$$\text{Div } I(x) = \sum_{y \sim x} I_{xy}; \qquad \text{div} I(x) = \frac{\text{Div } I(x)}{\mu_x}. \tag{2.4}$$

$\text{Div } I(x)$ is the *net flux out of* x, while div is the density of Div with respect to μ. For a flow I and $A \subset \mathbb{V}$ define the *net flux out of the set* A by

$$\text{Flux}(I; A) = \sum_{x \in A} \text{Div } I(x). \tag{2.5}$$

Lemma 2.3 *For* $f \in C(\mathbb{V})$, $\text{div} \nabla f(x) = \Delta f(x)$.

Proof By (2.3) and (2.4)

$$\text{div} \nabla f(x) = \mu_x^{-1} \sum_{y \sim x} (\nabla f)_{xy} = \mu_x^{-1} \sum_{y \sim x} \mu_{xy}(f(y) - f(x)) = \Delta f(x). \qquad \square$$

The behaviour of the currents in a network is described by Kirchhoff's two laws (see [Kir1, Kir2]). The first (the 'current law') states that if $x \in \mathbb{V}$, and no external current flows into or out of x, then

$$\text{Div } I(x) = \sum_{y \sim x} I_{xy} = 0.$$

The second (the 'voltage law') is that, given a cycle $x_0, x_1, \ldots, x_n = x_0$ in $\mathbb{V}$, then,

$$\sum_{i=1}^{n} \mu_{x_{i-1},x_i}^{-1} I_{x_{i-1},x_i} = 0. \tag{2.6}$$

It follows easily from this condition (a discrete analogue of curl(I) = 0) that there exists a function (or 'potential') f on $\mathbb{V}$ such that $I = \nabla f$; f is unique up to an additive constant. The proof is straightforward: let $x_0 \in \mathbb{V}$ and set $f(x_0) = 0$. Suppose f has been defined on a set $A \subset \mathbb{V}$. If $x \in A$ and $y \sim x$ with $y \in A^c$ then let $f(y) = f(x) + \mu_{xy}^{-1} I_{xy}$. The condition (2.6) ensures that f is well defined.

Plainly, if we just have a bunch of wires connected together, nothing will happen – that is, no currents will flow and the whole network will remain at the same potential. Electrical engineers therefore allow, as well as wires between x and y, the possibility of voltage or current sources. However, from our viewpoint, which is to use electrical networks to explore the properties of a graph Γ, it is better to fix the 'passive' network Γ (or a subset of it) and then allow 'external' sources (voltage or current) to be imposed in various ways.

Let $D \subset \mathbb{V}$. Suppose we impose a potential f_0 on $\mathbb{V} - D$ (by adding 'external' inputs of currents), and then allow currents to flow in D according to Kirchhoff's laws. Let $f \in C(\mathbb{V})$ be the resulting potential; by Ohm's law the current from x to y is $(\nabla f)_{xy}$. For $x \in D$ conservation of current then gives

$$\mathrm{div}\nabla f(x) = 0,$$

so that f is harmonic in D. Thus f is the solution to the following Dirichlet problem:

$$\begin{aligned} f(x) &= f_0(x) \text{ on } \mathbb{V} - D, \\ \Delta f(x) &= 0 \ \text{ on } D. \end{aligned} \tag{2.7}$$

Of course it is only the values of f_0 on ∂D which will affect f inside D.

Lemma 2.4 *Let f be any solution of the Dirichlet problem (2.7). Then $I = \nabla f$ satisfies Kirchhoff's laws.*

Proof Since I is the gradient of a function, (2.6) holds. For $x \in D$ we have, by Lemma 2.3, $\mathrm{Div}\, I(x) = \mu_x \Delta(x) = 0$. □

Assume now that f_0 is bounded, recall the definition of the exit time τ_D from (1.4), and set

$$\varphi(x) = \mathbb{E}^x\big(f_0(X_{\tau_D})\mathbf{1}_{\{\tau_D < \infty\}}\big). \tag{2.8}$$

Theorem 2.5 *Let D and φ be as above.*

(a) φ is a solution to (2.7).
(b) If D is finite then φ is the unique solution to (2.7).
(c) If $\mathbb{P}^x(\tau_D < \infty) = 1$ for all $x \in \mathbb{V}$ then φ is the unique bounded solution to (2.7).

Proof (a) If $x \in D$ then by the Markov property

$$\begin{aligned} \varphi(x) &= \sum_y \mathbb{E}^x(f_0(X_{\tau_D}); X_1 = y, \tau_D < \infty) \\ &= \sum_y \mathscr{P}(x, y)\mathbb{E}^y(f_0(X_{\tau_D}); \tau_D < \infty) = P\varphi(x). \end{aligned}$$

So $\Delta\varphi(x) = (P - I)\varphi(x) = 0$.

(b) Let φ' be another solution, and $h = \varphi - \varphi'$. Then h is harmonic in D, and is zero on $\mathbb{V} - D$. Then using either the maximum principle or the Gauss–Green formula, it is easy to prove that $h = 0$. By the maximum principle (Theorem 1.37) h attains its minimum (and maximum) on ∂D; hence $h = 0$ and $\varphi = \varphi'$.

Alternatively, note that $h(x) = 0$ for $x \in \mathbb{V} - D$, while $\Delta h(x) = 0$ if $x \in D$. As $h \in C_0(\mathbb{V})$ we can apply (1.20) and

$$\mathscr{E}(h, h) = \langle -\Delta h, h \rangle = 0,$$

proving that h is constant and therefore zero.
(c) Since f_0 is bounded, φ is bounded, and so if φ' is another bounded solution then $h = \varphi - \varphi'$ is bounded. Thus $M_n = h(X_n)$ is a bounded martingale; applying the optional sampling theorem at τ_D we have

$$h(x) = \mathbb{E}^x M_0 = \mathbb{E}^x M_{\tau_D} = 0,$$

so h is identically zero. □

Remark If $\mathbb{V}$ is transient and D is infinite then we may have $h_D(x) = \mathbb{P}^x(\tau_D = \infty) > 0$ for some $x \in D$. In this case uniqueness will fail for (2.7) since, for any λ, the function $\varphi + \lambda h_D$ is also a solution of (2.7). Thus the Dirichlet problem (2.7) is not enough to specify the potential f and current I arising from a potential f_0 on ∂D. We will see that if we impose the additional condition that f has 'minimal energy', then we do have uniqueness.

Lemma 2.6 *Let I be a flow on $\mathbb{V}$ with finite support. Then*

$$\sum_{x \in \mathbb{V}} \operatorname{Div} I(x) = 0.$$

Proof Let A be the set of vertices x such that $I_{xy} \neq 0$ for some $y \sim x$. Since A is finite,

$$\sum_{x \in \mathbb{V}} \operatorname{Div} I(x) = \sum_{x \in A} \operatorname{Div} I(x) = \sum_{x \in A} \sum_{y \in A} I_{xy} = -\sum_{x \in A} \sum_{y \in A} I_{yx} = -\sum_{y \in A} \operatorname{Div} I(y).$$

□

Remark If I has infinite support then the interchange of sums above may not be valid, and the result is false in general. Consider for example the graph $\mathbb{N}$ with flow $I_{n,n+1} = 1$ for all n; $\operatorname{Div} I(n) = 0$ for all $n \geq 2$ but $\operatorname{Div} I(1) = 1$.

2.2 Transience and Recurrence

Let B_0, B_1 be disjoint non-empty subsets of $\mathbb{V}$, and assume that $D = \mathbb{V} - (B_0 \cup B_1)$ is finite. Let $f_0 = \mathbf{1}_{B_1}$, and let φ be the unique solution, given by (2.8), to the Dirichlet problem (2.7). Let $\mathbb{V} - D = B_0 \cup B_1$ (with $B_0 \cap B_1 = \emptyset$), and let $f_0 = 0$ on B_0 and $f_0 = 1$ on B_1. Then

$$\varphi(x) = \mathbb{P}^x(T_{B_1} < T_{B_0}). \tag{2.9}$$

Let $I = \nabla\varphi$ be the associated current. As $0 \le \varphi \le 1$ and $\operatorname{Div} I(x) = \mu_x \Delta\varphi(x)$, we have

$$\operatorname{Div} I(x) \ge 0, \quad x \in B_0, \qquad \operatorname{Div} I(y) \le 0, \quad y \in B_1.$$

Using Lemma 2.6 we deduce that the net current flowing between B_0 and B_1 is given by

$$\operatorname{Flux}(I; B_0) = \sum_{x \in B_0} \operatorname{Div} I(x) = -\sum_{x \in B_1} \operatorname{Div} I(x) = -\operatorname{Flux}(I; B_1).$$

Definition 2.7 Let $\mathbb{V}$ be finite. The *(finite network) effective resistance between B_0 and B_1*, denoted $R_{\mathrm{eff}}^{\mathrm{FN}}(B_0, B_1)$, is $\operatorname{Flux}(I; B_0)^{-1}$, where I is the current flowing from B_0 to B_1 when B_0 is held at potential 0, and B_1 at potential 1.

Theorem 2.8 *Let $B_0 \subset A \subset \mathbb{V}$, with A finite. Then*

$$\sum_{x \in B_0} \mathbb{P}^x(\tau_A < T_{B_0}^+)\mu_x = \frac{1}{R_{\mathrm{eff}}^{\mathrm{FN}}(B_0, \mathbb{V} - A)}. \tag{2.10}$$

Further, for $x_0 \in A$,

$$R_{\mathrm{eff}}^{\mathrm{FN}}(x_0, \mathbb{V} - A) = g_A(x_0, x_0). \tag{2.11}$$

Proof Let $D = A - B_0$, $B_1 = \mathbb{V} - A$, φ be given by (2.9), and $I = \nabla\varphi$. Let $x \in B_0$. Then $\varphi(x) = 0$ and

$$\begin{aligned}\operatorname{Div} I(x) &= \sum_{y \sim x} \varphi(y)\mu_{xy} = \mu_x \sum_{y \sim x} \mathcal{P}(x, y)\mathbb{P}^y(T_{B_1} < T_{B_0}) \\ &= \mu_x \mathbb{P}^x(T_{B_1} < T_{B_0}^+).\end{aligned}$$

Thus the left side of (2.10) is $\operatorname{Flux}(I; B_0)$, and (2.10) follows from the definition of effective resistance.

Taking $B_0 = \{x_0\}$ we deduce that

$$\mathbb{P}^{x_0}(\tau_A < T_{x_0}^+)\mu_{x_0} = \frac{1}{R_{\text{eff}}^{\text{FN}}(x_0, \mathbb{V} - A)}, \tag{2.12}$$

and comparing (2.12) with (1.29) gives (2.11). □

Definition 2.9 Let (Γ, μ) be an infinite weighted graph, let $B_0 \subset \mathbb{V}$, and let $A_n \uparrow\uparrow \mathbb{V}$ with $B_0 \subset A_1$. We define

$$R_{\text{eff}}(B_0, \infty) = \lim_n R_{\text{eff}}^{\text{FN}}(B_0, A_n^c). \tag{2.13}$$

Lemma 2.10 *(a) If $B_0 \subset D_1 \subset D_2$, and D_2 is finite, then*

$$R_{\text{eff}}^{\text{FN}}(B_0, D_1^c) \le R_{\text{eff}}^{\text{FN}}(B_0, D_2^c).$$

(b) $R_{\text{eff}}(B_0, \infty)$ does not depend on the sequence (A_n) used.

Proof As $\tau_{D_1} \le \tau_{D_2}$ (a) is immediate from (2.10). (We will see a better proof later.)
(b) Let A'_n be another sequence, and denote the limits in (2.13) by R and R'. Given n there exists m_n such that $A_n \subset A'_{m_n}$. So, by (a),

$$R_{\text{eff}}^{\text{FN}}(B_0, A_n^c) \le R_{\text{eff}}^{\text{FN}}(B_0, (A'_{m_n})^c) \le \lim_m R_{\text{eff}}^{\text{FN}}(B_0, (A'_m)^c) = R'.$$

Thus $R \le R'$ and similarly $R' \le R$. □

Theorem 2.11 *Let (Γ, μ) be an infinite weighted graph.*

(a) For $x \in \mathbb{V}$

$$\mu_x \mathbb{P}^x(T_x^+ = \infty) = \frac{1}{R_{\text{eff}}(x, \infty)}.$$

(b) (Γ, μ) is recurrent if and only if $R_{\text{eff}}(x, \infty) = \infty$.
(c) (Γ, μ) is transient if and only if $R_{\text{eff}}(x, \infty) < \infty$.

Proof This is immediate from Theorem 2.8 and the definitions of $R_{\text{eff}}(x, \infty)$, transience, and recurrence. □

Remark 2.12 The connection between the probabilistic property of transience or recurrence, and resistance to infinity, is striking. Its usefulness comes from two sources. First, R_{eff} has monotonicity properties which follow from its characterisation via variational problems, but are not evident from a probabilistic viewpoint. Second, there exist techniques of 'circuit reduction', which allow (some) complicated networks to be reduced to simple ones. These include the well known formulae for calculating the effective resistance due to resistors in series or parallel, and the less familiar $Y - \nabla$ transform.

Examples (1) $\mathbb{Z}$ is recurrent since $R_{\text{eff}}^{\text{FN}}(\{0\}, [-n, n]^c) = \frac{1}{2}(n + 1)$.

(2) For $\mathbb{Z}^2$ one reduces effective resistance if one 'shorts' boxes around 0. (A precise mathematical justification for this operation will be given later in this chapter.) For $x = (x_1, x_2) \in \mathbb{Z}^2$, define $\|x\|_\infty = \max\{|x_1|, |x_2|\}$, and note that $S_n = \{x : \|x\|_\infty = n\}$ is a square with centre 0 and side $2n$. Each side of S_n contains $2n + 1$ vertices, and so the total number of edges between S_n and S_{n+1} is $4(2n + 1)$.

If we 'short' all the edges between vertices in S_n, for each n, then we obtain a new network $\tilde{\Gamma}$ with 'vertices' $S_0, S_1, \dots$, such that S_n is only connected to S_{n-1} and S_{n+1}. Using the formulae for resistors in parallel and series we have $\tilde{R}_{\text{eff}}^{\text{FN}}(S_n, S_{n+1}) = \frac{1}{4}(2n + 1)^{-1}$, and hence

$$R_{\text{eff}}^{\mathbb{Z}^2}(0, \infty) \geq \tilde{R}_{\text{eff}}(S_0, \infty) = \lim_{N \to \infty} \tilde{R}_{\text{eff}}^{\text{FN}}(S_0, S_N)$$

$$= \lim_{N \to \infty} 4^{-1} \sum_{k=0}^{N-1} (2k + 1)^{-1} = \infty.$$

(3) A similar argument shows that the triangular lattice is recurrent.

(4) Consider the graph (Γ, μ) with vertex set $\mathbb{Z}_+$ and weights $\mu_{n,n+1} = a_n$. Then $R_{\text{eff}}^{\text{FN}}(0, n) = \sum_{r=0}^{n-1} a_r^{-1}$, so that (Γ, μ) is transient if and only if $\sum_n a_n^{-1} < \infty$. In particular the graph considered in Examples 1.4(6) with $a_n = \alpha^n$ is transient if and only if $\alpha > 1$.

Remark 2.13 Let A be finite, $x_0 \in A$, and consider the flow I associated with the potential $g_A(\cdot, x_0)$. Taking $\varphi(x) = \mathbb{P}^x(T_{x_0} < \tau_A)$ as in Theorem 2.8 we have

$$g_A(x, x_0) = \varphi(x) g_A(x_0, x_0).$$

By (1.31) and Lemma 2.3 we have

$$\operatorname{Div} I(x_0) = \mu_x \Delta g_A^{x_0}(x_0) = -1,$$

so that I is a unit flow from A^c to x_0.

2.3 Energy and Variational Methods

In the examples above we used shorting and cutting techniques to calculate resistance without justification. But facts like 'shorts decrease resistance' should be provable from the definition, so we will now go on to discuss in more detail the mathematical background to effective resistance. In addition, we wish to study infinite networks, where not surprisingly some additional care is needed.

Let (Γ, μ) be a weighted graph, $B_0, B_1 \subset \mathbb{V}$ with $B_0 \cap B_1 = \emptyset$. Set

$$\mathscr{V}(B_0, B_1) = \{f \in H^2(\mathbb{V}) : f|_{B_1} = 1, f|_{B_0} = 0\}, \tag{2.14}$$

and consider the variational problem:

$$C_{\text{eff}}(B_0, B_1) = \inf\{\mathscr{E}(f, f) : f \in \mathscr{V}(B_0, B_1)\}. \tag{VP1}$$

We call $C_{\text{eff}}(B_0, B_1)$ the 'effective conductance' between B_0 and B_1, and will (of course) find that it is the reciprocal of $R_{\text{eff}}^{\text{FN}}(B_0, B_1)$ in those cases when $R_{\text{eff}}^{\text{FN}}(B_0, B_1)$ has been defined. In general we may have $C_{\text{eff}}(B_0, B_1) = \infty$. We call a function f a *feasible* function for (VP1) if it satisfies the constraints; that is, $f \in \mathscr{V}(B_0, B_1)$. We write $C_{\text{eff}}(x, y)$ for $C_{\text{eff}}(\{x\}, \{y\})$. Whenever we consider variational problems of this kind we will assume that B_0 and B_1 are non-empty.

Remark 2.14 If $B_0 \cap B_1 \neq \emptyset$ then $\mathscr{V}(B_0, B_1) = \emptyset$, so with the usual convention that $\inf \emptyset = \infty$ we can take $C_{\text{eff}}(B_0, B_1) = \infty$.

An immediate consequence of the definition is:

Lemma 2.15 *(a)* $C_{\text{eff}}(B_0, B_1) = C_{\text{eff}}(B_1, B_0)$.
(b) $C_{\text{eff}}(B_0, B_1) < \infty$ *if either* B_0 *or* B_1 *is finite and* $B_0 \cap B_1 = \emptyset$.
(c) $C_{\text{eff}}(B_0, B_1)$ *is increasing in each of* B_0, B_1.

Remark If instead we had used $H_0^2(\mathbb{V})$ in (2.14) then (a) would fail in general.

We will make frequent use of the following theorem from the theory of Hilbert spaces.

Theorem 2.16 *Let* $\mathscr{H}$ *be a Hilbert space, and* $A \subset \mathscr{H}$ *be a non-empty closed convex subset.*

(a) There exists a unique element $g \in A$ *with minimal norm.*
(b) If f_n *is any sequence in* A *with* $||f_n|| \to ||g||$ *then* $f_n \to g$ *in* $\mathscr{H}$.
(c) If $h \in \mathscr{H}$ *and there exists* $\delta > 0$ *such that* $g + \lambda h \in A$ *for all* $|\lambda| < \delta$ *then* $\langle h, g\rangle = 0$.

Proof (a) Let

$$M = \inf\{||f||^2 : f \in A\}.$$

Suppose $g_1, g_2 \in A$ with $||g_i||^2 \le M + \varepsilon$, where $\varepsilon \ge 0$. As A is convex, $h = \frac{1}{2}(g_1 + g_2) \in A$, and so

$$4M \le \langle 2h, 2h\rangle = ||g_1||^2 + 2\langle g_1, g_2\rangle + ||g_2||^2 \le 2M + 2\varepsilon + 2\langle g_1, g_2\rangle,$$

which implies that $-2\langle g_1, g_2\rangle \le -2M + 2\varepsilon$. So

$$\langle g_1 - g_2, g_1 - g_2\rangle \le 2(M + \varepsilon) - 2\langle g_1, g_2\rangle \le 4\varepsilon, \tag{2.15}$$

and it follows that the minimal element is unique.
(b) Let f_n be a sequence in A with $||f_n||^2 \to M$, and set $\delta_n = ||f_n||^2 - M$. Then by (2.15) we have $||f_n - f_m||^2 \le 4(\delta_n \vee \delta_m)$, so that (f_n) is Cauchy and converges to a limit f; since $||f||^2 = M$, then $f = g$.
(c) We have

$$0 \le \langle g + \lambda h, g + \lambda h\rangle - \langle g, g\rangle = 2\lambda\langle g, h\rangle + \lambda^2\langle h, h\rangle,$$

and as this holds for $|\lambda| < \delta$ we must have $\langle g, h\rangle = 0$. □

Theorem 2.17 *Suppose that* $C_{\text{eff}}(B_0, B_1) < \infty$.

(a) There exists a unique minimiser f for (VP1).
(b) f is a solution of the Dirichlet problem:

$$\begin{aligned} f(x) &= 0 \text{ on } B_0, \\ f(x) &= 1 \text{ on } B_1, \\ \Delta f(x) &= 0 \text{ on } \mathbb{V} - (B_0 \cup B_1). \end{aligned} \tag{2.16}$$

Proof (a) Consider the Hilbert space $H^2(\mathbb{V})$ where we choose the base point $x_0 \in B_0$. Clearly $\mathscr{V} = \mathscr{V}(B_0, B_1)$ is a convex subset of the Hilbert space $H^2(\mathbb{V})$. Further, if $f_n \in \mathscr{V}$ and $f_n \to f$ in $H^2(\mathbb{V})$, then (since $f_n \to f$ pointwise), $f \in \mathscr{V}$, so $\mathscr{V}$ is closed. Hence, by Theorem 2.16 $\mathscr{V}$ has a unique element f which minimises $\|g\|_{H^2}$, and so $\mathscr{E}(g, g)$, over $g \in \mathscr{V}$.
(b) It is clear that f satisfies the boundary conditions in (2.16). Let $x \in \mathbb{V} - (B_0 \cup B_1)$, and let $h(y) = \mathbf{1}_{\{x\}}(y)$. Then for any $\lambda \in \mathbb{R}$ the function $f + \lambda h \in \mathscr{V}(B_0, B_1)$, so by Theorem 2.16 $\mathscr{E}(f, h) = 0$. Since $h \in C_0(\mathbb{V})$,

$$0 = \mathscr{E}(f, h) = \langle \Delta f, h\rangle = \mu_x \Delta f(x),$$

so that f is harmonic in $\mathbb{V} - (B_0 \cap B_1)$. □

Proposition 2.18 *Let $\mathbb{V}_n \uparrow\uparrow \mathbb{V}$. Write $C_{\text{eff}}(\cdot, \cdot; \mathbb{V}_n)$ for effective conductance in the subgraph $\mathbb{V}_n$. Then if $B_0, B_1 \subset \mathbb{V}$,*

$$\lim_n C_{\text{eff}}(B_0 \cap \mathbb{V}_n, B_1 \cap \mathbb{V}_n; \mathbb{V}_n) = C_{\text{eff}}(B_0, B_1). \tag{2.17}$$

Proof Note first that

$$\begin{aligned} C_{\text{eff}}(B_0 \cap \mathbb{V}_n, B_1 \cap \mathbb{V}_n; \mathbb{V}_n) &\leq C_{\text{eff}}(B_0 \cap \mathbb{V}_n, B_1 \cap \mathbb{V}_n; \mathbb{V}_{n+1}) \\ &\leq C_{\text{eff}}(B_0 \cap \mathbb{V}_{n+1}, B_1 \cap \mathbb{V}_{n+1}; \mathbb{V}_{n+1}) \\ &\leq C_{\text{eff}}(B_0 \cap \mathbb{V}_{n+1}, B_1 \cap \mathbb{V}_{n+1}) \\ &\leq C_{\text{eff}}(B_0, B_1). \end{aligned}$$

The first and third inequalities hold because for any $f \in C(\mathbb{V})$ we have $\mathscr{E}_{\mathbb{V}_n}(f, f) \leq \mathscr{E}_{\mathbb{V}_{n+1}}(f, f) \leq \mathscr{E}(f, f)$, while the second and fourth are immediate by the monotonicity given by Lemma 2.15. Thus the limit on the left side of (2.17) exists, and we have proved the upper bound on this limit.

Let φ_n be the optimal function for $C_{\text{eff}}(B_0 \cap \mathbb{V}_n, B_1 \cap \mathbb{V}_n; \mathbb{V}_n)$; we extend φ_n to $\mathbb{V}$ by taking it to be zero outside $\mathbb{V}_n$. Let $C = \lim_n C_{\text{eff}}(B_0 \cap \mathbb{V}_n, B_1 \cap \mathbb{V}_n; \mathbb{V}_n)$. If $C = \infty$ there is nothing to prove. Otherwise $\mathscr{E}_{\mathbb{V}_n}(\varphi_n, \varphi_n)$ are uniformly bounded, and by Proposition 1.21(h) there exists a subsequence n_k such that φ_{n_k} converges pointwise in $\mathbb{V}$ to a function φ with

$$\mathscr{E}(\varphi, \varphi) \leq \liminf_n \mathscr{E}_{\mathbb{V}_n}(\varphi_n, \varphi_n) = C.$$

We have $\varphi = k$ on B_k, $k = 0, 1$ and so φ is feasible for (VP1), and therefore $C_{\text{eff}}(B_0, B_1) \leq \mathscr{E}(\varphi, \varphi) \leq C$, proving equality in (2.17). □

Remark 2.19 Since the function φ satisfies $\mathscr{E}(\varphi, \varphi) = C_{\text{eff}}(B_0, B_1)$, φ is the unique minimiser of (VP1). Further, since any subsequence of φ_n will have a sub-subsequence converging to φ, we have $\varphi_n \to \varphi$ pointwise, and deduce that

$$\varphi(x) = \lim_n \mathbb{P}^x(T_{B_1 \cap \mathbb{V}_n} < T_{B_0 \cap \mathbb{V}_n}).$$

We will shortly introduce a second variational problem in terms of flows. To formulate this, we need to define a Hilbert space for flows. Let $\vec{E}$ be the set of oriented edges in $\mathbb{V} \times \mathbb{V}$:

$$\vec{E} = \{(x, y) : \{x, y\} \in E\}.$$

Given flows I and J write

$$E(I, J) = \tfrac{1}{2} \sum_x \sum_{y \sim x} \mu_{xy}^{-1} I_{xy} J_{xy}$$

for the inner product in the space $L^2(\vec{E}, \mu^{-1})$. We call $E(I, I)$ the *energy dissipation* of the current I. Note that if $f, g \in H^2$ then

$$E(\nabla f, \nabla g) = \mathscr{E}(f, g). \tag{2.18}$$

Lemma 2.20 *Let I be a flow. Then $||\mathrm{div} I||_2^2 \le 2E(I, I)$.*

Proof We have

$$\begin{aligned} ||\mathrm{div} I||_2^2 &= \sum_x \mu_x \Big(\sum_{y\sim x} \mu_x^{-1} I_{xy}\Big)^2 = \sum_x \mu_x^{-1} \Big(\sum_{y\sim x} \mu_{xy}^{-1/2} I_{xy} \mu_{xy}\Big)^2 \\ &\le \sum_x \mu_x^{-1} \Big(\sum_{y\sim x} \mu_{xy}\Big)\Big(\sum_{y\sim x} \mu_{xy}^{-1} I_{xy}^2\Big) = 2E(I, I). \end{aligned}$$

□

Lemma 2.21 *Let $f \in C(\mathbb{V})$, and I be a flow. Then if $I \in L^2(\vec{E})$ and $f \in L^2$,*

$$E(I, \nabla f) \le E(I, I)^{1/2} \mathscr{E}(f, f)^{1/2}, \tag{2.19}$$

$$\langle -\mathrm{div} I, f\rangle \le \langle \mathrm{div} I, \mathrm{div} I\rangle^{1/2} \langle f, f\rangle^{1/2}. \tag{2.20}$$

Further we have

$$E(I, \nabla f) = -\sum_x \mathrm{div} I(x) f(x) \mu_x \tag{2.21}$$

in the following cases:

(a) if either I or f has finite support,
(b) if $I \in L^2(\vec{E})$ and $f \in L^2$,
(c) if $I \in L^2(\vec{E})$, $\mathrm{div} I \in C_0$, and $f \in H_0^2$.

Proof The inequalities (2.19) and (2.20) are immediate from Cauchy–Schwarz.
(a) Suppose that either I or f has support in a finite set A. Set $A_1 = \overline{A}$. Then

$$\begin{aligned} E(I, \nabla f) &= \tfrac{1}{2} \sum_{x\in A_1} \sum_{y\in A_1} I_{xy}(f(y) - f(x)), \\ &= \tfrac{1}{2} \sum_{x\in A_1} \sum_{y\in A_1} I_{xy} f(y) - \tfrac{1}{2} \sum_{x\in A_1} \sum_{y\in A} I_{xy} f(x) \\ &= -\sum_{x\in A_1} \sum_{y\in A_1} I_{xy} f(x) \\ &= -\sum_{x\in A_1} f(x) \mathrm{div} I(x) \mu_x = \langle -\mathrm{div} I, f\rangle. \end{aligned}$$

(b) Let $f_n \in C_0$ with $f_n \to f$ in L^2. Thus $f_n \to f$ in H^2 also. Then

$$E(I, \nabla f_n) = \langle -\mathrm{div} I, f_n \rangle \tag{2.22}$$

by (a), and using (2.19) and (2.20) we obtain (2.21).
(c) Let $f_n \in C_0$ with $f_n \to f$ in H^2. Then again (2.22) holds and we have

$$|E(I, \nabla f_n - \nabla f)|^2 \le E(I, I)\mathcal{E}(f - f_n, f - f_n) \to 0 \quad \text{as } n \to \infty,$$

so the left side of (2.22) converges to $E(I, \nabla f)$. Since $f_n \to f$ pointwise, and $\mathrm{div} I$ has finite support, the right side of (2.22) also converges. □

Remark 2.22 See Remark 2.33 for examples which show that (2.21) can fail if $f \in H^2$ and $I \in L^2(\vec{E})$ with $\mathrm{div} I \in C_0$. In addition, if $I \in L^2(\vec{E})$ and $f \in H_0^2$ then the sum on the right side of (2.21) may not converge absolutely.

The next result connects C_{eff} with our previous definition of $R_{\mathrm{eff}}^{\mathrm{FN}}$.

Proposition 2.23 *Let $\mathbb{V}$ be finite, and $B_0 \cap B_1 = \emptyset$.*

(a) The unique minimiser for (VP1) is given by

$$\varphi(x) = \mathbb{P}^x(T_{B_1} < T_{B_0}).$$

(b) The flow $I = \nabla\varphi$ satisfies

$$\mathrm{Flux}(I; B_0) = \sum_{x \in B_0} \mathrm{Div}\, I(x) = \mathcal{E}(\varphi, \varphi) = C_{\mathrm{eff}}(B_0, B_1).$$

(c) $C_{\mathrm{eff}}(B_0, B_1) = R_{\mathrm{eff}}^{\mathrm{FN}}(B_0, B_1)^{-1}$.

Proof (a) By Theorem 2.17 a unique minimiser exists, and satisfies (2.16). Since φ also satisfies (2.16), and by Theorem 2.5 this Dirichlet problem has a unique solution, we deduce that φ is this minimiser.
(b) Let $I = \nabla\varphi$. Note that $\mathrm{div} I(x) \ge 0$ if $x \in B_0$, $\mathrm{div} I(x) \le 0$ if $x \in B_1$, and $\mathrm{div} I(x) = 0$ for $x \in \mathbb{V} - B$. We apply Lemma 2.21 with $f = \varphi$ and $I = \nabla\varphi$, to obtain

$$\mathscr{E}(\varphi,\varphi) = -\sum_x \varphi(x)\mathrm{div} I(x)\mu_x$$
$$= -\sum_{x\in B_1} \mathrm{Div}\, I(x) = -\mathrm{Flux}(I;\, B_1) = \mathrm{Flux}(I;\, B_0).$$

(c) By definition $R^{\mathrm{FN}}_{\mathrm{eff}}(B_0, B_1) = \mathrm{Flux}(I;\, B_0)^{-1}$. □

Definition 2.24 We now define for $B_0, B_1 \subset \mathbb{V}$, with $B_0 \cap B_1 = \varnothing$,

$$R_{\mathrm{eff}}(B_0, B_1) = C_{\mathrm{eff}}(B_0, B_1)^{-1}.$$

By Proposition 2.2.3(c), this agrees with our previous definition when $\mathbb{V} - (B_0 \cup B_1)$ is finite, and we will now drop the temporary notation $R^{\mathrm{FN}}_{\mathrm{eff}}$.

Lemma 2.25 *For $f \in H^2$, $x, y \in \mathbb{V}$*

$$|f(x) - f(y)|^2 \le \mathscr{E}(f, f)R_{\mathrm{eff}}(x, y). \tag{2.23}$$

Proof This is immediate from the definition of R_{eff} via (VP1). If $f(x) = f(y)$ there is nothing to prove; otherwise set

$$g(\cdot) = \frac{f(\cdot) - f(x)}{f(y) - f(x)}.$$

Then $g(x) = 0$, $g(y) = 1$ so g is feasible for (VP1). Hence

$$R_{\mathrm{eff}}(x, y)^{-1} \le \mathscr{E}(g, g) = |f(y) - f(x)|^{-2}\mathscr{E}(f, f);$$

rearranging then proves (2.23). □

Definition 2.26 ('Shorts' and 'cuts') Let (Γ, μ) be a weighted graph, and $e_0 = \{x, y\}$ be an edge in Γ. The graph $(\Gamma^{(c)}, \mu^{(c)})$ obtained by *cutting the edge* e_0 is the weighted graph with vertex set $\mathbb{V}$, edge set $E^{(c)} = E - \{e_0\}$, and weights given by

$$\mu^{(c)}_e = \begin{cases} \mu_e, & e \ne e_0, \\ 0, & e = e_0. \end{cases}$$

(Note that this graph may not be connected.)

The graph $(\Gamma^{(s)}, \mu^{(s)})$ obtained by *shorting the edge* $\{x, y\}$ is the weighted graph obtained by identifying the vertices x and y: that is, by collapsing the set $\{x, y\}$ to a single point a, as in Definition 1.5(2).

Corollary 2.27 *Let μ, μ' be weights on Γ, and write R_{eff}, R'_{eff} for effective resistance in the weighted graphs (Γ, μ) and (Γ, μ').*

(a) If $\mu \le c_1 \mu'$ *then*

$$C_{\text{eff}}(B_0, B_1) \le c_1 C'_{\text{eff}}(B_0, B_1), \qquad R_{\text{eff}}(B_0, B_1) \ge c_1^{-1} R'_{\text{eff}}(B_0, B_1).$$

(b) Cuts increase $R_{\text{eff}}(B_0, B_1)$.
(c) Shorts decrease $R_{\text{eff}}(B_0, B_1)$.

Proof (a) Since $\mathcal{E} \le c_1 \mathcal{E}'$, this is immediate from the definition of C_{eff}.
Now fix an edge $e_0 = \{x, y\}$, and write $\mathcal{E}_\lambda(f, f)$ for the energy of f in the graph (Γ, μ^λ), where $\mu^\lambda_e = \mu_e$ if $e \ne e_0$ and $\mu^\lambda_{e_0} = \lambda$. Write $C_{\text{eff}}(\cdot, \cdot, \lambda)$ for the effective conductance in the graph (Γ, μ^λ), and $C^{(c)}_{\text{eff}}(\cdot, \cdot)$ and $C^{(s)}_{\text{eff}}(\cdot, \cdot)$ for effective conductance in $(\Gamma^{(c)}, \mu^{(c)})$ and $(\Gamma^{(s)}, \mu^{(s)})$.
(b) We have $C^{(c)}_{\text{eff}}(B_0, B_1) = C_{\text{eff}}(B_0, B_1, 0)$, and the result follows from (a).
(c) We continue with the notation from (a). Define $F : \mathbb{V} \to \mathbb{V}'$ by taking F to be the identity on $\mathbb{V} - \{x, y\}$, and $F(x) = F(y) = a$, and let $B'_j = F(B_j)$ for $j = 0, 1$. If B'_0 and B'_1 are disjoint then

$$C^{(s)}_{\text{eff}}(B'_0, B'_1) = \lim_{\lambda \to \infty} C_{\text{eff}}(B_0, B_1, \lambda), \tag{2.24}$$

so again the monotonicity follows from (a). If $B'_0 \cap B'_1 \ne \emptyset$ then the right side of (2.24) is infinite, and by Remark 2.14 we should also take the left side to be infinite. □

We can also regard $C_{\text{eff}}(B_0, B_1)$ as a capacity, or, following [Duf] as the *modulus* of a family of curves. The connection with capacity will be made in Chapter 7 – see Proposition 7.9. To see the connection with moduli, let $\mathcal{P}$ be a family of (finite) paths in $\mathbb{V}$. For a path $\gamma = (z_0, z_1, \dots, z_n) \in \mathcal{P}$ and $f \in C(\mathbb{V})$ set

$$O_\gamma(f) = \sum_{i=1}^{n} |f(z_i) - f(z_{i-1})|,$$

and let

$$\mathcal{F}(\mathcal{P}) = \big\{ f \in C(\mathbb{V}) : O_\gamma(f) \ge 1 \text{ for all } \gamma \in \mathcal{P} \big\}.$$

Define the modulus of $\mathcal{P}$ by

$$M(\mathcal{P}) = \inf\{\mathcal{E}(f, f) : f \in \mathcal{F}(\mathcal{P})\}.$$

The quantity $M(\mathcal{P})^{-1}$ is the *extremal length* of the family of curves $\mathcal{P}$. (This concept has a long history in complex analysis – see for example [Ah].)

Proposition 2.28 *Let B_0, B_1 be disjoint subsets of $\mathbb{V}$ and $\mathscr{P}$ be the set of all finite paths with one endpoint in B_0 and the other in B_1. Then $M(\mathscr{P}) = C_{\mathrm{eff}}(B_0, B_1)$.*

Proof Let φ be an optimal solution for (VP1). Then if $\gamma = (x_0, \dots, x_n)$ is any path in $\mathscr{P}$ with $x_0 \in B_0$, $x_n \in B_1$,

$$O_\gamma(\varphi) = \sum_{i=1}^n |\varphi(x_i) - \varphi(x_{i-1})| \geq \sum_{i=1}^n (\varphi(x_i) - \varphi(x_{i-1})) = \varphi(x_n) - \varphi(x_0) = 1.$$

So, $\varphi \in \mathscr{F}(\mathscr{P})$, and hence $M(\mathscr{P}) \leq \mathscr{E}(\varphi, \varphi) = C_{\mathrm{eff}}(B_0, B_1)$.

To prove the other inequality, let $f \in \mathscr{F}(\mathscr{P})$. Define

$$g(x) = \min \big\{ O_\gamma(f) : \gamma \text{ is a path from a point } x_0 \in B_0 \text{ to } x \big\}.$$

Clearly $g(x) = 0$ if $x \in B_0$, and (as $f \in \mathscr{F}(\mathscr{P})$), $g(x) \geq 1$ for $x \in B_1$.

Now let $x \sim y$, and suppose that $g(x) \geq g(y)$. Then since a path to y can be extended by one step to give a path to x, $g(x) \leq g(y) + |f(x) - f(y)|$, proving that

$$|g(x) - g(y)| \leq |f(x) - f(y)|, \qquad \text{whenever } x \sim y.$$

Let $\psi = g \wedge 1$; ψ is feasible for (VP1). Then from the above and Lemma 1.27

$$C_{\mathrm{eff}}(B_0, B_1) \leq \mathscr{E}(\psi, \psi) \leq \mathscr{E}(g, g) \leq \mathscr{E}(f, f),$$

and hence $C_{\mathrm{eff}}(B_0, B_1) \leq M(\mathscr{P})$. □

Now we consider a second variational problem, given by minimising the energy of flows in $\mathbb{V}$. Recall from (2.5) the definition of $\mathrm{Flux}(I; A)$.

Definition 2.29 Let B_0, B_1 be disjoint non-empty subsets of $\mathbb{V}$. Let $\mathscr{I}_0(B_0, B_1)$ be the set of flows I with finite support such that

$$\mathrm{div} I(x) = 0, \quad x \in \mathbb{V} - (B_0 \cup B_1), \tag{2.25}$$

$$\mathrm{Flux}(I; B_0) = 1, \tag{2.26}$$

$$\mathrm{Flux}(I; B_1) = -1. \tag{2.27}$$

Let $\overline{\mathscr{I}}_0(B_0, B_1)$ be the closure of $\mathscr{I}_0(B_0, B_1)$ with respect to $E(\cdot, \cdot)$. By taking a unit flow along a finite path between $x_0 \in B_0$ and $x_1 \in B_1$ we see that $\mathscr{I}_0(B_0, B_1)$ and $\overline{\mathscr{I}}_0(B_0, B_1)$ are non-empty. The conditions (2.26) and (2.27) are not preserved by L^2 convergence, so need not be satisfied by flows in $\overline{\mathscr{I}}_0(B_0, B_1)$; see the examples in Remark 2.33.

It is easy to check that:

Lemma 2.30 *Let $I \in \overline{\mathscr{I}}_0(B_0, B_1)$ and let $I_n \in \mathscr{I}_0(B_0, B_1)$ with $E(I - I_n, I - I_n) \to 0$. Then $(I_n)_{xy} \to I_{xy}$ for each x, y, and I satisfies (2.25). In addition, if $B_0 \cup B_1$ is finite then I satisfies (2.26) and (2.27).*

The variational problem is:

$$\text{minimise } E(I, I) \text{ over } I \in \overline{\mathscr{I}}_0(B_0, B_1). \tag{VP2}$$

As before, we say that a flow is *feasible* for (VP2) if it is in $\overline{\mathscr{I}}_0(B_0, B_1)$.

Theorem 2.31 (Thompson's principle) *Let B_0 and B_1 be disjoint non-empty subsets of $\mathbb{V}$. Then (VP2) has a unique minimiser I^*, and*

$$R_{\text{eff}}(B_0, B_1) = E(I^*, I^*) = \inf\big\{E(I, I) : I \in \overline{\mathscr{I}}_0(B_0, B_1)\big\}. \tag{2.28}$$

If $C_{\text{eff}}(B_0, B_1) < \infty$ let φ be the unique minimiser for (VP1). Then

$$I^* = R_{\text{eff}}(B_0, B_1)\nabla\varphi. \tag{2.29}$$

If $C_{\text{eff}}(B_0, B_1) = \infty$ then $I^* = 0$.

Proof Since $\overline{\mathscr{I}}_0$ is closed and convex, a unique minimiser for (VP2) exists by Theorem 2.16. If $J \in \mathscr{I}_0(B_0, B_1)$ then by Lemma 2.21(a)

$$E(J, \nabla\varphi) = \langle -\text{div}J, \varphi\rangle = -\sum_{x \in B_1} \text{div}J(x)\mu_x = 1.$$

Hence if $C_{\text{eff}}(B_0, B_1) < \infty$ then

$$1 = E(J, \nabla\varphi) \le E(J, J)^{1/2}\mathscr{E}(\varphi, \varphi)^{1/2},$$

and therefore $E(J, J) \ge R_{\text{eff}}(B_0, B_1)$. This last inequality is trivial if $C_{\text{eff}}(B_0, B_1) = \infty$.

If $J' \in \overline{\mathscr{I}}_0(B_0, B_1)$ then there exists a sequence $J_n \in \mathscr{I}_0(B_0, B_1)$ with $E(J' - J_n, J' - J_n) \to 0$. Since $E(J_n, J_n) \ge R_{\text{eff}}(B_0, B_1)$ for each n, we have $E(J', J') \ge R_{\text{eff}}(B_0, B_1)$, and we have proved that

$$R_{\text{eff}}(B_0, B_1) \le \inf\{E(J, J) : J \in \overline{\mathscr{I}}_0(B_0, B_1)\}. \tag{2.30}$$

Now let I^* be defined by (2.29). Then since $E(\nabla\varphi, \nabla\varphi) = \mathscr{E}(\varphi, \varphi)$,

$$E(I^*, I^*) = R_{\text{eff}}(B_0, B_1)^2\mathscr{E}(\varphi, \varphi) = R_{\text{eff}}(B_0, B_1). \tag{2.31}$$

It remains to prove that $I^* \in \overline{\mathscr{I}}_0(B_0, B_1)$. Suppose first that $\mathbb{V}$ is finite. Then by (2.31)

$$1 = E(I^*, \nabla\varphi) = \langle -\text{div}I^*, \varphi\rangle = -\sum_{x \in B_1} \text{div}I^*(x)\mu_x = -\text{Flux}(I^*, B_1).$$

Then $\text{Flux}(I^*; B_0) = 1$, and so $I^* \in \mathscr{I}_0(B_0, B_1)$.

Now suppose $\mathbb{V}$ is infinite. Let $\mathbb{V}_n \uparrow\uparrow \mathbb{V}$, and let I_n be the optimum flow for (VP2) between $B_0 \cap \mathbb{V}_n$ and $B_1 \cap \mathbb{V}_n$ in the graph $\mathbb{V}_n$. Then each $I_n \in \mathcal{I}_0(B_0, B_1)$, and by Proposition 2.18

$$E(I_n, I_n) = R_{\text{eff}}(B_0 \cap \mathbb{V}_n, B_1 \cap \mathbb{V}_n; \mathbb{V}_n) \to R_{\text{eff}}(B_0, B_1) = E(I^*, I^*).$$

By Theorem 2.16, $E(I_n - I^*, I_n - I^*) \to 0$, and so $I^* \in \overline{\mathcal{I}}_0(B_0, B_1)$. □

Remark 2.32 (1) The two variational principles (VP1) and (VP2) allow one to estimate $R_{\text{eff}}(B_0, B_1)$ in cases when an exact calculation may not be possible. If f and I are feasible functions and flows for (VP1) and (VP2), respectively, then

$$\mathcal{E}(f, f)^{-1} \le R_{\text{eff}}(B_0, B_1) \le E(I, I).$$

It is usually easier to find a 'good' function f than a 'good' flow I.

(2) Let $\mathbb{B}$ be the rooted binary tree given in Examples 1.4, and recall that $\mathbb{B}_n$ is the set of 2^n points at a distance n from the root o. Let I be the flow which sends a flux of 2^{-n} into each point x in $\mathbb{B}_n$ from its ancestor $a(x)$. Then

$$E(I, I) = \sum_{n=1}^{\infty} \sum_{x \in \mathbb{B}_n} (2^{-n})^2 = \sum_{n=1}^{\infty} 2^{-n} = 1.$$

Using the flow I we have

$$R_{\text{eff}}(o, \mathbb{B}_n) \le \sum_{k=1}^{n} 2^{-k} = 1 - 2^{-n}.$$

In fact equality holds, as can be seen by considering the function $f(x) = (1 - 2^{-d(o,x)})/(1 - 2^{-n})$.

Remark 2.33 The following examples of infinite networks show why in (VP2) we take the space of flows to be $\overline{\mathcal{I}}_0(B_0, B_1)$.

(1) Let $\mathbb{V} = \{0, 1\} \times \mathbb{Z}_+$, and let $\Gamma = (\mathbb{V}, E)$ be the subgraph of $\mathbb{Z}^2$ induced by $\mathbb{V}$; Γ is like a ladder. Since it is a subgraph of $\mathbb{Z}^2$, Γ is recurrent using Corollary 2.39 in Section 2.4. Let $B_j = \{j\} \times \mathbb{Z}_+$. If I_n is the flow which sends current $1/n$ from $(0, k)$ to $(1, k)$ for $k = 1, \ldots, n$ then $E(I_n, I_n) = 1/n$. The flows I_n converge to $I^* = 0$, which is therefore in $\overline{\mathcal{I}}_0(B_0, B_1)$. Hence the infimum in (VP2) is zero, but there is clearly no minimising flow which satisfies the flux conditions (2.26) and (2.27).

(2) In the same graph, let J be the flow which sends current $1/n$ from $(0, n)$ to $(1, n)$ for each $n \ge 1$. Then $E(J, J) = \sum(1/n^2) < \infty$. If $f = 1$ then, since Γ is recurrent, we have $f \in H_0^2$ by Theorem 2.37. Then since

$\nabla f = 0$ we have $E(J, \nabla f) = 0$, while the sum in the expression for $\langle -\mathrm{div} J, f \rangle$ does not converge absolutely.

(3) Let Γ consist of two copies of a rooted binary tree, denoted $\mathbb{V}_0$, $\mathbb{V}_1$, with roots x_0 and x_1, connected by a path $A = (x_0, y_1, y_2, x_1)$ of length 3. Then if $B_j = \{x_j\}$ it is clear that $C_{\mathrm{eff}}(B_0, B_1) = \frac{1}{3}$. However, one can build a flow I (going from x_0 to infinity in $\mathbb{V}_0$, coming from infinity to x_1 in $\mathbb{V}_1$) with $E(I, I) = 2$. Of course I is not in $\overline{\mathscr{I}_0}$.

Let f be the function which equals 0 on $\mathbb{V}_0$ and 1 on $\mathbb{V}_1$, and is linear between x_0 and x_1. Then I and ∇f have disjoint supports, but

$$\langle -\mathrm{div} I, f \rangle = -\sum_{i=0}^{1} \mathrm{div} I(x_0) f(x_0) = -\mathrm{div} I(x_1) = -1,$$

so that (2.21) fails.

This example suggests that to handle flows on general infinite graphs one needs to impose the correct boundary behaviour at infinity. For more on this see [LP, So2, Z].

2.4 Resistance to Infinity

We now take B_1 to be a finite set, and B_0 to be 'infinity'. We consider the two variational problems

$$C(B_1, \infty) = \inf\{\mathscr{E}(f, f) : f|_{B_1} = 1, f \in H_0^2\} \qquad \text{(VP1')}$$

and

$$R(B_1, \infty) = \inf\{E(I, I) : \mathrm{Flux}(I; B_1) = -1,\ \mathrm{div} I(x) = 0,\ x \in \mathbb{V} - B_1\}. \qquad \text{(VP2')}$$

Set

$$\varphi(x) = \mathbb{P}^x(T_{B_1} < \infty). \qquad (2.32)$$

Recall from Definition 2.9 the definition of $R_{\mathrm{eff}}(B_1, \infty)$. We set

$$C_{\mathrm{eff}}(B_1, \infty) = R_{\mathrm{eff}}(B_1, \infty)^{-1}.$$

Proposition 2.34 *(a) The function $\varphi \in H_0^2(\mathbb{V})$, and attains the minimum in (VP1′). We have*

$$\mathscr{E}(\varphi, \varphi) = C(B_1, \infty) = C_{\mathrm{eff}}(B_1, \infty).$$

(b) If $f \in H_0^2$ with $f = 1$ on B_1 then

$$\mathscr{E}(\varphi, f) = \langle -\Delta\varphi, \mathbf{1}_{B_1} \rangle = R_{\mathrm{eff}}(B_1, \infty)^{-1}. \qquad (2.33)$$

Proof (a) By Theorem 2.16 there exists a unique minimiser $\widetilde{\varphi}$ for (VP1′). Let $\widetilde{\varphi}_n \in C_0$ with $\widetilde{\varphi}_n \to \widetilde{\varphi}$ in H^2. Then $\mathscr{E}(\widetilde{\varphi}_n, \widetilde{\varphi}_n) \to C(B_1, \infty)$. Since B_1 is finite and $\widetilde{\varphi}_n$ converge pointwise to $\widetilde{\varphi}$, we can assume that $\widetilde{\varphi}_n = 1$ on B_1.

Choose $x_0 \in B_1$, and let $A_n = \mathrm{supp}(\widetilde{\varphi}_k, k \le n)$, and $\mathbb{V}_n = B(x_0, n) \cup A_n$, so that $\mathbb{V}_n \uparrow\uparrow \mathbb{V}$. Let $\varphi_n = \mathbb{P}^x(T_{B_1} < \tau_{\mathbb{V}_n})$. By Proposition 2.23 φ_n is the minimiser for (VP1) for the sets $\partial\mathbb{V}_n$ and B_1 in the finite graph $\mathbb{V}_n \cup \partial\mathbb{V}_n$. Since $\widetilde{\varphi}_n$ is feasible for this problem, we have $\mathscr{E}(\varphi_n, \varphi_n) \le \mathscr{E}(\widetilde{\varphi}_n, \widetilde{\varphi}_n)$. Since also φ_n is feasible for (VP1′),

$$\mathscr{E}(\widetilde{\varphi}, \widetilde{\varphi}) = C(B_1, \infty) \le \mathscr{E}(\varphi_n, \varphi_n) \le \mathscr{E}(\widetilde{\varphi}_n, \widetilde{\varphi}_n).$$

By Theorem 2.16 we have $\varphi_n \to \widetilde{\varphi}$ in H_0^2 and therefore pointwise. However, we also have $\varphi_n(x) \uparrow \varphi(x)$ for all x, so $\widetilde{\varphi} = \varphi$ and $\varphi \in H_0^2$. By the definition of $C_{\mathrm{eff}}(B_1, \infty)$ we have

$$C_{\mathrm{eff}}(B_1, \infty) = \lim_n C_{\mathrm{eff}}(B_1, V_n^c) = \lim_n \mathscr{E}(\varphi_n, \varphi_n) = \mathscr{E}(\varphi, \varphi),$$

completing the proof of (a).
(b) Since B_1 is finite, using Proposition 2.23(b)

$$\langle -\Delta\varphi, \mathbf{1}_{B_1} \rangle = \lim_n \langle -\Delta\varphi_n, \mathbf{1}_{B_1} \rangle = \lim_n C_{\mathrm{eff}}(B_1, \mathbb{V}_n^c) = C_{\mathrm{eff}}(B_1, \infty).$$

Let $f \in H_0^2$ with $f = 1$ on B_1. Let $f_n \in C_0(\mathbb{V})$ with $f_n \to f$ in $H^2(\mathbb{V})$; as in (a) we can assume that $f_n = 1$ on B_1. Then

$$\mathscr{E}(\varphi, f_n) = \langle -\Delta\varphi, f_n \rangle = \langle -\Delta\varphi, \mathbf{1}_{B_1} \rangle,$$

and letting $n \to \infty$ gives (2.33). □

Lemma 2.35 *Let I be any feasible flow for (VP2′). Then $E(I, I) \ge R_{\mathrm{eff}}(B_1, \infty)$.*

Proof Let $\mathbb{V}_n \uparrow\uparrow \mathbb{V}$, and φ_n be as in Proposition 2.34. By Lemma 2.21

$$E(I, I)^{1/2} \mathscr{E}(\varphi_n, \varphi_n)^{1/2} \ge E(I, \nabla\varphi_n) = \langle -\mathrm{div} I, \varphi_n \rangle = -\operatorname{Flux}(I; B_1) = 1,$$

so $E(I, I) \ge R_{\mathrm{eff}}(B_1, \mathbb{V}_n^c)$. □

We can now summarise the behaviour of these variational problems; it is easier to deal separately with the recurrent and transient cases.

Theorem 2.36 *Let (Γ, μ) be recurrent, and $B_1 \subset \mathbb{V}$ be finite. Then:*

(a) $1 \in H_0^2$,
(b) $H_0^2 = H^2$,

(c) $R_{\text{eff}}(B_1, \infty) = \infty$,
(d) $R(B_1, \infty) = \infty$.

Proof As $\varphi \equiv 1$, (a) is immediate from Proposition 2.34, (b) follows by Proposition 1.29, (c) holds by Theorem 2.11, and (d) follows from Lemma 2.35. □

Theorem 2.37 *Let (Γ, μ) be transient, and $B_1 \subset \mathbb{V}$ be finite. Then:*

(a) the function φ given by (2.32) attains the minimum in (VP1′),
(b) $\mathscr{E}(\varphi, \varphi) = C_{\text{eff}}(B_1, \infty) = R_{\text{eff}}(B_1, \infty)^{-1}$,
(c) $R_{\text{eff}}(B_1, \infty) < \infty$,
(d) $1 \notin H_0^2$,
(e) $I^* = R_{\text{eff}}(B_1, \infty)\nabla\varphi$ attains the minimum in (VP2′) and

$$E(I^*, I^*) = R(B_1, \infty) = R_{\text{eff}}(B_1, \infty) < \infty.$$

Proof (a) and (b) have already been proved in Proposition 2.34, and (c) was proved in Theorem 2.8. For (d), if $1 \in H_0^2$ then the infimum in (VP1′) would be zero, which contradicts (c). For (e) we have

$$\text{Flux}(I^*; B_1) = \sum_{B_1} \text{div} I^*(x)\mu_x = R_{\text{eff}}(B_1, \infty)\langle \Delta\varphi, \mathbf{1}_{B_1}\rangle = -1,$$

by Proposition 2.34. Thus I^* is feasible for (VP2′). Since

$$E(I^*, I^*) = R_{\text{eff}}(B_1, \infty)^2 \mathscr{E}(\varphi, \varphi) = R_{\text{eff}}(B_1, \infty),$$

by Lemma 2.35 the flow I^* attains the minimum in (VP2′). □

Corollary 2.38 (See [Ly1]) *(Γ, μ) is transient if and only if there exists $x_1 \in \mathbb{V}$ and a flow J from x_1 to infinity with finite energy.*

Proof Theorem 2.36 and Theorem 2.37 together imply that (Γ, μ) is transient if and only if $R(B_1, \infty) < \infty$ for every finite set $B_1 \subset \mathbb{V}$. □

Corollary 2.39 *Let Γ be a subgraph of Γ'.*

(a) If Γ is transient then Γ' is transient.
(b) If Γ' is recurrent then Γ is recurrent.

Proof (a) is clear from Corollary 2.38 since if J is a flow from x_1 to infinity with finite energy in Γ then J is also such a flow in Γ'. (b) follows immediately from (a). □

The following criterion often provides an easy way to prove recurrence.

Corollary 2.40 (Nash–Williams criterion for recurrence [NW]) *Let (Γ, μ) be an infinite weighted graph. Suppose there exist $A_k \uparrow\uparrow \mathbb{V}$ with $\overline{A}_k \subset A_{k+1}$ such that*

$$\sum_{k=1}^{\infty} R_{\text{eff}}(A_k, A_k^c) = \infty.$$

Then Γ is recurrent.

Proof Let E_k be the set of edges $e = \{x, y\}$ with $x \in A_k$ and $y \in A_k^c$. The condition on A_k implies that the edge sets E_k are disjoint. We have

$$R_{\text{eff}}(A_k, A_k^c)^{-1} = C_{\text{eff}}(A_k, A_k^c) = \sum_{e \in E_k} \mu_e.$$

If I is any unit flow from A_1 to infinity then $\sum_{e \in E_k} |I(e)| \geq 1$. So,

$$1 \leq \sum_{e \in E_k} |I_e| \mu_e^{-1/2} \mu_e^{1/2} \leq \left(\sum_{e \in E_k} I_e^2 \mu_e^{-1} \right)^{1/2} \left(\sum_{e \in E_k} \mu_e \right)^{1/2}$$
$$= \left(\sum_{e \in E_k} I_e^2 \mu_e^{-1} \right)^{1/2} (R_{\text{eff}}(A_k, A_k^c))^{-1/2}.$$

Hence

$$E(I, I) \geq \sum_k \sum_{e \in E_k} \mu_e^{-1} I_e^2 \geq \sum_k R_{\text{eff}}(A_k, A_k^c) = \infty,$$

and so Γ is recurrent by Corollary 2.38. □

Examples 2.41 (1) $\mathbb{Z}^d$, $d \geq 3$ is transient. Since $\mathbb{Z}^3$ can be regarded as a subgraph of $\mathbb{Z}^d$ if $d > 3$, it is enough to prove this for $d = 3$.

Note first that a continuum approximation suggests that a finite energy flow from zero to infinity should exist. Let I be the symmetric flow on $\mathbb{R}^3$ from the unit ball to infinity. Then, at a distance r from the origin, I has density $(4\pi r^2)^{-1}$, so the energy of the flow I is:

$$\int_1^{\infty} (4\pi r^2)^{-2} 4\pi r^2 dr = (4\pi)^{-1} \int_1^{\infty} r^{-2} dr < \infty.$$

To prove that $\mathbb{Z}^3$ is transient we exhibit the flow in $\mathbb{Z}^3_+$ given in [DS]. For $(x, y, z) \in \mathbb{Z}^3_+$ set $A = x + y + z$, and let

$$I_{(x,y,z),(x+1,y,z)} = \frac{x+1}{(A+1)(A+2)(A+3)},$$

with $I_{(x,y,z),(x,y+1,z)}$ and $I_{(x,y,z),(x,y,z+1)}$ defined in a symmetric fashion. Then it is straightforward to verify that $\operatorname{div} I = 0$ except at 0, and $E(I, I) < \infty$.

(2) (Feller's problem.) The following problem is given on p. 425 of [Fe]: 'Let $\Gamma = (G, E)$ be a recurrent graph: Prove that any subgraph of Γ is also recurrent.' Discussions following Peter Doyle's solution of this problem using resistance techniques led to the book [DS].

The solution is given by Corollary 2.39. Note, however, that we do not need the full strength of the results such as this: one can also give a simple argument using just Corollary 2.27 and Theorem 2.11.

(3) In Remark 2.33(2) we constructed a flow I on the rooted binary tree with $E(I, I) = 1$; it follows that the binary tree is transient.

The following reformulation of Theorem 2.37, using the language of Sobolev inequalities rather than resistance, is useful and connects with the ideas in Chapter 3.

Theorem 2.42 *Let $x \in \mathbb{V}$. (Γ, μ) is transient if and only if there exists $c_1 = c_1(x) < \infty$ such that*

$$|f(x)|^2 \leq c_1 \mathscr{E}(f, f) \quad \textit{for all } f \in C_0(\mathbb{V}). \tag{2.34}$$

Proof Suppose first that (Γ, μ) is transient. Let $f \in C_0(\mathbb{V}) \subset H^2_0(\mathbb{V})$. If $f(x) = 0$ there is nothing to prove. Otherwise, set $g = f/f(x)$. Then g is feasible for (VP1$'$), so

$$\mathscr{E}(f, f) = f(x)^2 \mathscr{E}(g, g) \geq f(x)^2 C_{\text{eff}}(x, \infty).$$

If (2.34) holds, let φ_n be as in the proof of Theorem 2.37. Then $\mathscr{E}(\varphi_n, \varphi_n) \geq c_1^{-1}$, so $\mathscr{E}(\varphi, \varphi) = \lim_n \mathscr{E}(\varphi_n, \varphi_n) \geq c_1^{-1}$. □

As remarked above, it is usually easier to find a good feasible function for (VP1$'$) than a good feasible flow for (VP2$'$). It is therefore of interest to find other ways of bounding $R_{\text{eff}}(.,.)$ from above. One method is described in [Bov, Sect. 6]. Let A be a set, and let q_e^a, $a \in A$, $e \in E$ satisfy

$$q_e^a \geq 0, \quad \sum_a q_e^a \leq 1.$$

Set $\mu_e^a = q_e^a \mu_e$, and let $\mathscr{E}_a$ be the quadratic form associated with μ^a. Write $\mathscr{H}$ for the set of functions in $C_0(\mathbb{V})$ which are feasible for (VP1$'$). Then

$$C(B_1, \infty) = \inf_{f \in \mathscr{H}} \mathscr{E}(f, f) \geq \inf_{f \in \mathscr{H}} \sum_a \mathscr{E}_a(f, f) \geq \sum_a \inf_{f \in \mathscr{H}} \mathscr{E}_a(f, f).$$

If one can choose q_e^a so that the final term can be calculated, or estimated, then this gives a lower bound on $C(B_1, \infty)$ and hence an upper bound on $R_{\text{eff}}(B_1, \infty)$. Note that this generalises the method of cuts; cutting the edge e' is equivalent to taking $A = \{a\}$ and $q_{e'}^a = 0$, $q_e^a = 1$ for $e \neq e'$.

We now describe another method, the technique of *unpredictable paths* introduced in [BPP]. Fix a vertex $o \in \mathbb{V}$, and let Π be the set of paths $\pi = (o, \pi_1, \pi_2, \dots)$ from o to infinity with the property that $d(o, \pi_n) = n$ for all $n \geq 1$. Let $\mathbb{P}$ be a probability measure on Π, and let π and π' be independent paths chosen according to $\mathbb{P}$. We regard π, π' as sets of edges.

Exercise Prove that Π is non empty.

Theorem 2.43 (See [BPP], Sect. 2) *Suppose that*

$$\mathbb{E}\Big(\sum_e \mu_e^{-1} \mathbf{1}_{(e \in \pi \cap \pi')}\Big) < \infty. \tag{2.35}$$

Then (Γ, μ) *is transient.*

Proof We construct a flow I as follows. Let $e = \{x, y\}$ be an edge, and assume that $d(o, x) \leq d(o, y)$. If $d(o, x) = d(o, y)$ we take $I_{xy} = 0$, and otherwise we set

$$I_e = I_{xy} = \mathbb{P}(e \in \pi).$$

It is easy to check that I is a flow from o to infinity. (Each path in Π gives a unit flow from o to infinity, and I is the flow obtained by averaging these with respect to the probability measure $\mathbb{P}$.) Then

$$E(I, I) = \sum_{e \in E} \mu_e^{-1} I_e^2 = \sum_{e \in E} \mu_e^{-1} \mathbb{E}(\mathbf{1}_{(e \in \pi)} \mathbf{1}_{(e \in \pi')}) = \mathbb{E}\Big(\sum_{e \in E} \mu_e^{-1} \mathbf{1}_{(e \in \pi \cap \pi')}\Big).$$

Thus I has finite energy and so (Γ, μ) is transient. □

If Γ has natural weights then the expression in (2.35) is just the mean number of edges in both π and π'.

Example 2.44 We can use this method to give another proof of the transience of $\mathbb{Z}^3$. Let θ be chosen uniformly on the sphere $\mathbb{S}^2$, and let R_θ be the ray from zero to infinity in the direction θ.

Then there exists C such that we can define a path $\pi = \pi(\theta)$ from zero to infinity in $\mathbb{Z}^3$ in the direction θ so that for each n we have $\inf_{r \in R_\theta} |r - \pi_n| \le C$. (In fact we can take $C = 1$.) It follows that $|\pi(\theta) \cap \pi(\theta')| \le c d_S(\theta, \theta')^{-1}$, where d_S is the usual metric on $\mathbb{S}^2$. The random variable $d_S(\theta, \theta')$ has a density $f(t)$ on $[0, \pi]$ with $f(t) \le ct$. Hence

$$\mathbb{E}|\pi \cap \pi'| \le \int_0^\pi ct^{-1} f(t) dt \le c',$$

and we deduce that $\mathbb{Z}^3$ is transient.

Exercise 2.45 Use a similar construction to show that there exists $C < \infty$ such that if $x, y \in \mathbb{Z}^d$, with $d \ge 3$, then $R_{\text{eff}}(x, y) \le C$.

2.5 Traces and Electrical Equivalence

A standard result in electrical network theory is that n unit resistors in series can be replaced by a single resistor of strength n. In this subsection we will see how circuit reduction results of this kind can be formulated and proved in our framework.

Let $(\mathbb{V}, E, \mu)$ be a weighted graph. Let $D \subset \mathbb{V}$ be finite, and let $B = \mathbb{V} - D$. If $f \in C(B)$ let $Z_D f$ denote the function in $C(\mathbb{V})$ given by

$$Z_D f(x) = \begin{cases} f(x) & \text{if } x \in B, \\ 0 & \text{if } x \in D. \end{cases}$$

Recall from (1.16) the definition of $\mathscr{E}_B$ and set

$$H^2(B) = \{f \in C(B) : \mathscr{E}_B(f, f) < \infty\}.$$

Since D is finite, $Z_D f \in H^2(\mathbb{V})$ if and only if $f \in H^2(B)$. For $f \in H^2(B)$ set

$$\widetilde{\mathscr{E}}_B(f, f) = \inf\{\mathscr{E}(u, u) : u \in H^2(\mathbb{V}), u|_B = f\}. \tag{2.36}$$

Remark 2.46 We could consider infinite D, but all our applications just use finite sets. See [Kum] for a more general situation.

Proposition 2.47 *The infimum in* (2.36) *is obtained by a unique function. Writing $H_B f$ for this function, so that*

$$\widetilde{\mathscr{E}}_B(f, f) = \mathscr{E}(H_B f, H_B f), \quad f \in H^2(B),$$

$H_B f$ is the unique solution to the Dirichlet problem

$$\begin{aligned} H_B f(x) &= f \text{ on } B, \\ \Delta H_B f(x) &= 0 \text{ on } D. \end{aligned} \tag{2.37}$$

Consequently the map $H_B : H^2(B) \to C(\mathbb{V})$ *is linear and*

$$H_B f(x) = \mathbb{E}^x f(X_{\tau_D}). \tag{2.38}$$

Further $\widetilde{\mathscr{E}}_B(f, f)$ *has the Markov property:*

$$\widetilde{\mathscr{E}}_B(1 \wedge f_+, 1 \wedge f_+) \le \widetilde{\mathscr{E}}_B(f, f) \text{ for } f \in H^2(B).$$

Proof The existence and uniqueness of the minimiser is immediate from Theorem 2.16. The proof that $H_B f$ satisfies the Dirichlet problem (2.37) is exactly as in Theorem 2.17(b); uniqueness follows since D is finite, and this also proves (2.38). This characterisation implies that H_B is linear.

To prove the Markov property, let $f \in H^2(B)$ and $g = 1 \wedge f_+$. If $u|_B = f$ then $v = 1 \wedge u_+$ satisfies $v|_B = g$. The Markov property for $\mathscr{E}$ (see Lemma 1.27) implies that $\mathscr{E}(v, v) \le \mathscr{E}(u, u)$, and it follows that $\widetilde{\mathscr{E}}_B(g, g) \le \widetilde{\mathscr{E}}_B(f, f)$. □

We now define

$$\widetilde{\mathscr{E}}_B(f, g) = \mathscr{E}(H_B f, H_B g), \quad \text{for } f, g \in H^2(B).$$

We call $\widetilde{\mathscr{E}}_B$ the *trace* of $\mathscr{E}$ on B – see [FOT], Sect. 6.2.

Let $E(B)$ be the set of edges $e \in E$ such that both endpoints are in B. Let $E_D^+(B)$ be the set of $\{x, y\}$ in B with $x \neq y$ such that there exists a path $\gamma = (x = z_0, z_1, \dots, z_k = y)$ in Γ such that $z_i \in D$, $i = 1, \dots, k-1$. That is, x and y can be connected by a path which lies in D, except for its endpoints. Let

$$E_D(B) = E(B) \cup E_D^+(B).$$

Proposition 2.48 *There exist (symmetric) conductances* $\widetilde{\mu}_{xy}$ *such that, for* $f, g \in H^2(B)$,

$$\widetilde{\mathscr{E}}_B(f, g) = \tfrac{1}{2} \sum_{x \in B} \sum_{y \in B} (f(x) - f(y))(g(x) - g(y)) \widetilde{\mu}_{xy}. \tag{2.39}$$

We have

$$\widetilde{\mu}_{xy} = \mu_x \mathbb{P}^x(X_{T_B^+} = y) \text{ if } x \neq y. \tag{2.40}$$

In particular $\widetilde{\mu}_{xy} = \mu_{xy}$ *unless* $x, y \in \partial_i B$.

Proof $\widetilde{\mathscr{E}}_B$ is a symmetric bilinear form on $C_0(\mathbb{V})$, and $\widetilde{\mathscr{E}}_B(f, f) \ge 0$ for all f. Further, for $f \in L^2(\mathbb{V})$,

$$\widetilde{\mathscr{E}}_B(f, f) \le \mathscr{E}(Z_D f, Z_D f) \le 2||Z_D f||_2^2 = 2||f||^2_{L^2(B)}.$$

Since $\widetilde{\mathscr{E}}_B$ has the Markov property, by Theorem A.33 there exist $(\widetilde{\mu}_{xy})$ with $\widetilde{\mu}_{xy} \geq 0$ for $x \neq y$ and $k_x \geq 0$ such that, for $f, g \in L^2(B)$,

$$\widetilde{\mathscr{E}}_B(f, g) = \tfrac{1}{2} \sum_{x \in B} \sum_{y \in B} (f(x) - f(y))(g(x) - g(y))\widetilde{\mu}_{xy} + \sum_{x \in B} f(x)g(x)k_x. \tag{2.41}$$

Let $x \in B$. Then for sufficiently large m we have $\overline{D} \subset B(x, m)$, and so $H_B \mathbf{1}_{B(x,m)} = 1$ on $B(x, m-1)$. Then as $H_B \mathbf{1}_x$ has support contained in $\{x\} \cup \overline{D}$,

$$\widetilde{\mathscr{E}}_B(\mathbf{1}_x, \mathbf{1}_{B(x,m)}) = \mathscr{E}(H_B \mathbf{1}_x, H_B \mathbf{1}_{B(x,m)}) = 0.$$

Consequently we have $k_x = 0$, and we obtain the representation (2.39).

Now let $x, y \in B$, with $x \neq y$. Then

$$\widetilde{\mu}_{xy} = -\widetilde{\mathscr{E}}_B(\mathbf{1}_x, \mathbf{1}_y) = -\mathscr{E}(H_B \mathbf{1}_x, H_B \mathbf{1}_y) = \langle H_B \mathbf{1}_x, \Delta H_B \mathbf{1}_y \rangle.$$

Since $H_B \mathbf{1}_x = 0$ on $B - \{x\}$, while $\Delta H_B \mathbf{1}_y = 0$ on D,

$$\widetilde{\mu}_{xy} = \mu_x \Delta H_B \mathbf{1}_y(x) = \sum_{z \sim x} \mu_{xz} H_B \mathbf{1}_y(z) = \sum_{z \sim x} \mu_{xz} \mathbb{P}^z(X_{T_B} = y),$$

and (2.40) then follows.

Finally, since $\widetilde{\mu}_{xy} = \mu_{xy}$ except for finitely many pairs (x, y), the representation (2.39) also holds for $f, g \in H^2(B)$. □

Note that Proposition 2.48 does not specify $\widetilde{\mu}_{xx}$. Summing (2.40) we have

$$\mu_x \mathbb{P}^x(X_{T_B^+} = x) = \mu_x - \sum_{y \in B, y \neq x} \widetilde{\mu}_{xy}.$$

It is therefore convenient to take

$$\widetilde{\mu}_{xx} = \mu_x \mathbb{P}^x(X_{T_B^+} = x),$$

so that $\widetilde{\mu}_x = \sum_{y \in B} \widetilde{\mu}_{xy} = \mu_x$. We then have that $(B, E_D(B), \widetilde{\mu})$ is a weighted graph.

Definition 2.49 We say that the two networks $(\mathbb{V}, E, \mu)$ and $(B, E_D(B), \widetilde{\mu})$ are *electrically equivalent* on B.

The intuition for this definition is that an electrical engineer who is only able to access the vertices in B (imposing voltages and measuring currents there) would be unable to distinguish between the two networks. Writing $\widetilde{\Gamma}$ for the new network, Proposition 2.50 gives that resistances between any sets outside D are the same in the old network Γ and the new network $\widetilde{\Gamma}$; hence if it is simpler we can do the calculations in the new network.

Proposition 2.50 *Let B_0, B_1 be disjoint non-empty subsets of B, and let B_2 be a finite subset of B. Write $\widetilde{R}_{\text{eff}}(B_0, B_1)$ and $\widetilde{R}_{\text{eff}}(B_2, \infty)$ for effective resistances in $(B, E_D(B), \widetilde{\mu})$. Then*

$$\widetilde{R}_{\text{eff}}(B_0, B_1) = R_{\text{eff}}(B_0, B_1), \qquad \widetilde{R}_{\text{eff}}(B_2, \infty) = R_{\text{eff}}(B_2, \infty).$$

Proof To prove the first inequality, let f be the function which attains the minimum for (VP1) in the network (Γ, μ), and set $g = f|_B$. Then since $D \cap (B_0 \cup B_1) = \emptyset$, f is harmonic in D and thus $f = H_B g$. Then

$$\widetilde{C}_{\text{eff}}(B_0, B_1) \le \widetilde{\mathscr{E}}_B(g, g) = \mathscr{E}(f, f) = R_{\text{eff}}(B_0, B_1)^{-1}.$$

On the other hand, let $u \in H^2(B)$ be the function which attains the minimum for (VP1) in $(B, E_D(B), \widetilde{\mu})$. Then, since $H_B u$ is feasible for (VP1) in (Γ, μ),

$$\widetilde{C}_{\text{eff}}(B_0, B_1) = \widetilde{\mathscr{E}}_B(u, u) = \mathscr{E}(H_B u, H_B u) \ge C_{\text{eff}}(B_0, B_1).$$

The second equality is proved in the same way. □

Example 2.51 (Resistors in series) Suppose Γ contains a subset $A = \{x_0, x_1, \dots, x_n\}$, with the property that for $i = 1, \dots, n-1$ the only neighbours of x_i are x_{i-1} and x_{i+1}. Let

$$\mu_{x_{i-1}, x_i} = R_i^{-1}, \quad i = 1, \dots, n.$$

A standard procedure in electrical network theory is to replace the resistors in A with a single wire (i.e. edge) with resistance $R = \sum_{i=1}^n R_i$. This can be justified using Proposition 2.50 as follows. Set $D = \{x_1, \dots, x_{n-1}\}$, and take $B = \mathbb{V} - D$.

If $g \in H^2(B)$ then it is straightforward to verify that the unique solution to the Dirichlet problem (2.37) is given by

$$H_B g(x_k) = g(x_0) + \big(g(x_1) - g(x_0)\big) \frac{\sum_{i=1}^k R_i}{\sum_{i=1}^n R_i}.$$

So, with R as above,

$$\sum_{k=1}^n R_k^{-1} (g(x_{k-1} - g(x_k))^2 = \sum_{k=1}^n R_k R^{-2} = R^{-1},$$

and thus

$$\widetilde{\mu}_{x_0, x_n} = \mu_{x_0, x_n} + R^{-1}.$$

(Note that we did not exclude the possibility that there was already an edge between x_0 and x_n.)

Remark 2.52 (Resistors in parallel) In electrical network theory one combines resistances in parallel by adding their conductances. In this book we have not treated graphs with multiple edges. If we did allow these, then the process X and Dirichlet form $\mathscr{E}$ would not be changed by replacing two edges between vertices x and y, with conductances μ_1 and μ_2, by a single edge with conductance $\mu_1 + \mu_2$.

Example 2.53 (Y–∇ or Wye–Delta transform; see Figure 2.1) Let Γ be a network, and suppose $\mathbb{V}$ contains a set $A = \{a_1, a_2, a_3, b\}$ such that the only neighbours of b are a_1, a_2, a_3 and

$$\mu_{a_j,b} = \beta_j, \quad j = 1, 2, 3.$$

Assume for simplicity that there is no edge between a_i and a_j for $i \neq j$. Let $D = \{b\}$. If we remove D we obtain a network $\widetilde{\Gamma}$ where there are now edges on the triangle $\{a_1, a_2, a_3\}$. Let

$$\alpha_{ij} = \widetilde{\mu}_{a_i,a_j}, \quad i \neq j.$$

Then

$$\alpha_{ij} = \frac{\beta_i \beta_j}{\beta_1 + \beta_2 + \beta_3}. \tag{2.42}$$

Equivalently if i, j, k are distinct then

$$\beta_k = \frac{\alpha_{12}\alpha_{23} + \alpha_{23}\alpha_{31} + \alpha_{31}\alpha_{12}}{\alpha_{ij}}.$$

To prove (2.42) take $f = \mathbf{1}_{a_1}$, $g = \mathbf{1}_{a_2}$. Then, writing $S = \beta_1 + \beta_2 + \beta_3$,

$$H_B f(b) = \frac{\sum_j \beta_j f(a_j)}{S} = \frac{\beta_1}{S},$$

and similarly $H_B g(b) = \beta_2 / S$. Then

$$\begin{aligned} -\alpha_{01} = \mathscr{E}(H_B f, H_B g) &= \sum_{i=1}^{3} \beta_i (f(a_i) - f(b))(g(a_i) - g(b)) \\ &= -\beta_1 g(b) - \beta_2 f(b) + f(b) g(b) S = -\frac{\beta_1 \beta_2}{S}. \end{aligned}$$

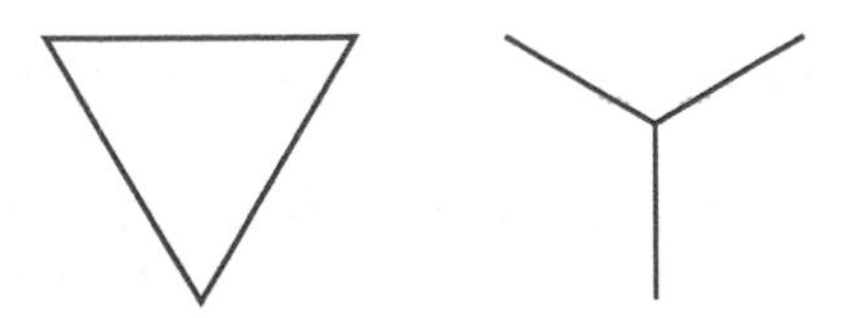

Figure 2.1. The ∇–Y transform.

Interchanging 'new' and 'old' networks, this transformation also allows a triangle in the original network to be replaced by a 'Y' in the new one.

A minor modification allows us to extend this to the case when $a_i \sim a_j$. Use the previous example to first add an extra vertex a_{ij} between a_i and a_j, then apply the Y–∇ transform, and then remove a_{ij}.

We shall see in the following section an application of this transform to the computation of effective resistances on the Sierpinski gasket graph.

Remark Any planar network can be reduced to a single resistor by a sequence of these three transformations – that is, the Y–∇ transform, and the formulae for resistors in series and parallel – see [Tru].

The 'trace operation' on the Dirichlet form $\mathscr{E}$ corresponds to time changing X so that the time spent in D is cut out. We continue to assume that D is finite, and $B = \mathbb{V} - D$. Define the stopping times

$$T_0 = 0,$$
$$T_{k+1} = \min\{n \geq T_k + 1 : X_n \in B\}.$$

(If X hits D for the first time at time n, and then stays in D until the time it hits B again at $n + m$, we would have $T_k = k$ for $k = 0, \ldots, n - 1$, $T_n = n + m$.) Set

$$\widetilde{X}_k = X_{T_k}, \quad k \geq 0.$$

Note that if $X_0 = x \in \partial_i B$ we will have, with positive probability, that $\widetilde{X}_1 = x$, so jumps from a point to itself are possible for $\widetilde{X}$.

Theorem 2.54 *Let $D, B, \widetilde{\mu}$ be as above. The process $\widetilde{X}$ is a SRW on the weighted graph $(B, E_D(B), \widetilde{\mu})$.*

Proof Since D is finite we have $\mathbb{P}^x(T_k < \infty) = 1$ for all k. The (strong) Markov property of X implies that

$$\mathbb{P}^w(\widetilde{X}_{k+1} = y | \widetilde{X}_k = x, X_0, X_1, \ldots, X_{T_k})$$

depends only on x and y. So it is enough to prove that, for $x, y \in B$,

$$\widetilde{\mathscr{P}}(x, y) = \mathbb{P}^x(\widetilde{X}_1 = y | \widetilde{X}_0 = x) = \frac{\widetilde{\mu}_{xy}}{\widetilde{\mu}_x}.$$

However, since $\widetilde{\mu}_x = \mu_x$, this is immediate from (2.40). □

Proposition 2.55 *Let $D \subset \mathbb{V}$ be finite, and let $\widetilde{\Gamma}$ be the graph obtained from $\Gamma = (\mathbb{V}, E)$ by collapsing D^c to a single point. Write $\tilde{R}_{\text{eff}}$ for effective resistance in $\widetilde{\Gamma}$. Then, for $x, y \in D$,*

$$g_D(x, x) + g_D(y, y) - 2g_D(x, y) = \tilde{R}_{\text{eff}}(x, y), \tag{2.43}$$
$$2g_D(x, y) = R_{\text{eff}}(x, D^c) + R_{\text{eff}}(y, D^c) - \tilde{R}_{\text{eff}}(x, y).$$

Proof Note that the Green's function $g_D(x, y)$ is the same in the original network Γ and the collapsed one $\widetilde{\Gamma}$. Since $g_D(x, x) = R_{\text{eff}}(x, D^c) = \tilde{R}_{\text{eff}}(x, D^c)$, the second formula follows easily from the first.

To prove (2.43), write d for the vertex D^c. Let I^x be the current in $\widetilde{\Gamma}$ associated with the potential g_D^x. Then, by Lemma 2.6,

$$\sum_{z \sim d} \widetilde{\mu}_{zd}(g_D^x(z) - g_D^x(d)) = \text{Div}\, I^x(d) = -\,\text{Div}\, I^x(x) = 1.$$

Similarly we have $\text{Div}\, I^y(d) = 1$, and so if $f = g_D^x - g_D^y$

$$\sum_{z \sim d} \widetilde{\mu}_{zd}(f(z) - f(d)) = 0,$$

so that f is harmonic at d. Since f is harmonic at all points $z \in D - \{x, y\}$, the associated flow $I^x - I^y$ is optimal for (VP2), and we have

$$\begin{aligned}\tilde{R}_{\text{eff}}(x, y) &= \widetilde{E}(I^x - I^y, I^x - I^y) = \widetilde{\mathscr{E}}(g_D^x - g_D^y, g_D^x - g_D^y)\\ &= g_D(x, x) + g_D(y, y) - 2g_D(x, y).\end{aligned}$$

□

2.6 Stability under Rough Isometries

We now consider the stability of transience/recurrence under rough isometries (of weighted graphs). First note that if (Γ, μ) has controlled weights (i.e. satisfies (H5)) and $x \sim y$, $y \sim z$, then

$$\mu_{xy} \geq C_2^{-1} \mu_y \geq C_2^{-1} \mu_{yz}.$$

Iterating, it follows that if $d(x, y) = k$ and $x \sim x'$, $y \sim y'$ then

$$\mu_{yy'} \leq C_2^{k+1} \mu_{xx'}. \tag{2.44}$$

Proposition 2.56 *Let (Γ_i, μ_i), $i = 1, 2$ be weighted graphs satisfying (H5), and let $\varphi : \mathbb{V}_1 \to \mathbb{V}_2$ be a rough isometry. There exists $K < \infty$ such that, for $f_2 \in C_0(\mathbb{V}_2)$,*

$$\mathscr{E}_1(f_2 \circ \varphi, f_2 \circ \varphi) \leq K \mathscr{E}_2(f_2, f_2).$$

Remark 2.57 It is easy to see that the converse is false; that is, that one cannot control $\mathcal{E}_2(f_2, f_2)$ in terms of $\mathcal{E}_1(f_2 \circ \varphi, f_2 \circ \varphi)$. Let for example $\Gamma_i = \mathbb{Z}$, $\varphi(n) = 2n$ and $f_2(n) = \mathbf{1}_{\{n=1\}}$. Then $f_2 \circ \varphi = 0$, while f_2 has non-zero energy.

Proof Let C_1, C_2, C_3 be the constants associated with φ in (1.5)–(1.7), and let C_5 be a constant so that (H5) holds with constant C_5 for both Γ_1 and Γ_2. We use the notation $\mu_i(x, y)$ to denote conductances in the graphs Γ_i. Set $f_1 = f_2 \circ \varphi$.

Let $x \sim y$ in $\mathbb{V}_1$, and $x' = \varphi(x)$, $y' = \varphi(y)$. Then

$$0 \le d_2(x', y') \le C_1(1 + C_2).$$

Let $k = \lceil C_1(1 + C_2) \rceil$, so there exists a path $x' = z_0, \dots, z_k = y'$ connecting x' and y'. Then

$$\begin{aligned} |f_1(x) - f_1(y)| &= |f_2(x') - f_2(y')| \\ &\le \sum_{i=1}^{k} |f_2(z_{i-1}) - f_2(z_i)| \\ &\le k^{1/2} \Big(\sum_{i=1}^{k} |f_2(z_{i-1}) - f_2(z_i)|^2 \Big)^{1/2}. \end{aligned}$$

Using (2.44) we have, for $1 \le i \le k$,

$$\mu_1(x, y) \le \mu_1(x) \le C_3 \mu_2(x') \le C_3 C_5 \mu_2(z_0, z_1) \le C_3 C_5^i \mu_2(z_{i-1}, z_i).$$

Hence,

$$|f_1(x) - f_1(y)|^2 \mu_1(x, y) \le (k C_3 C_5^k) \sum_{i=1}^{k} |f_2(z_{i-1}) - f_2(z_i)|^2 \mu_2(z_{i-1}, z_i). \tag{2.45}$$

Summing over $x \sim y \in \mathbb{V}_1$, the left hand side gives $\mathcal{E}_1(f_1, f_1)$, while the right hand side is less than $k C_3 C_5^k M \mathcal{E}_2(f_2, f_2)$, where M is the maximum number of pairs $x \sim y \in \mathbb{V}_1$ such that the same edge $w \sim w'$ in $\mathbb{V}_2$ occurs in a sum of the form (2.45).

It remains to prove that $M < \infty$. Fix an edge $\{w, w'\}$ in Γ_2. If $\{w, w'\}$ is on a path connecting $\varphi(x')$ and $\varphi(y')$, where $x' \sim y'$ are in $\mathbb{V}_1$, then $d_2(\varphi(x'), w) \le k$. So it remains to bound $|A|$, where $A = \varphi^{-1}(B_2(w, k))$. Let $z_1, z_2 \in A$. Then by (1.5)

$$C_1^{-1}(d_1(z_1, z_2) - C_2) \le d_2(\varphi(z_1), \varphi(z_2)) \le 2k,$$

so $d(z_1, z_2) \le c_1 = c_1(C_1, C_2)$. Hence $|A| \le c_2 = c_2(C_1, C_2)$, and $M \le c_2^2$. $\square$

Theorem 2.58 *Transience and recurrence of weighted graphs is stable under rough isometries.*

Proof Let (Γ_i, μ_i), $i = 1, 2$, be as in Proposition 2.56, and let $\varphi : \mathbb{V}_1 \to \mathbb{V}_2$ be a rough isometry. Suppose that (Γ_1, μ_1) is transient, and let $f_2 \in C_0(\mathbb{V}_2)$, and $f_1 = f_2 \circ \varphi$. Then if $x_1 \in \mathbb{V}_1$, $x_2 = \varphi(x_1)$, using Theorem 2.42,

$$|f(x_1)|^2 = |f_2(x_2)|^2 \le C\mathscr{E}_2(f_2, f_2) \le KC\mathscr{E}_1(f_1, f_1).$$

Thus, by Theorem 2.42, (Γ_2, μ_2) is transient. □

Proposition 2.59 *Let $\mathscr{G}$ be a finitely generated infinite group, and let Λ_1, Λ_2 be two sets of generators. Let Γ_1, Γ_2 be the associated Cayley graphs. Then Γ_1 and Γ_2 are roughly isometric.*

Proof Let $\varphi(x) = x$ for $x \in \mathscr{G}$. Let d_i be the metric for the Cayley graph of $(\mathscr{G}, \Lambda_i)$. The property (1.6) of rough isometries is immediate, so it remains to prove (1.5). For this it is enough to prove that there exists $C_{21} \ge 1$ such that

$$d_2(x, y) \le C_{21} d_1(x, y), \quad x, y \in \mathscr{G}, \tag{2.46}$$

since interchanging d_1 and d_2 in the proof we also have that there exists $C_{12} \ge 1$ such that $d_1(x, y) \le C_{12} d_2(x, y)$ for all x, y. Taking $C_1 = C_{12} \vee C_{21}$ and $C_2 = 0$ then gives (1.5).

To prove (2.46) recall that given a set of generators Λ we set $\Lambda^* = \{g, g^{-1}, g \in \Lambda\}$. Let

$$\Lambda_2(k) = \left\{ \prod_{j=1}^{k'} x_j : x_j \in \Lambda_2^*, k' \le k \right\}$$

be the set of all products of k or fewer elements of Λ_2^*. It is easy to check that $d_2(x, y) \le k$ if and only if $x^{-1}y \in \Lambda_2(k)$, and that if $w_1 \in \Lambda_2(k_1)$ and $w_2 \in \Lambda_2(k_2)$ then $w_1 w_2 \in \Lambda_2(k_1 + k_2)$. Since the group generated by Λ_2^* is the whole of $\mathscr{G}$, for each $g \in \Lambda_1^*$ there exists $k(g)$ such that $g \in \Lambda_2(k(g))$. Set $C_{21} = \max\{k(g), g \in \Lambda_1\}$.

Now let $x, y \in \mathscr{G}$ with $d_1(x, y) = m$. Then $x^{-1}y = z_1 \dots z_m$, where $z_j \in \Lambda_1^*$, and therefore in $\Lambda_2(C_{21})$. By the above, $x^{-1}y \in \Lambda_2(mc_{21})$, and thus $d_2(x, y) \le C_{21} m$, proving (2.46). □

Corollary 2.60 *The type of a Cayley graph does not depend on the choice of the (finite) set of generators.*

Proof This is immediate from Proposition 2.59 and Theorem 2.58. □

Remark A group $\mathcal{G}$ has recurrent Cayley graphs if and only if it is a finite extension of $\mathbb{Z}$ or $\mathbb{Z}^2$. The 'if' part is immediate from Corollary 2.60. The 'only if' part requires a deep theorem on the structure of groups due to Gromov [Grom1], and is beyond the scope of this book.

2.7 Hitting Times and Resistance

In this section we show how the resistance between points and sets gives information on hitting times of a random walk. We begin by extending Theorem 2.8 to an infinite graph.

Lemma 2.61 *Let $A \subset \mathbb{V}$ and $x \in \mathbb{V} - A$. Then*

$$\frac{1}{\mu_x R_{\text{eff}}(x, A)} - \mathbb{P}^x(T_x^+ = T_A = \infty) \le \mathbb{P}^x(T_x^+ > T_A) \le \frac{1}{\mu_x R_{\text{eff}}(x, A)}. \tag{2.47}$$

Proof Recall that by $\mathbb{P}^x(T_x^+ > T_A)$ we mean $\mathbb{P}^x(T_x^+ > T_A, T_A < \infty)$. Let $V_n \uparrow\uparrow \mathbb{V}$ with $x \in V_1$. Let $X^{(n)}$ be the random walk on V_n, and write $T_\cdot^{(n)}$ for hitting times of $X^{(n)}$. We can define X and $X^{(n)}$ on the same probability space so that they agree until $S_n = T_{\partial_i V_n} = T_{\partial_i V_n}^{(n)}$. Thus

$$\{T_x^+ > T_A, T_A < S_n\} = \{T_x^{(n,+)} > T_A^{(n)}, T_A^{(n)} < S_n\}.$$

Since $\{T_A < \infty\} = \cup_{n=1}^\infty \{T_A < S_n\}$, we have

$$\begin{aligned}
\mathbb{P}^x(T_x^+ > T_A) &= \lim_n \mathbb{P}^x(T_x^{(n,+)} > T_A^{(n)}, T_A^{(n)} < S_n) \\
&= \lim_n \Big(\mathbb{P}^x(T_x^{(n,+)} > T_A^{(n)}) - \mathbb{P}^x(T_x^{(n,+)} > T_A^{(n)}, T_A^{(n)} \ge S_n)\Big).
\end{aligned}$$

Dropping the second term, we have

$$\mathbb{P}^x(T_x^+ > T_A) \le \lim_n \frac{1}{\mu_x R_{\text{eff}}(x, A \cap V_n : V_n)},$$

which gives the upper bound in (2.47) by Proposition 2.18.

For the lower bound, we have

$$\mathbb{P}^x(T_x^{(n,+)} > T_A^{(n)}, T_A^{(n)} \ge S_n) \le \mathbb{P}^x(T_A \ge S_n, T_x^+ \ge S_n),$$

and these probabilities have limit $\mathbb{P}^x(T_A = T_x^+ = \infty)$. □

Lemma 2.62 *Let $x \notin A \cup B$.*

(a) If $\mathbb{P}^x(T_{A\cup B} < \infty) = 1$ then

$$\mathbb{P}^x(T_A < T_B) \le \frac{R_{\text{eff}}(x, A\cup B)}{R_{\text{eff}}(x, A)} \le \frac{R_{\text{eff}}(x, B)}{R_{\text{eff}}(x, A)}.$$

(b) If $\mathbb{P}^x(T_A < \infty) = 1$ then

$$\mathbb{P}^x(T_A < T_B) \ge \frac{R_{\text{eff}}(x, A\cup B)}{R_{\text{eff}}(x, A)} - \frac{R_{\text{eff}}(x, A\cup B)}{R_{\text{eff}}(x, B)}.$$

Proof We have

$$\mathbb{P}^x(T_A < T_B) = \mathbb{P}^x(T_A < T_B, T_{A\cup B} < T_x^+) + \mathbb{P}^x(T_A < T_B, T_{A\cup B} > T_x^+).$$

By the Markov property at T_x^+ the second term is

$$\mathbb{P}^x(T_A < T_B)\mathbb{P}^x(T_{A\cup B} > T_x^+),$$

so

$$\mathbb{P}^x(T_A < T_B) = \frac{\mathbb{P}^x(T_A < T_B, T_{A\cup B} < T_x^+)}{\mathbb{P}^x(T_{A\cup B} < T_x^+)} \le \frac{\mathbb{P}^x(T_A < T_x^+)}{\mathbb{P}^x(T_{A\cup B} < T_x^+)},$$

and using Lemma 2.61 gives the upper bound.

For the lower bound,

$$\begin{aligned}\mathbb{P}^x(T_A < T_B, T_{A\cup B} < T_x^+) &\ge \mathbb{P}^x(T_A < T_x^+, T_B > T_x^+)\\ &\ge \mathbb{P}^x(T_A < T_x^+) - \mathbb{P}^x(T_B < T_x^+),\end{aligned}$$

and again using Lemma 2.61 completes the proof. □

The following result, called the *commute time identity*, was discovered in 1989 – see [CRR]. Other proofs can be found in [Tet, BL].

Theorem 2.63 *Let (Γ, μ) be a weighted graph with $\mu(\mathbb{V}) < \infty$. Then*

$$\mathbb{E}^{x_0} T_{x_1} + \mathbb{E}^{x_1} T_{x_0} = R_{\text{eff}}(x_0, x_1)\mu(\mathbb{V}).$$

Proof Let $\varphi(x) = \mathbb{P}^x(T_{x_1} < T_{x_0})$, $I = \nabla\varphi$, and write $R = R_{\text{eff}}(x_0, x_1)$. Then

$$\begin{aligned}\Delta\varphi(x) &= 0, \quad x \in \mathbb{V} - \{x_0, x_1\},\\ \operatorname{div} I(x_1) &= \mu_{x_1}\Delta\varphi(x_1) = -R^{-1}.\end{aligned}$$

Let $f_0(x) = \mathbb{E}^x T_{x_0}$. Then $f_0(x_0) = 0$ and if $x \ne x_0$ then

$$\Delta f_0(x) = \mathbb{E}^x f_0(X_1) - f_0(x) = -1.$$

By the self-adjointness of Δ we have

$$\langle \varphi, -\Delta f_0 \rangle = \langle -\Delta\varphi, f_0 \rangle. \tag{2.48}$$

We calculate each side of (2.48). The left side is

$$\sum_{x \neq x_0} \varphi(x)\mu_x + \varphi(x_0)(-\Delta\varphi(x_0))\mu_{x_0} = \sum_x \varphi(x)\mu_x,$$

since $\varphi(x_0) = 0$. The right side is

$$\Delta\varphi(x_0)(-f_0(x_0))\mu_{x_0} + \Delta\varphi(x_1)(-f_0(x_1))\mu_{x_1} = R^{-1} f_0(x_1).$$

Thus

$$\mathbb{E}^{x_1} T_{x_0} = R_{\text{eff}}(x_0, x_1) \sum_x \varphi(x)\mu_x. \tag{2.49}$$

If $\varphi'(x) = \mathbb{P}^x(T_{x_1} > T_{x_0})$ then $\varphi(x) + \varphi'(x) = 1$ and

$$\mathbb{E}^{x_0} T_{x_1} = R_{\text{eff}}(x_0, x_1) \sum_x \varphi'(x)\mu_x.$$

Adding these two equations completes the proof of the theorem. □

Remark (1) In an unweighted graph $\mu(\mathbb{V}) = 2|E|$.
(2) See Theorem 7.10 for the commute time between two sets.

Theorem 2.64 (See [Ki], Th. 1.6) *Let Γ be a graph (finite or infinite). Then R_{eff} is a metric on $\mathbb{V}$.*

Proof Symmetry is clear, so it remains to prove the triangle inequality. Let $x, y, z \in \mathbb{V}$. If $\mu(\mathbb{V}) < \infty$ then by the Markov property

$$\mathbb{E}^x T_y + \mathbb{E}^y T_z \geq \mathbb{E}^x T_z,$$
$$\mathbb{E}^z T_y + \mathbb{E}^y T_x \geq \mathbb{E}^z T_x.$$

Adding, one deduces that $R_{\text{eff}}(x, y) + R_{\text{eff}}(y, z) \geq R_{\text{eff}}(x, z)$. If $\mathbb{V}$ is infinite, let $\mathbb{V}_n \uparrow\uparrow \mathbb{V}$, with $\{x, y, z\} \subset \mathbb{V}_1$. Then for each n

$$R_{\text{eff}}(x, y; \mathbb{V}_n) + R_{\text{eff}}(y, z; \mathbb{V}_n) \leq R_{\text{eff}}(x, z; \mathbb{V}_n). \tag{2.50}$$

Taking the limit and using Proposition 2.18 completes the proof. □

Remark 2.65 An alternative proof of (2.50) is to use electrical equivalence. Using the Y–∇ relation, the inequality is clear for a graph with three points. It then follows, for a general finite graph, by considering the network on $B = \{x, y, z\}$ which is equivalent to the original network.

The following corollary was proved in [T1], shortly before the discovery of Theorem 2.63.

Corollary 2.66 *Let* (Γ, μ) *be a graph, and* $A \subset \mathbb{V}$. *Then*

$$\mathbb{E}^x \tau_A \le R_{\text{eff}}(x_0, A^c)\mu(A). \tag{2.51}$$

Proof If $\mu(A) = \infty$ there is nothing to prove. If $\mu(A) < \infty$ this follows from (2.49) on collapsing A^c to a single point. □

Example Suppose (Γ, μ) is a transient weighted graph, $\{x_0, x_1\}$ is an edge in Γ, and there is a finite subgraph $\mathbb{V}_1 = \{x_1, \dots, x_n\} \subset \mathbb{V}$ 'hanging' from x_1. More precisely, if $y \in \mathbb{V}_1$ then every path from y to x_0 passes through x_1. Then we can ask how much the existence of $\mathbb{V}_1$ 'delays' the SRW X, that is, what is $E^{x_1} T_{x_0}$?

Let $\mathbb{V}_0 = \{x_0, \dots, x_n\}$, and work with the subgraph generated by $\mathbb{V}_0$, except we eliminate any edges between x_0 and itself. Write μ^0 for the measure on $\mathbb{V}_0$ given by (1.1), and for clarity denote the hitting times for the SRW on $\mathbb{V}_0$ by T_y^0. Then, by Theorem 2.63,

$$E^{x_0} T^0_{x_1} + E^{x_1} T^0_{x_0} = R_{\text{eff}}(x_0, x_1)\mu^0(\mathbb{V}_0).$$

Since $E^{x_0} T^0_{x_1} = 1$, $R_{\text{eff}}(x_0, x_1) = \mu^{-1}_{x_0,x_1}$, $\mu_0(\mathbb{V}_0) = \mu(\mathbb{V}_1) + \mu^{-1}_{x_0,x_1}$, and $E^{x_1} T^0_{x_0} = E^{x_1} T_{x_0}$, we obtain

$$E^{x_1} T_{x_0} = \frac{\mu(\mathbb{V}_1)}{\mu_{x_0,x_1}}.$$

2.8 Examples

Example 2.67 Consider the graph Γ obtained by identifying the origins of $\mathbb{Z}_+$ and $\mathbb{Z}^3$. Since $\mathbb{Z}^3$ is transient, Γ is transient, so by Theorem 2.37 the identity function 1 is not in H_0^2. Let f_n be zero on $\mathbb{Z}^3 - \{0\}$ and define f_n on $\mathbb{Z}_+$ by

$$f_n(k) = \begin{cases} 1 - k/n & \text{if } k \le n, \\ 0 & \text{if } k > n. \end{cases}$$

Let $f = \mathbf{1}_{\mathbb{Z}_+}$ be the pointwise limit of the f_n. Then $f_n \in C_0$ and $\mathscr{E}(f - f_n, f - f_n) = 1/n$, so that $f_n \to f$ in H^2 and thus $f \in H_0^2$. So 1 is in H^2 but not H_0^2, while f is in H_0^2 but not L^2, showing that in general all three spaces are distinct.

In addition, this graph satisfies the Liouville property but not the strong Liouville property. To show the latter, let

$$h(x) = \mathbb{P}^x(T_0 < \infty), \quad x \in \mathbb{Z}^3.$$

Then h is harmonic in $\mathbb{Z}^3 - \{0\}$. Let $a > 0$ and define h on $\mathbb{Z}_+$ by $h(n) = 1 + an$; clearly h is also harmonic in $\mathbb{Z}_+ - \{0\}$. Let $p_0 > 0$ be the probability that a SRW on $\mathbb{Z}^3$ started at any neighbour of 0 does not return to 0. Then

$$\begin{aligned}\Delta h(0) &= (2d+1)^{-1}\Big([(1+a)-1] + 2d[(1-p_0)-1]\Big)\\ &= (2d+1)^{-1}(a - 2dp_0).\end{aligned}$$

So if $a = 2dp_0$ then h is a positive non-constant harmonic function on Γ.

To prove the Liouville property, let h be a bounded harmonic function on Γ; we can assume $h(0) = 1$. Then h must be of the form $h(n) = 1 + bn$ on $\mathbb{Z}_+$, for some $b \in \mathbb{R}$. But since h is bounded, $b = 0$, and h is constant on $\mathbb{Z}_+$. As h is harmonic at 0, writing $A = \{x \in \mathbb{Z}^3 : x \sim 0\}$,

$$1 = h(0) = (2d+1)^{-1}\Big(1 + \sum_{x \in A} h(x)\Big),$$

and therefore

$$h(0) = (2d)^{-1} \sum_{x \in A} h(x).$$

So the restriction of h to $\mathbb{Z}^3$ is harmonic in $\mathbb{Z}^3$, and by the Liouville property for $\mathbb{Z}^3$ (Corollary 1.52) it follows that h is constant.

Example 2.68 (Binary tree with weights) Recall from Examples 1.4(3) the definition of the rooted binary tree. Let $(r_n, n \geq 0)$ be strictly positive, and for an edge $\{x, y\}$ with $x \in \mathbb{B}_n$, $y \in \mathbb{B} + n + 1$ set

$$\mu_{xy} = \frac{1}{r_n}.$$

Thus $R_{\text{eff}}(\mathbb{B}_n, \mathbb{B}_{n+1}) = 2^{-n} r_n$, and the weighted tree is transient if and only if

$$R_{\text{eff}}(o, \infty) = \lim_n R_{\text{eff}}(o, \mathbb{B}_n) = \sum_{k=0}^{\infty} 2^{-k} r_k < \infty.$$

Example 2.69 (Spherically symmetric trees) Again we start from the rooted binary tree and a sequence $r_n, n \geq 0$ of positive integers. The tree is obtained by replacing each edge $\{x, y\}$ with $x \in \mathbb{B}_n$, $y \in \mathbb{B}_{n+1}$ by a sequence of r_k edges. Again we have that the tree is transient if and only if $\sum 2^{-k} r_k < \infty$.

2.9 The Sierpinski Gasket Graph

The 'true' Sierpinski gasket is a compact fractal subset of $\mathbb{R}^2$, which can be constructed by a procedure analogous to that for the Cantor set. Starting with the (closed) unit triangle K_0 in $\mathbb{R}^2$, one removes the open downward facing triangle whose vertices are the midpoints of the sides of K_0. The resulting set, K_1, consists of three closed triangles of side 2^{-1}. Repeating this procedure, after n steps one obtains a set K_n consisting of 3^n triangles each of side 3^{-n} (see Figure 2.2). The Sierpinski gasket is the set $K_\infty = \cap_n K_n$; it is a compact connected subset of $\mathbb{R}^2$ with Hausdorff dimension $\log 3/\log 2$ – see [Fa].

A more formal description of the sets K_n can be given using an iterated function scheme – see p. 113 of [Fa]. Let $V_0 = \{(0,0), (1,0), (1/2, \sqrt{3}/2)\} = \{a_1, a_2, a_3\}$ be the vertices of the unit triangle in $\mathbb{R}^2$, and let K_0 be the closed convex hull of V_0. Define the functions $\psi_i : \mathbb{R}^2 \to \mathbb{R}^2$ by

$$\psi_i(x) = \tfrac{1}{2}(x - a_i).$$

Given a compact set $K \subset \mathbb{R}$ let

$$\Psi(K) = \cup_{i=1}^3 \psi_i(K).$$

It is easy to check that setting $K_n = \Psi(K_{n-1})$ for $n \geq 1$ defines a decreasing sequence of compact sets with the properties described above.

Using this construction as a guide, we can define an infinite graph with a similar structure. Note that $K_n = \Psi^n(V_0)$ consists of 3^n triangles each of side 2^{-n}. We define the set $\mathcal{T}_n$ to be the set of these triangles, that is

Figure 2.2. The set K_4.

$$\mathscr{T}_n = \{\psi_{i_1} \circ \cdots \circ \psi_{i_n}(V_0) : (i_1, \dots, i_n) \in \{1, 2, 3\}^n\}.$$

Let $U_n = \Psi^n(V_0)$ be the set of vertices of these triangles, and for $x, y \in U_n$ let $x \overset{n}{\sim} y$ if there exists a triangle $T \in \mathscr{T}_n$ with $x, y \in T$. Define a graph $\mathscr{G}_n = (V_n, E_n)$ by

$$V_n = 2^n U_n = \{2^n x : x \in U_n\}, \quad E_n = \{\{2^n x, 2^n y\} : x \overset{n}{\sim} y, \ x, y \in U_n\}.$$

The vertices in V_n are all contained in $2^n K_0$, and V_{n+1} consists of V_n together with two copies of V_n, which meet at the vertices $2^n a_1$ and $2^n a_2$. We define $V = \cup_n V_n$, and for $x, y \in V$ we let $\{x, y\} \in E$ if $\{x, y\} \in E_m$ for all sufficiently large m. The graph $\Gamma_{SG} = (V, E)$ is the *Sierpinski gasket graph* (Figure 2.3). We call a subset A of V a *triangle of side* 2^n if it is isometric to V_n. The *extreme points of* A are the three vertices in A such that A is contained in the convex hull of these vertices. We say that two triangles of side 2^n are *adjacent* if they share an extreme point. Let D_n be the set of extreme points of vertices of triangles of side 2^n.

The following lemma gives some basic properties of Γ_{SG}. We write $B(x, r)$ for balls in Γ_{SG} in the usual graph metric $d(x, y)$, and set $\alpha = \log 3/\log 2$. Write $B_E(x, r) = \{y \in \mathbb{R}^2 : |x - y| \le r\}$ for (closed) balls in the usual Euclidean metric in $\mathbb{R}^2$.

Lemma 2.70 *(a) If A is a triangle of side 2^n then $|A| = \frac{1}{2}(3^{n+1} + 3)$.*

(b) If $x \in V$ then $B_E(x, 2^{n-2})$ intersects at most two triangles of side 2^n, and these triangles are adjacent.

Figure 2.3. Part of the Sierpinski gasket graph Γ_{SG}.

(c) *We have, for* $x, y \in V$,

$$|x - y| \le d(x, y) \le c_1 |x - y|.$$

(d) *If* $x \in V$ *and* $r \ge 1$ *then*

$$|B(x, r)| \asymp \mu(B(x, r)) \asymp r^{\alpha}. \tag{2.52}$$

Proof (a) is easy on noting that A contains 3^n edges and that all but three of the vertices in A are in two triangles of side 1.
(b) Let $x \in V$, and let H be a triangle of side 2^{n-1} containing x. Write w_i, $i = 1, 2, 3$ for the extreme points of H. Then the set $\cup_{i=1}^{3} B_E(w_i, 2^{n-2})$ intersects at most two triangles of side 2^n, and it follows that the same holds for $B_E(x, 2^{n-2})$.
(c) Since Γ_{SG} is embedded in the plane and all edges are of length 1, it is clear that $d(x, y) \ge |x - y|$. It is easy to prove by induction that if A is a triangle of side 2^n and z is an extreme point of A then $d(x, z) \le 2^n$ for all $x \in A$. Hence if x, y are in two adjacent triangles of side 2^n and w is the common extreme point of the two triangles, then $d(x, y) \le 2.2^n$; using (b), (c) then follows.
(d) If $x \in V$ then $\mu_x = 4$ unless $x = 0$, when we have $\mu_0 = 2$. So $|B(x, r)| \asymp \mu(B(x, r))$. If $r \ge 1$ let m be the smallest integer such that $r \le 2^m$. Then $r > 2^{m-1}$, $B(x, r)$ is contained in two triangles of side 2^{m+1}, and so $|B(x, r)| \le 3^{m+2} + 3 \le c r^{\alpha}$.

For the lower bound, x is contained in at least one triangle A of side 2^{m-2}; let z be an extreme point of this triangle. Then for $y \in A$ we have $d(x, y) \le d(x, z) + d(y, z) \le 2^{m-1} \le r$, so that $A \subset B(x, r)$ and thus $|B(x, r)| \ge c' r^{\alpha}$. □

Lemma 2.71 *The SRW on* Γ_{SG} *is recurrent.*

Proof As Γ_{SG} is a subgraph of the triangular lattice in $\mathbb{R}^2$, this is immediate from Corollary 2.39. It is also easy to give a direct proof using the Nash–Williams criterion, since for each n we have $|\partial V_n| = 4$. □

Theorem 2.72 *Let* $\beta = \log 5 / \log 2$. *For* $x, y \in V$

$$c_1 d(x, y)^{\beta - \alpha} \le R_{\text{eff}}(x, y) \le c_2 d(x, y)^{\beta - \alpha}. \tag{2.53}$$

Proof It is sufficient to prove this in the finite graph $\mathcal{G}_m$ for large enough m.

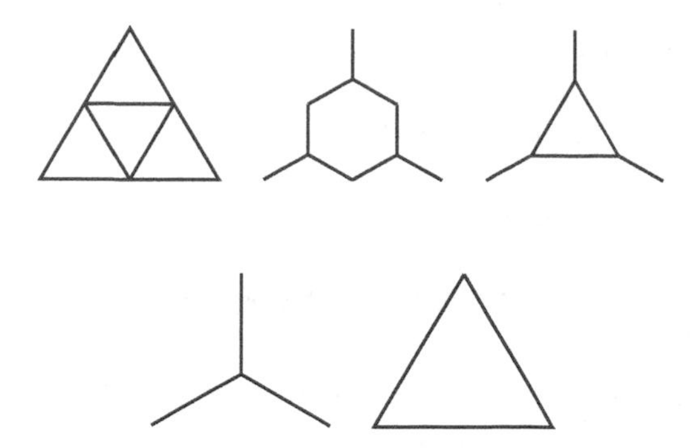

Figure 2.4. Sequence of Y–∇ transforms for the Sierpinski gasket.

If $d(x, y) = 1$ then the unit flow from x to y has energy 1, so $R_{\text{eff}}(x, y) \le 1$. If $f = \mathbf{1}_y$ then $R_{\text{eff}}(x, y)^{-1} \le \mathscr{E}(f, f) \le 4$, so we deduce that

$$\tfrac{1}{4} \le R_{\text{eff}}(x, y) \le 1.$$

Now consider a triangle A of side 2 in Γ_{SG}, with vertices z_1, z_2, z_3. Then by a sequence of Y–∇ transformations (see Figure 2.4) we can remove the vertices inside A, and replace the edges $\{z_i, z_j\}$ by edges which (a short calculation shows) have conductance $5/3$.

Repeating this operation on every triangle of side 2, we can replace $\mathscr{G}_m$ by $\mathscr{G}_{m-1}$, but with every edge having conductance $5/3$. Continuing in this way, we can replace $\mathscr{G}_m$ by $\mathscr{G}_{m-k}$ with edge conductances of $(5/3)^k$. Thus if x, y are distinct extreme points of a triangle with side 2^k, then by (2.9) we have

$$\tfrac{1}{4}(5/3)^k \le R_{\text{eff}}(x, y) \le (5/3)^k. \tag{2.54}$$

Let A_n be a triangle of side n, with extreme points w_1, w_2, w_3. This calculation also shows that the unique harmonic function in A_n with boundary values given by $h_n(w_1) = 1$, $h_n(w_2) = h_n(w_3) = 0$ satisfies $\mathscr{E}_{A_n}(h_n, h_n) = 2(3/5)^n$.

Now let A_k be a triangle of side 2^k, and $y \in A_k$. Choose z_k to be an extreme point of A_k closest to y. Then we can find a sequence of triangles A_j, $0 \le j \le k$ and extreme points z_j of A_j such that $z_0 = y$ and for each j both z_j and z_{j-1} are extreme points of A_{j-1}. (A_j is a triangle of side 2^j.) Since R_{eff} is a metric it follows that

$$R_{\text{eff}}(y, z_k) \le \sum_{j=1}^{k} R_{\text{eff}}(z_{j-1}, z_j) \le \sum_{j=1}^{k} (5/3)^{j-1} \le (3/2)(5/3)^k;$$

the upper bound in (2.53) then follows.

For the lower bound, suppose that $d(x, y) \geq 2^k$. Then there exist disjoint triangles of side 2^{k-2}, A_x, and A_y say, with $x \in A_x$ and $y \in A_y$. There are either two or three triangles of side 2^{k-2} adjacent to A_x; call the union of these triangles A. Consider the function f which is 1 on A_x, harmonic in A^o, and zero on $\partial(A \cup A_x)$. As A_y is disjoint from A^o, we have $f(x) = 1$, $f(y) = 0$. The function f has energy $2(3/5)^{k-2}$ in each of the triangles of side 2^{k-2} in A, so we have $\mathscr{E}(f, f) \leq 6(3/5)^{k-2}$. Hence

$$C_{\text{eff}}(x, y) \leq \mathscr{E}(f, f) \leq c(3/5)^k,$$

proving the lower bound in (2.53). □

We will see in Chapter 6 that these calculations for the resistance metric on Γ_{SG} are the input needed to be able to obtain bounds on the heat kernel for the SRW.

3

Isoperimetric Inequalities and Applications

3.1 Isoperimetric Inequalities

The study of the close connections between random walks and isoperimetric inequalities was opened up by Varopoulos – see in particular [V1, V2]. We continue to assume that (Γ, μ) is a locally finite connected weighted graph.

Definition 3.1 For $A, B \subset \mathbb{V}$ set

$$\mu_E(A; B) = \sum_{x \in A} \sum_{y \in B} \mu_{xy}.$$

Let $\psi : \mathbb{R}_+ \to \mathbb{R}_+$ be increasing. We say that (Γ, μ) satisfies the ψ-*isoperimetric inequality* (I_ψ) if there exists $C_0 < \infty$ such that

$$\frac{\mu_E(A; \mathbb{V} - A)}{\psi(\mu(A))} \geq C_0^{-1} \quad \text{for every finite non-empty } A \subset \mathbb{V}. \qquad (I_\psi)$$

For $\alpha \in [1, \infty)$ we write (I_α) for (I_ψ) with $\psi(t) = t^{1-1/\alpha}$, and (I_∞) for (I_ψ) with $\psi(t) = t$. (I_∞) is also called the *strong isoperimetric inequality*.

Examples 3.2 (1) The Euclidean lattice $\mathbb{Z}^d$ satisfies (I_d).
(2) The binary tree satisfies (I_∞) with $C_0 = 3$.
(3) The Sierpinski gasket graph does not satisfy (I_α) for any $\alpha > 1$.
(4) Any infinite connected graph (with natural weights) satisfies (I_1).
(5) If (Γ, μ) satisfies $(I_{\alpha+\delta})$ then it satisfies (I_α).

(1) If $A \subset \mathbb{Z}^d$ is finite, construct a subset $D \subset \mathbb{R}^d$ by replacing each $x \in A$ with a unit cube. Then, writing Λ_k for the k-dimensional Lebesgue measure, $\mu(A) = 2d\Lambda_d(D)$, and $\mu_E(A; A^c) = \Lambda_{d-1}(\partial D)$; thus the isoperimetric inequality for $\mathbb{Z}^d$ follows from that for $\mathbb{R}^d$.

However, while intuitive, the isoperimetric inequality for $\mathbb{R}^d$ is not elementary. There is a very extensive literature on this topic, much of it related to determining the optimal constant and proving that the sets for which equality is attained are balls. See [Ga] for an easy proof via the Brunn–Minkowski inequality, and also the same article for a (fairly) quick proof of the Brunn–Minkowski inequality.

Alternatively, see Theorem 3.26 in Section 3.3 for a proof which does not give the optimal constant.

(2) To prove (I_∞) for the binary tree, it is enough to consider connected sets A. The inequality is then easy by induction on $|A|$; each point added increases $\mu(A)$ by 3 and $\mu_E(A; A^c)$ by at least 1.

(3) If A_n is the triangle of side 2^n containing the origin, then $\mu(A_n) = 4 + 2.3^n$, while $\mu_E(A_n; A_n^c) = 4$. So $\mu_E(A_n; A_n^c)/\mu(A_n)^{1-1/\alpha} \leq c3^{-n(1-1/\alpha)}$.

(4) If A is a finite set then $\mu_E(A; \mathbb{V} - A) \geq 1$, and so (I_ψ) with $\psi(t) = 1$ and $C_0 = 1$ holds.

(5) By Lemma 3.3(a) we have $\mu(A) \geq C_0^{-(\alpha+\delta)}$ for all (non-empty) subsets A. Hence $\mu(A)^{-1/(\alpha+\delta)} \geq C_0^{-\delta/\alpha} \mu(A)^{-1/\alpha}$, and (I_α) follows.

The condition (I_α) gives a lower bound sizes of balls.

Lemma 3.3 *Let (Γ, μ) satisfy (I_α).*

(a) For any $x \in \mathbb{V}$,

$$\mu_x \geq C_0^{-\alpha}. \tag{3.1}$$

(b) If $\alpha \in [1, \infty)$ then

$$\mu(B(x,n)) \geq c_1(\alpha) n^\alpha, \quad n \geq 1.$$

(c) If $\alpha = \infty$ then $\mu(B(x,n)) \geq (1 + C_0^{-1})^n \mu_x$, so (Γ, μ) has (at least) exponential volume growth.

Proof (a) Taking $A = \{x\}$ we have $\mu(A) = \mu_E(A; A^c) = \mu_x$, and (3.1) is then immediate.

(b), (c) Let $B_n = B(x_0, n)$, and $b_n = \mu(B_n)$. Then

$$\mu(\partial B_n) = \sum_{x \in \partial B_n} \mu_x \geq \sum_{x \in \partial B_n} \sum_{y \sim x, y \in B_n} \mu_{xy} = \mu_E(B_n; B_n^c).$$

So,

$$\begin{aligned} b_{n+1} = \mu(B_{n+1}) &= \mu(B_n) + \mu(\partial B_n) \\ &\geq \mu(B_n) + C_0^{-1}\mu(B_n)^{1-1/\alpha} = b_n(1 + C_0^{-1}b_n^{-1/\alpha}). \end{aligned}$$

If $\alpha = \infty$ then it follows immediately that $b_n \geq (1 + C_0^{-1})^n b_0$.

If $\alpha \in [1, \infty)$ then we have $1 + (2^\alpha - 1)t \geq (1 + t)^\alpha$ for $t \in [0, 1]$. Choose $A \in (0, C_0^{-\alpha})$ small enough so that $\frac{1}{2}C_0^{-1}A^{-1/\alpha} \geq 2^\alpha - 1$. We have $b_1 \geq C_0^{-\alpha} \geq A$. We now prove (b) with $c(\alpha) = A$ by induction. Suppose that $b_n \geq An^\alpha$. If $b_n \geq A(n + 1)^\alpha$ then $b_{n+1} \geq b_n \geq A(n + 1)^\alpha$ and we are done. If not then $b_n \leq A(n + 1)^\alpha \leq 2^\alpha An^\alpha$, and hence

$$b_{n+1} \geq An^\alpha(1 + \tfrac{1}{2}C_0^{-1}A^{-1/\alpha}n^{-1}) \geq An^\alpha(1 + n^{-1})^\alpha = A(n + 1)^\alpha.$$

□

Theorem 3.4 *The property (I_α) is weight stable, and stable under rough isometries.*

Proof Stability under bounded perturbation of weights is immediate from the definition.

To prove stability under rough isometries, let (Γ_i, μ_i) be weighted graphs, and let $\varphi : \mathbb{V}_1 \to \mathbb{V}_2$ be a rough isometry. Suppose that (Γ_1, μ_1) satisfies (I_α). If $A_2 \subset \mathbb{V}_2$ let $f_2 = \mathbf{1}_{A_2}$, $A_1 = \varphi^{-1}(A_2)$, and $f_1 = \mathbf{1}_{A_1} = f_2 \circ \varphi$. Then $\mathscr{E}_i(f_i, f_i) = \mu_E(A_i; \mathbb{V}_i - A_i)$ for $i = 1, 2$. Since φ is a rough isometry of weighted graphs, $\mu_1(A_1) \asymp \mu_2(A_2)$, while Proposition 2.56 gives that

$$\mu_E(A_2; \mathbb{V}_2 - A_2) \geq K^{-1}\mu_E(A_1; \mathbb{V}_1 - A_1).$$

Combining these inequalities we have that (Γ_2, μ_2) also satisfies (I_α). □

Exercise 3.5 Let Γ_i, $i = 1, 2$, be graphs, with natural weights, which satisfy (I_{α_i}). Then the join of Γ_1 and Γ_2 satisfies $(I_{\alpha_1 \wedge \alpha_2})$.

Definition 3.6 Now define

$$||\nabla f||_p^p = \frac{1}{2}\sum_x \sum_y \mu_{xy}|f(y) - f(x)|^p.$$

We will be mainly interested in the cases $p = 1$ and $p = 2$; of course, $||\nabla f||_2^2 = \mathscr{E}(f, f)$. Note that if $f = \mathbf{1}_A$ then

$$||\nabla f||_p^p = \sum_{x \in A} \sum_{y \in A^c} \mu_{xy} = \mu_E(A; A^c). \tag{3.2}$$

We introduce the following inequalities:

$$||f||_{\alpha p/(\alpha-p)} \leq C||\nabla f||_p \quad \text{for } f \in C_0(\mathbb{V}), \qquad (S_\alpha^p)$$

$$||f||_p \leq C\mu(D)^{1/\alpha}||\nabla f||_p, \quad \text{if } \operatorname{supp}(f) \subset D. \qquad (F_\alpha^p)$$

The inequality (S_α^p) only makes sense when $p < \alpha$, and is the graph analogue of a classical Sobolev inequality in $\mathbb{R}^d$. The inequalities (F_α^p) were introduced by Coulhon in [Co1], and it is proved there (in the context of manifolds) that when $\alpha > p$ the inequalities (S_α^p) and (F_α^p) are equivalent. This equivalence also holds in the graph context, but we will just discuss the cases $p = 1$ and $p = 2$. The connection with the Sobolev inequalities (S_α^p) is included mainly for historical interest; as we will see the (F_α^p) inequalities are on the whole easier to use, and contain the same information. For $\alpha = \infty$ both inequalities above take the form

$$||f||_p \leq C||\nabla f||_p, \quad \text{if } f \in C_0(\mathbb{V}). \qquad (F_\infty^p)$$

Theorem 3.7 *Let $\alpha \in (1, \infty]$. The following are equivalent:*

(a) (Γ, μ) satisfies (I_α) with constant C_0.
(b) (Γ, μ) satisfies (S_α^1) with constant C_0.
(c) (Γ, μ) satisfies (F_α^1) with constant C_0.

In addition (I_1) and (F_1^1) are equivalent.

Proof (c) $\Rightarrow$ (a) Let $\alpha \in [1, \infty]$. Given a finite subset $A \subset \mathbb{V}$ set $f = \mathbf{1}_A$. Then $\|f\|_1 = \mu(A)$, while $||\nabla f||_1 = \mu_E(A; A^c)$. So by (F_α^1)

$$\mu(A) \leq C_0\mu(A)^{1/\alpha}\mu_E(A; A^c),$$

which is (I_α).
(b) $\Rightarrow$ (c) This is immediate from Hölder's inequality; note that $p = \alpha/(\alpha-1)$ and $p' = \alpha$ are conjugate indices. If f has support D then

$$||f||_1 = ||f\mathbf{1}_D||_1 \leq ||f||_{\alpha/(\alpha-1)}||\mathbf{1}_D||_\alpha \leq C_0||\nabla f||_1\mu(D)^{1/\alpha}.$$

The final implication, that (a) implies (b), requires a bit more work. First define

$$\Lambda_t(f) = \{x : f(x) \geq t\}.$$

Lemma 3.8 (Co-area formula) *Let $f : \mathbb{V} \to \mathbb{R}_+$. Then*

$$||\nabla f||_1 = \int_0^\infty \mu_E(\Lambda_t(f); \Lambda_t(f)^c)dt.$$

Proof We have

$$\begin{aligned}||\nabla f||_1 &= \sum_x \sum_y \mu_{xy}(f(y) - f(x))_+ \\ &= \sum_x \sum_y \mu_{xy} \int_0^\infty \mathbf{1}_{(f(y) \geq t > f(x))} dt \\ &= \int_0^\infty dt \sum_x \sum_y \mu_{xy} \mathbf{1}_{(f(y) \geq t > f(x))} \\ &= \int_0^\infty dt \mu_E(\Lambda_t(f); \Lambda_t(f)^c).\end{aligned}$$

□

Proof of the remainder of Theorem 3.7 We begin by proving that (a) $\Rightarrow$ (b). First, note that it is enough to consider $f \geq 0$. For, if $g = |f|$ then $||g||_p = ||f||_p$ and $||\nabla g||_1 \leq ||\nabla f||_1$.

The idea of the proof is to use (I_α) on the sets $\Lambda_t(f)$. However, if we do what comes naturally we find we need to use the Hardy–Littlewood–Polya inequality: for a non-negative decreasing function f,

$$\int_0^\infty p t^{p-1} f(t)^p dt \leq \Big(\int_0^\infty f(t) dt \Big)^p.$$

We can avoid this by using a trick from Rothaus [Ro]. First let $1 < \alpha < \infty$. Let $p = \alpha/(\alpha - 1)$, and $p' = \alpha$. Let $g \in L_+^{p'}(\mathbb{V}, \mu)$ with $||g||_{p'} = 1$. Then

$$\begin{aligned}C_0||\nabla f||_1 &= \int_0^\infty C_0 \mu_E(\Lambda_t(f); \Lambda_t(f)^c)dt \geq \int_0^\infty \mu(\Lambda_t(f))^{1/p} dt \\ &= \int_0^\infty dt ||\mathbf{1}_{\Lambda_t(f)}||_p \geq \int_0^\infty dt ||g \mathbf{1}_{\Lambda_t(f)}||_1 \\ &= \int_0^\infty dt \sum_x g(x) \mu_x \mathbf{1}_{\Lambda_t(f)}(x) = \sum_x g(x) \mu_x \int_0^\infty dt \mathbf{1}_{(f(x) \geq t)} \\ &= \sum_x f(x) g(x) \mu_x = ||fg||_1.\end{aligned}$$

So,

$$||f||_p = \sup\{||fg||_1 : ||g||_{p'} = 1\} \leq C_0 ||\nabla f||_1.$$

For $\alpha = \infty$ take $p = 1$, $g = 1$ in the calculation above.

Finally, we need to show that (I_1) implies (F_1^1). Let $f \geq 0$ have finite support D and let $M = \max f$. Then

$$||\nabla f||_1 = \int_0^M \mu_E(\Lambda_t(f); \Lambda_t(f)^c)dt \geq C_0^{-1} M \geq C_0^{-1}\mu(D)^{-1}||f||_1. \quad \square$$

Lemma 3.9 *Let $\alpha \in [1, \infty]$. Then (F_α^1) implies (F_α^2).*

Proof As before it is enough to prove this for $f \geq 0$. Let f have support D, and set $g = f^2$. Applying (F_α^1) to g, and writing C_0 for the constant in (F_α^1), we obtain

$$||f||_2^2 = ||g||_1 \leq C_0 \mu(D)^{1/\alpha} ||\nabla g||_1. \tag{3.3}$$

If we were working with functions on $\mathbb{R}^d$ we would have

$$||\nabla g||_1 = \int |\nabla g| = \int |\nabla f^2| = 2\int |f||\nabla f| \leq 2\Big(\int |\nabla f|^2\Big)^{1/2} ||f||_2.$$

However, in a discrete space one cannot use the Leibnitz rule to obtain $|\nabla f^2| = 2|f||\nabla f|$. This is not a serious difficulty, since

$$\begin{aligned} ||\nabla g||_1 &= \tfrac{1}{2}\sum_{x,y} \mu_{xy}|f(x)^2 - f(y)^2| \\ &= \tfrac{1}{2}\sum_{x,y} \mu_{xy}|f(x) - f(y)||f(x) + f(y)| \\ &\leq \sum_{x,y} \mu_{xy}|f(x) - f(y)||f(x)| \leq 2^{1/2}\mathcal{E}(f, f)^{1/2}||f||_2. \end{aligned} \tag{3.4}$$

Combining (3.3) and (3.4) gives (F_α^2) with constant $2^{1/2}C_0$. $\square$

3.2 Nash Inequality

We now introduce another inequality, which following [CKS] is called a *Nash inequality*. Nash used inequalities of this type in his 1958 paper [N] to prove Hölder continuity of solutions of divergence form PDEs – see the first inequality on p. 936 of [N]. In Section 4.1, which closely follows the ideas of [CKS], we see that Nash inequalities are closely related to the behaviour of the heat kernel on (Γ, μ).

Definition 3.10 Let $\alpha \in [1, \infty)$. Γ satisfies the Nash inequality (N_α) if for all $f \in L^1 \cap L^2$

$$\mathcal{E}(f, f) \geq C_N ||f||_2^{2+4/\alpha} ||f||_1^{-4/\alpha}. \tag{N_α}$$

Remark 3.11 (1) Suppose we wish to prove (N_α), and have proved it for all $f \in C_{0,+}(\mathbb{V})$. Then if $f \in C_0(\mathbb{V})$, writing $f = f_+ - f_-$, we have $||f||_p^p = ||f_+||_p^p + ||f_-||_p^p$, and $\mathscr{E}(f, f) \geq \mathscr{E}(f_+, f_+) + \mathscr{E}(f_-, f_-)$ by Lemma 1.27. Hence we obtain (N_α) (with a different constant) for all $f \in C_0(\mathbb{V})$. If now $f \in L^1 \cap L^2$, let $f_n \in C_0(\mathbb{V})$ with $f_n \to f$ in L^1 and L^2. Then by (1.17) $f_n \to f$ in $H^2(\mathbb{V})$, and hence (N_α) holds for f. It is also enough just to prove (N_α) for f with $||f||_1 = 1$.

(2) Since $\mathscr{E}(f, f) \leq 2||f||_2^2$ by Proposition 1.21, (N_α) also implies that $||f||_2 \leq c_1||f||_1$ for all f.

(3) Suppose (N_α) holds. Let $A \subset \mathbb{V}$. Applying (N_α) and using (3.2) gives

$$\mu_E(A, \mathbb{V} - A) \geq C_N \mu(A)^{1-2/\alpha},$$

so that $(I_{\alpha/2})$ holds. Taking $A = \{x\}$ we have

$$\mu_x \geq \sum_{y \neq x} \mu_{xy} \geq C_N \mu_x^{1-2/\alpha},$$

so that $\mu_x \geq C_N^{\alpha/2}$. Thus the Nash inequality gives a global lower bound on μ_x.

Exercise 3.12 Let (Γ, μ) be a weighted graph, let $\lambda > 0$, and let $\mu^\lambda = \lambda\mu$. If (Γ, μ) satisfies (N_α) with constant C_N then (Γ, μ^λ) satisfies (N_α) with constant $\lambda^{2/\alpha} C_N$.

The following simple proof of the Nash inequality in $\mathbb{Z}^d$ is based on an argument of Stein – see p. 935 of [N].

Lemma 3.13 *The Nash inequality (N_d) holds in $\mathbb{Z}^d$.*

Proof Note that since we have $\mu_{xy} = 1$ when $x \sim y$, we have $\mu_x = 2d$ for each $x \in \mathbb{Z}^d$, so that expressions for $||f||_p$ will differ from familiar ones by powers of $2d$.

Write $I = [-\pi, \pi]^d$, and let $f \in L^1(\mathbb{Z}^d)$. Then $f \in L^2(\mathbb{Z}^d)$ by Lemma 1.20. For $\theta \in \mathbb{R}^d$ let $\widehat{f}(\theta)$ be the Fourier transform of f:

$$\widehat{f}(\theta) = \sum_{x \in \mathbb{Z}^d} e^{i\theta \cdot x} f(x).$$

Clearly $|\widehat{f}(\theta)| \leq (2d)^{-1}||f||_1$, while by Parseval's formula

$$(2d)^{-1}||f||_2^2 = (2\pi)^{-d} \int_I |\widehat{f}(\theta)|^2 d\theta.$$

Let e_j denote the unit vectors in $\mathbb{Z}^d$ and $h_j(x) = f(x+e_j) - f(x)$. Then

$$\widehat{h}_j(\theta) = \sum_x e^{i\theta\cdot(x-e_j)} f(x) - \widehat{f}(\theta) = (e^{-i\theta\cdot e_j} - 1)\widehat{f}(\theta).$$

Since there exists a constant c_0 such that $|1 - e^{i\theta\cdot e_j}|^2 \geq c_0|\theta_j|^2$,

$$(2d)^{-1}||h_j||_2^2 = (2\pi)^{-d}\int_I |\widehat{h}_j(\theta)|^2 d\theta \geq (2\pi)^{-d}\int_I c_0|\theta_j|^2|\widehat{f}(\theta)|^2 d\theta,$$

and hence

$$\begin{aligned}\mathscr{E}(f,f) &= \sum_{j=1}^d \sum_x (f(x+e_j) - f(x))^2 = (2d)^{-1}\sum_{j=1}^d ||h_j||_2^2 \\ &\geq (2\pi)^{-d} c_0 \int_I |\theta|^2 |\widehat{f}(\theta)|^2 d\theta.\end{aligned}$$

Let $r > 0$, and let C_d be the volume of the unit ball $B(0,1)$ in $\mathbb{R}^d$. Then

$$\begin{aligned}(2\pi)^d \|f\|_2^2 &= \int_{B(0,r)} |\widehat{f}(\theta)|^2 d\theta + \int_{I-B(0,r)} |\widehat{f}(\theta)|^2 d\theta \\ &\leq C_d(2d)^{-1} r^d \|f\|_1^2 + \int_{I-B(0,r)} |\theta/r|^2 |\widehat{f}(\theta)|^2 d\theta \\ &\leq C_d(2d)^{-1} r^d \|f\|_1^2 + r^{-2} c_0^{-1} (2\pi)^d \mathscr{E}(f,f).\end{aligned}$$

If we choose r so that the last two terms are equal up to constants then $r^{d+2} = \mathscr{E}(f,f)\|f\|_1^{-2}$, and we obtain

$$\|f\|_2^2 \leq C'_d \mathscr{E}(f,f)^{d/(d+2)} \|f\|_1^{4/(d+2)},$$

which is (N_d). (Note that the proof works even if r is large enough so that $I - B(0,r)$ is empty.) □

Theorem 3.14 *Let $\alpha \in [1,\infty)$. The following are equivalent:*

(a) (Γ,μ) satisfies (F_α^2).
(b) (Γ,μ) satisfies (N_α).

In addition, if $\alpha > 2$ then (a) and (b) are equivalent to:

(c) (Γ,μ) satisfies (S_α^2).

Proof If $\alpha = \infty$ then for $f \in C_0(\mathbb{V})$ all three inequalities take the form $||f||_2^2 \leq c\mathscr{E}(f,f)$. An easy approximation argument then gives the general equivalence.

Now suppose that $\alpha < \infty$. The implication (b) $\Rightarrow$ (a) is easy. Let $\mathrm{supp}(f) = D$. Then

$$||f||_1 = ||f\mathbf{1}_D||_1 \le ||f||_2||\mathbf{1}_D||_2 = ||f||_2\mu(D)^{1/2}.$$

So by (N_α)

$$C_N||f||_2^{2+4/\alpha} \le \left(||f||_2\mu(D)^{1/2}\right)^{4/\alpha}\mathscr{E}(f,f),$$

and (F_α^2) follows immediately.

For the implication (a) $\Rightarrow$ (b) we use an argument of Grigor'yan [Gg2] – see also [Co2]. By Remark 3.11 we can suppose that $f \in C_0^+(\mathbb{V})$. For $\lambda \ge 0$ we have

$$f^2 \le 2\lambda f\mathbf{1}_{(f\le 2\lambda)} + 4(f-\lambda)_+^2\mathbf{1}_{(f>2\lambda)} \le 2\lambda f + 4(f-\lambda)_+^2. \tag{3.5}$$

Applying (F_α^2) to $g = (f-\lambda)_+$, and using Lemma 1.27(a) and Markov's inequality,

$$||g||_2^2 \le C^2\mu(\{f>\lambda\})^{2/\alpha}\mathscr{E}(g,g) \le C^2\mathscr{E}(f,f)\Big(\frac{||f||_1}{\lambda}\Big)^{2/\alpha}.$$

Let $\lambda = ||f||_2^2/4||f||_1$. Then using (3.5)

$$\begin{aligned}||f||_2^2 &\le 2\lambda||f||_1 + 4||g||_2^2\\ &\le \tfrac{1}{2}||f||_2^2 + 4C^2\mathscr{E}(f,f)(||f||_1\lambda^{-1})^{2/\alpha}\\ &= \tfrac{1}{2}||f||_2^2 + 4^{1+2/\alpha}C^2\mathscr{E}(f,f)||f||_1^{4/\alpha}||f||_2^{-4/\alpha},\end{aligned}$$

which is (N_α).

Now let $\alpha > 2$. The implication (c) $\Rightarrow$ (a) is easy. Let $D = \mathrm{supp}(f)$ and $p = \alpha/(\alpha-2)$, so that the conjugate index $p' = 2/\alpha$. Then using (S_α^2)

$$\begin{aligned}||f||_2^2 &= ||f^2\mathbf{1}_D||_1 \le ||f^2||_p||\mathbf{1}_D||_{p'}\\ &= ||f||_{2\alpha/(\alpha-2)}^2\mu(D)^{\alpha/2} \le C\mathscr{E}(f,f)\mu(D)^{\alpha/2}.\end{aligned}$$

That (a) implies (c) needs a bit more work. We include the argument for completeness, but will not need it. For $k \in \mathbb{Z}$ define $\Phi_k : \mathbb{R}_+ \to \mathbb{R}_+$ by

$$\Phi_k(t) = (t-2^k)_+ \wedge 2^k = \begin{cases} 2^k & \text{if } t \ge 2^{k+1},\\ t-2^k & \text{if } 2^k \le t < 2^{k+1},\\ 0 & \text{if } t < 2^k.\end{cases}$$

So if n is such that $2^n \le t < 2^{n+1}$ then

$$\Phi_k(t) = (t-2^n)\mathbf{1}_{(k=n)} + 2^k\mathbf{1}_{(k<n)},$$

and thus we have $t = \sum_{k \in \mathbb{Z}} \Phi_k(t)$ for $t \geq 0$. If $s, t \geq 0$ then

$$(t-s)^2 = \Big(\sum_k (\Phi_k(t) - \Phi_k(s))\Big)^2 \geq \sum_k (\Phi_k(t) - \Phi_k(s))^2.$$

Now let $f \in C_{0,+}(\mathbb{V})$, and set $f_k = \Phi_k \circ f$. Then by the above

$$\mathscr{E}(f, f) \geq \sum_{k \in \mathbb{Z}} \mathscr{E}(f_k, f_k).$$

Let $\gamma_k = \mu(\{f > 2^k\})$ and $p = 2\alpha/(\alpha - 2)$. We have

$$\begin{aligned} ||f||_p^p &= \sum_x \sum_k f(x)^p \mathbf{1}_{(2^k < f(x) \leq 2^{k+1})} \mu_x \\ &\leq \sum_k 2^{p(k+1)} \sum_x \mathbf{1}_{(2^k < f(x) \leq 2^{k+1})} \mu_x \leq \sum_k 2^{p(k+1)} \gamma_k. \end{aligned}$$

For $q \geq 2$

$$||f_k||_q^q \geq 2^{qk} \mu(\{f > 2^{k+1}\}) = 2^{qk} \gamma_{k+1}. \tag{3.6}$$

Applying (F_α^2) to f_k gives

$$||f_k||_2^2 \leq C \mu(\{f > 2^k\})^{2/\alpha} \mathscr{E}(f_k, f_k) = C \gamma_k^{2/\alpha} \mathscr{E}(f_k, f_k),$$

and using (3.6), first with $q = 2$ and then with $q = p$,

$$2^{2k} \gamma_{k+1} \leq ||f_k||_2^2 \leq C \gamma_k^{2/\alpha} \mathscr{E}(f_k, f_k) \leq C \big(||f_{k-1}||_p^p 2^{-(k-1)p}\big)^{2/\alpha} \mathscr{E}(f_k, f_k).$$

Therefore since $p - 2p/\alpha = 2$ and $|f_k| \leq |f|$,

$$\begin{aligned} ||f||_p^p &\leq 2^{2p} \sum_k 2^{pk} \gamma_{k+1} \\ &\leq 2^{2p + 2p/\alpha} C \sum_k 2^{(p-2)k} ||f||_p^{2p/\alpha} 2^{-2kp/\alpha} \mathscr{E}(f_k, f_k) \\ &\leq C' ||f||_p^{2p/\alpha} \sum_k \mathscr{E}(f_k, f_k) \leq C ||f||_p^{p-2} \mathscr{E}(f, f). \end{aligned}$$

Since $f \in C_0(\mathbb{V})$ we have $||f||_p < \infty$, and dividing the last equation by $||f||_p^{p-2}$ gives (S_α^2). □

Corollary 3.15 *Let $\alpha \in [1, \infty)$. If (Γ, μ) satisfies (I_α) then it satisfies (N_α).*

Proof Combining Theorem 3.7, Lemma 3.9, and Theorem 3.14, we have the following chain of implications: $(I_\alpha) \Rightarrow (F_\alpha^1) \Rightarrow (F_\alpha^2) \Rightarrow (N_\alpha)$. □

Remark 3.16 In number theory a constant in an equation or inequality is called *effective* if it could in principle be calculated. Most proofs of inequalities in analysis and probability yield constants which are effective, but occasionally a 'pure existence' proof will show that some inequality holds without giving any control on the constants.

If one inspects the proof of Theorem 3.14 then one sees that the constants in the results are effective in the sense that they depend only on the constants in the 'input data'. So, if for example (Γ, μ) satisfies (F^2_α) with a constant C_F then the proof of the implication (a) $\Rightarrow$ (b) in Theorem 3.14 shows that (Γ, μ) satisfies (N_α) with a constant C which depends only on α and C_F. The same holds for the other implications, and, although we do not remark on this again, it holds for the remaining results in this book.

Corollary 3.17 *Let (Γ, μ) satisfy (I_α) with $\alpha \in (2, \infty]$. Then (Γ, μ) is transient.*

Proof Let $x_0 \in \mathbb{V}$, $p = 2\alpha/(\alpha - 2)$. Then using (S^α_2)

$$|f(x_0)|^2 \le \mu_{x_0}^{-2/p} ||f||_p^2 \le c\,\mathscr{E}(f, f),$$

and so (Γ, μ) is transient by Theorem 2.42. □

Lemma 3.18 *(a) Let $\alpha \in (2, \infty)$. If (Γ, μ) satisfies (S^2_α) then (Γ, μ) satisfies $(I_{\alpha/2})$.*
(b) If (Γ, μ) satisfies (S^2_∞) then (Γ, μ) satisfies (I_∞).

Proof (a) Let $A \subset \mathbb{V}$ and $f = \mathbf{1}_A$. Then $||f||^2_{2\alpha/(\alpha-2)} = \mu(A)^{(\alpha-2)/\alpha}$, and by (3.2) $\mathscr{E}(f, f) = \mu_E(A; A^c)$. So by (S^2_α),

$$\mu_E(A; A^c) \ge c\mu(A)^{(\alpha-2)/\alpha} = c\mu(A)^{1-2/\alpha},$$

giving $I_{\alpha/2}$.
(b) Again taking $f = \mathbf{1}_A$, we have $||f||^2 = \mu(A)$, so that (I_∞) is immediate. □

Remark 3.19 $\mathbb{Z}^2$ satisfies (S^1_α) with $\alpha = 2$. One might hope that one would obtain the $\alpha \downarrow 2$ limit of (S^2_α), that is,

$$||f||^2_\infty \le C_1 \mathscr{E}(f, f), \quad f \in C_0(\mathbb{Z}^2), \tag{3.7}$$

but since $\mathbb{Z}^2$ is recurrent, using Theorem 2.42 one sees that (3.7) cannot hold. There can be a loss of information in passing from (S^1_α) to (S^2_α).

Exercise 3.20 For $i = 1, 2$ let Γ_i satisfy (N_{α_i}). Then the join of Γ_1 and Γ_2 satisfies $(N_{\alpha_1 \wedge \alpha_2})$.

3.3 Poincaré Inequality

We now introduce a second kind of isoperimetric inequality, and begin by considering finite graphs.

Definition 3.21 A finite weighted graph (Γ, μ) satisfies a *relative isoperimetric inequality* (with constant C_R) if, for any $A \subset \mathbb{V}$ satisfying $\mu(A) \le \frac{1}{2}\mu(\mathbb{V})$, one has

$$\frac{\mu_E(A; \mathbb{V} - A)}{\mu(A)} \ge C_R. \tag{3.8}$$

We write $R_I = R_I(\Gamma) = R_I(\Gamma, \mu)$ for the largest constant C_R such that (3.8) holds, and call R_I the *relative isoperimetric constant* of Γ.

This is closely related to the *Cheeger constant* for a finite graph, defined by

$$\chi(\Gamma) = \min_{A \ne \varnothing, \mathbb{V}} \frac{\mu(\mathbb{V})\mu_E(A; \mathbb{V} - A)}{\mu(A)\mu(\mathbb{V} - A)}. \tag{3.9}$$

Since for any A with $0 < \mu(A) \le \frac{1}{2}\mu(\mathbb{V})$ one has

$$\frac{\mu_E(A; \mathbb{V} - A)}{\mu(A)} \le \frac{\mu(\mathbb{V})\mu_E(A; \mathbb{V} - A)}{\mu(A)\mu(\mathbb{V} - A)} \le 2\frac{\mu_E(A; \mathbb{V} - A)}{\mu(A)},$$

it follows that $R_I(\Gamma) \le \chi(\Gamma) \le 2R_I(\Gamma)$.

Proposition 3.22 *Let $\Gamma = (\mathbb{V}, E)$ be a finite graph. The following are equivalent:*

(a) (Γ, μ) satisfies a relative isoperimetric inequality with constant C_R.
(b) For any $f : \mathbb{V} \to \mathbb{R}$,

$$\min_a \sum_{y \in \mathbb{V}} |f(y) - a| \mu_y \le C_R^{-1} ||\nabla f||_1. \tag{3.10}$$

Consequently

$$R_I = \inf \left\{ \frac{||\nabla f||_1}{\min_a \sum_{y \in \mathbb{V}} |f(y) - a| \mu_y} : f \in C(\mathbb{V}),\ f \text{ non-constant} \right\}.$$

Proof Note first that the left side of (3.10) is minimised by choosing a to be a median of f, that is, so that $\mu(\{x : f(x) > a\})$ and $\mu(\{x : f(x) < a\})$ are both less than or equal to $\frac{1}{2}\mu(\mathbb{V})$.

(a) $\Rightarrow$ (b) Choose a as above, and let $g = f - a$. Since $|g(x) - g(y)| = |g_+(x) - g_+(y)| + |g_-(x) - g_-(y)|$, we have $||\nabla g||_1 = ||\nabla g_+||_1 + ||\nabla g_-||_1$. Using (3.8) and the co-area formula Lemma 3.8,

$$\begin{aligned}\sum_{x \in \mathbb{V}} g_+(x)\mu_x &= \int_0^\infty \mu(\{x : g_+(x) > t\})dt \\ &\le C_R^{-1} \int_0^\infty \mu_E(\{g_+ > t\}; \{g_+ \le t\})dt \\ &= C_R^{-1}||\nabla g_+||_1.\end{aligned}$$

Adding this and a similar inequality for g_- gives (3.10).
(b) $\Rightarrow$ (a) Let $\mu(A) \le \frac{1}{2}\mu(\mathbb{V})$ and $f = \mathbf{1}_A$. Then f has median zero, and (3.10) gives $\mu(A) \le C_R^{-1}\mu_E(A; \mathbb{V} - A)$, which is the relative isoperimetric inequality with constant C_R.

The final characterisation of R_I is immediate from the equivalence of (a) and (b), and the definition of R_I. □

Given a (finite) graph Γ it is often easy to give a good upper bound for $R_I(\Gamma)$ by exhibiting a suitable set A. For lower bounds one can use a dual argument which finds a good (i.e. well spread out) family of paths on the graph.

Definition 3.23 (See [DiS, SJ]) A family of paths $\mathcal{M}$ *covers* Γ if for each pair of distinct points $x, y \in \mathbb{V}$ there exists a path $\gamma = \gamma(x, y) \in \mathcal{M}$ such that γ is a path from x to y. Given such a family of paths, set

$$\kappa(\mathcal{M}) = \max_{e \in E} \mu_e^{-1} \sum_{\{(x,y): e \in \gamma(x,y)\}} \mu_x \mu_y.$$

We wish to choose the family of paths $\mathcal{M}$ so that $\kappa(\mathcal{M})$ is small.

Theorem 3.24 *Let Γ be a finite graph and $\mathcal{M}$ be a covering family of paths. Then*

$$R_I(\Gamma) \ge \frac{\mu(\mathbb{V})}{\kappa(\mathcal{M})}.$$

Proof Let $f : \mathbb{V} \to \mathbb{R}$. Then

$$\begin{aligned}\mu(\mathbb{V}) \min_\lambda \sum_x |f(x) - \lambda|\mu_x &= \sum_y \mu_y \min_\lambda \sum_x |f(x) - \lambda|\mu_x \\ &\le \sum_y \mu_y \sum_x |f(x) - f(y)|\mu_x.\end{aligned}$$

But since, for any $x, y \in \mathbb{V}$,

$$f(y) - f(x) = \sum_{i=1}^{k} \big(f(z_i) - f(z_{i-1})\big),$$

where $(z_0, z_1, \dots, z_k)$ is the path $\gamma(x, y)$, we have

$$\begin{aligned} \sum_y \sum_x |f(x) - f(y)| \mu_x \mu_y &\le \sum_y \sum_x \mu_x \mu_y \sum_{e \in \gamma(x,y)} |\nabla f(e)| \\ &= \sum_{e \in E} |\nabla f(e)| \mu_e \sum_{x,y} \mu_e^{-1} \mathbf{1}_{\{(x,y): e \in \gamma(x,y)\}} \mu_x \mu_y \\ &\le \kappa(\mathscr{M}) ||\nabla f||_1. \end{aligned}$$

Using Proposition 3.22 completes the proof. □

Theorem 3.25 *Let Γ be the cube $Q = \{1, \dots, R\}^d$ in $\mathbb{Z}^d$. Then $R_I(\Gamma) \ge c_d/R$.*

Proof Note that $R^d \le \mu(Q) \le (2dR)^d$. For each pair $x, y \in Q$ let $\gamma(x, y)$ be the path which first matches the first coordinate, then the second, and so on.

Now let $e = \{z, z + e_k\}$ be an edge in Γ, where $1 \le k \le d$. If $e \in \gamma(x, y)$ then the first $k - 1$ coordinates of x must already have been set to those of y, while coordinates $k+1, \dots, d$ will not yet have been altered. So we must have $z_i = y_i$, $1 \le i \le k - 1$, and $z_i = x_i$ for $k + 1 \le i \le d$. Thus x lies in a set of size at most R^k and y in a set of size at most $R^{d-(k-1)}$. So,

$$|\{(x, y) : e \in \gamma(x, y)\}| \le R^k R^{d-k+1} = R^{d+1}.$$

Thus $\kappa(\mathscr{M}) \le (2d)^2 R^{d-1}$ and hence

$$R_I(\Gamma) \ge \frac{\mu(Q)}{(2d)^2 R^{d+1}} \ge \frac{1}{(2d)^2 R}.$$ □

This leads easily to a proof of the isoperimetric inequality in $\mathbb{Z}^d$.

Theorem 3.26 *Let A be a finite subset of $\mathbb{Z}^d$. Then*

$$\mu_E(A; \mathbb{Z}^d - A) \ge C_d \mu(A)^{(d-1)/d}.$$

Proof Choose r such that $\frac{1}{4} r^d \le \mu(A) \le \frac{1}{2} r^d$. Let $(Q_n, n \in \mathbb{N})$ be a tiling of $\mathbb{Z}^d$ by disjoint cubes each containing r^d points, and let $A_n = A \cap Q_n$. Set $a_n = \mu(A_n)$, $b_n = \mu_E(A_n; Q_n - A_n)$. Then $\mu(A) = \sum a_n$, while $\mu_E(A; \mathbb{Z}^d -$

$A) \geq \sum b_n$. Since $\mu(A_n) \leq \frac{1}{2}\mu(Q_n)$, by Theorem 3.25 we have $b_n \geq (c/r)a_n$ for each n. Hence

$$\mu(A) = \sum a_n \leq cr \sum b_n \leq cr\mu_E(A; \mathbb{Z}^d - A),$$

and as r has been chosen so that $r^d \asymp \mu(A)$, the inequality (I_d) follows. □

Proposition 3.27 *Let* $\Gamma = (\mathbb{V}, E)$ *be a finite graph. Then for any* $f : \mathbb{V} \to \mathbb{R}$

$$\min_\lambda \sum_{y \in \mathbb{V}} |f(y) - \lambda|^2 \mu_y \leq 2R_I^{-2}\mathscr{E}(f, f). \tag{3.11}$$

Proof Note that the left side of (3.11) is minimised by taking λ to be the average of f, that is,

$$\lambda = \overline{f}_{\mathbb{V}} = \mu(\mathbb{V})^{-1} \sum_{x \in \mathbb{V}} f(x)\mu_x.$$

Let $f : \mathbb{V} \to \mathbb{R}$. By subtracting a constant we can assume that the median of f is zero. Set $g(x) = f(x)^2 \mathrm{sgn}(f(x))$, where $\mathrm{sgn}(x) = \mathbf{1}_{(x>0)} - \mathbf{1}_{(x<0)}$ is the sign of x. Then g also has median zero, so by Proposition 3.22

$$||f||_2^2 = \sum_{x \in \mathbb{V}} |g(x)|\mu_x \leq R_I^{-1}\tfrac{1}{2} \sum_{x \in \mathbb{V}} \sum_{y \in \mathbb{V}} |g(x) - g(y)|\mu_{xy}.$$

It is not hard to check that, for any $a, b \in \mathbb{R}$,

$$|a^2\mathrm{sgn}(a) - b^2\mathrm{sgn}(b)| \leq |a - b|(|a| + |b|).$$

Hence

$$\begin{aligned} ||f||_2^2 &\leq R_I^{-1}\tfrac{1}{2} \sum_{x \in \mathbb{V}} \sum_{y \in \mathbb{V}} |f(x) - f(y)|(|f(x)| + |f(y)|)\mu_{xy} \\ &\leq R_I^{-1}\mathscr{E}(f, f)^{1/2}(2||f||_2^2)^{1/2} = \sqrt{2}\, R_I^{-1}\mathscr{E}(f, f)^{1/2}||f||_2. \end{aligned}$$

Dividing by $||f||_2$ gives (3.11). □

We now consider infinite graphs.

Definition 3.28 We say (Γ, μ) satisfies a (weak) *Poincaré inequality* (PI) if there exist $C_P < \infty$ and $\lambda \geq 1$ such that, for all $x \in \mathbb{V}$, $R \geq 1$, if $B = B(x, R)$, $B^* = B(x, \lambda R)$, and $f : B^* \to \mathbb{R}$ then

$$\begin{aligned} \sum_{y \in B} (f(y) - \overline{f}_B)^2 \mu_y &\leq C_P R^2 \tfrac{1}{2} \sum_{x \in B^*} \sum_{y \in B^*} \mu_{xy}(f(x) - f(y))^2 \\ &= C_P R^2 \mathscr{E}_{B^*}(f, f). \end{aligned}$$

Here

$$\overline{f}_B = \mu(B)^{-1} \sum_{y \in B} f(y)\mu_y$$

is the real number a which minimises $\sum_{y \in B}(f(y) - a)^2 \mu_y$. Note that the PI is not a single inequality such as the Nash inequality; rather it gives a family of inequalities which hold for all balls in Γ. We call λ the *expansion factor* of the PI.

We say (Γ, μ) satisfies a *strong Poincaré inequality* if the weak PI holds with $\lambda = 1$. If we just use the term 'Poincaré inequality' then this will refer to the weak PI, but we will sometimes refer to 'weak PI' in contexts where we wish to emphasise the distinction between the weak and strong versions of the PI.

Given Proposition 3.27 a natural way to prove that a graph Γ satisfies a PI is to prove a suitable relative isoperimetric inequality for balls in Γ. However, μ is not the measure which arises from the μ_e when we restrict to the subgraph of Γ generated by a subset A. For $D \subset \mathbb{V}$ write

$$\mu_x^D = \sum_{y \in D} \mu_{xy}, \quad \mu^D(A) = \sum_{x \in A} \mu_x^D.$$

Lemma 3.29 *Suppose Γ satisfies (H5) with constant C_H, and that D is a finite subset of $\mathbb{V}$. Then for any $f : D \to \mathbb{R}$*

$$\min_{\lambda} \sum_{y \in D} (f(y) - \lambda)^2 \mu_y \le 2 C_H R_I(D)^{-2} \mathscr{E}_D(f, f).$$

Proof The hypothesis (H5) implies that $\mu_x \le C_H \mu_x^D$, so

$$\min_{\lambda} \sum_{y \in D} (f(y) - \lambda)^2 \mu_y \le C_H \min_{\lambda} \sum_{y \in D} (f(y) - \lambda)^2 \mu_y^D,$$

and the conclusion follows by Proposition 3.27. □

Corollary 3.30 *$\mathbb{Z}^d$ satisfies the (weak) PI.*

Proof Let $B = B(x, R)$ be a ball in $\mathbb{Z}^d$ and $Q(x, R) = \{y \in \mathbb{Z}^d : ||x - y||_\infty \le R\}$ be the cube with the same centre. Thus

$$B(x, R) \subset Q(x, R) \subset B(x, d^{1/2} R).$$

By Theorem 3.25 we have $R_I(Q(x, R)) \geq cR^{-1}$, and so if $f : B(x, d^{1/2}R) \to \mathbb{R}$ then

$$\min_b \sum_{x \in B(x,R)} |f(x) - b|^2 \mu_x \leq \min_b \sum_{x \in Q(x,R)} |f(x) - b|^2 \mu_x$$
$$\leq c_1 R^2 \mathcal{E}_{Q(x,R)}(f, f) \leq c_1 R^2 \mathcal{E}_{B(x,d^{1/2}R)}(f, f),$$

which is the PI with $\lambda = d^{1/2}$. □

Remark This corollary shows why it is useful to consider the weak PI: the relative isoperimetric inequality is most easily proved for cubes, but this then yields the strong PI for cubes in $\mathbb{Z}^d$ rather than balls. In Appendix A.9 it is proved that under some fairly mild additional conditions the weak PI implies the strong one – see Corollary A.51. (These conditions do hold for $\mathbb{Z}^d$.)

The following identity is often helpful in handling Poincaré inequalities.

Lemma 3.31 *Let $A \subset \mathbb{V}$ be finite and $f : A \to \mathbb{R}$. Then*

$$\sum_{x \in A} (f(x) - \overline{f}_A)^2 \mu_x = \tfrac{1}{2} \mu(A)^{-1} \sum_{x \in A} \sum_{y \in A} (f(x) - f(y))^2 \mu_{xy}.$$

Proof We can assume that $\overline{f} = 0$. Then expanding the right hand side, and noting that $\sum_{y \in A} f(y) \mu_y = 0$, gives the left hand side. □

Example 3.32 The weak PI fails on the join of two copies of $\mathbb{Z}^d$ if $d \geq 2$. This is easy to see if $d \geq 3$. Suppose the copies are joined at their origins, denoted 0 and 0′; let $B = B(0, R)$ and let f be 1 on one copy and -1 on the other. Then

$$\sum_{x \in B} (f - \overline{f}_B)^2 \mu_x \asymp R^d, \text{ and } R^2 \mathcal{E}_B(f, f) = R^2.$$

For $d = 2$ one needs to look at functions which decay like $\log(R/|x|)$ away from the origins.

Proposition 3.33 *The (weak) Poincaré inequality is stable under rough isometries of weighted graphs.*

Proof Let $(\Gamma_1, \mu^{(1)})$ and $(\Gamma_2, \mu^{(2)})$ be weighted graphs, $\varphi : \mathbb{V}_1 \to \mathbb{V}_2$ be a rough isometry, and C_i be as in Definition 1.14. Suppose that $(\Gamma_1, \mu^{(1)})$ satisfies the PI, with expansion factor λ_1.

Let $x_2 \in \mathbb{V}_2$, $R \geq 1$. Then there exists $x_1 \in \mathbb{V}_1$ such that $d_2(\varphi(x_1), x_2)) \leq C_2$, and c_1 such that $\varphi^{-1}(B(x_2, R)) \subset B_1(x_1, c_1 R)$. Choose λ_2 large enough so that $\varphi(B(x_1, c_1\lambda_1 R)) \subset B_2(x_2, \lambda_2 R)$. Write $B_1 = B_1(x_1, c_1)$, $B_1^* = B_1(x_1, c_1\lambda_1 R)$, $B_2 = B_2(x_2, R)$, and $B_2^* = B_2(x_2, \lambda_2)$.

Now let $f_2 : B_2^* \to \mathbb{R}$. Set $A = \varphi(\mathbb{V}_1) \cap B_2$. Let $a \in \mathbb{R}$. Since every point in B_2 is within a distance C_2 of a point in A, an argument similar to that in Proposition 2.56 gives that

$$\sum_{x \in B_2} (f_2(x) - a)^2 \mu_x^{(2)} \leq \sum_{x \in A} (f_2(x) - a)^2 \mu_x^{(2)} + c_2 \mathscr{E}_{B_2^*}^{(2)}(f_2, f_2).$$

Then taking $f_1 = f_2 \circ \varphi$, and a to be the mean of f_1 on B_1,

$$\sum_{x \in A} (f_2(x) - a)^2 \mu_x^{(2)} \leq c_3 \sum_{y \in B_1} (f_1(y) - a)^2 \mu_y^{(1)} \leq c_4 \mathscr{E}_{B_1^*}^{(1)}(f_1, f_1).$$

As in Proposition 2.56

$$\mathscr{E}_{B_1^*}^{(1)}(f_1, f_1) \leq c_5 \mathscr{E}_{B_2^*}^{(2)}(f_2, f_2),$$

and combining the chain of inequalities completes the proof. □

3.4 Spectral Decomposition for a Finite Graph

Recall that for a (real) symmetric $N \times N$ matrix $A = (A_{ij})$ we have

$$A = U \Lambda U^T, \tag{3.12}$$

where U is unitary (i.e. $UU^T = U^T U = I$) and Λ is a diagonal matrix with $\Lambda_{ii} = \rho_i$, the eigenvalues of A. (Both U and Λ are real.)

Now let (Γ, μ) be a weighted graph with $|\mathbb{V}| = N$. Define a symmetric matrix A by

$$A_{ij} = \frac{\mu_{ij}}{(\mu_i \mu_j)^{1/2}} = \frac{\mu_i \mathscr{P}(i, j)}{(\mu_i \mu_j)^{1/2}} = \frac{\mu_i^{1/2} \mathscr{P}(i, j)}{\mu_j^{1/2}}.$$

In matrix terms

$$A = MPM^{-1},$$

where $P_{ij} = \mathscr{P}(i, j)$ and M is a diagonal matrix with $M_{ii} = \mu_i^{1/2}$. Let $\rho_1 \geq \rho_2 \geq \cdots \geq \rho_N$ be the eigenvalues of P. Since M is invertible, P and A have the same eigenvalues, and, as $||P||_{2\to 2} \leq 1$ by Proposition 1.16, we have $|\rho_i| \leq 1$ for each $i = 1, \ldots, N$. Let U and Λ be as in (3.12); then $AU = U\Lambda$, and if we write $u^{(k)}$ for the kth column of U this gives

$$Au^{(k)} = \rho_k u^{(k)}, \quad k = 1, \ldots, N.$$

Setting $v^{(k)} = M^{-1}u^{(k)}$ we have

$$Pv^{(k)} = \rho_k v^{(k)}, \quad k = 1, \dots, N. \tag{3.13}$$

We now rewrite (3.13) using the notation of this book. Let $\varphi_k(x) = v_x^{(k)}, x \in \mathbb{V}$.

Lemma 3.34 *The functions φ_k are an orthonormal basis for $L^2(\mathbb{V}, \mu)$, $\rho_k \in [-1, 1]$ for each k, and*

$$P\varphi_k = \rho_k \varphi_k, \quad k = 1, \dots, N, \tag{3.14}$$

$$\Delta\varphi_k = -(1 - \rho_k)\varphi_k, \quad k = 1, \dots, N. \tag{3.15}$$

Proof Equations (3.14) and (3.15) are immediate from (3.13). To prove that (φ_i) are orthonormal, we make the computation

$$\begin{aligned}
\langle \varphi_i, \varphi_j \rangle &= \sum_{x \in \mathbb{V}} \varphi_i(x)\varphi_j(x)\mu_x \\
&= \sum_{x \in \mathbb{V}} \mu_x^{-1/2} u_x^{(i)} \mu_x^{-1/2} u_x^{(j)} \mu_x \\
&= \sum_{x \in \mathbb{V}} U_{ix} U_{jx} = \sum_{x \in \mathbb{V}} U_{ix} U^T_{xj} = (UU^T)_{ij} = \mathbf{1}_{(i=j)}.
\end{aligned}$$

□

Remark Since $\sum_y \mathscr{P}(x, y) = 1$ the function $f \equiv 1$ is an eigenfunction of P with eigenvalue 1. If Γ is bipartite then there exists $A \subset \mathbb{V}$ such that all edges are of the form $e = \{x, y\}$, with $x \in A$ and $y \in \mathbb{V} - A$. We have $P(\mathbf{1}_A - \mathbf{1}_{\mathbb{V}-A}) = -(\mathbf{1}_A - \mathbf{1}_{\mathbb{V}-A})$, and hence -1 is also an eigenvalue.

Corollary 3.35 *We have*

$$p_n(x, y) = \sum_{i=1}^{N} \varphi_i(x)\varphi_i(y)\rho_i^n, \quad n \geq 0. \tag{3.16}$$

Proof Since φ_i are orthonormal, for any f we have

$$f = \sum_{i=1}^{N} \langle f, \varphi_i \rangle \varphi_i. \tag{3.17}$$

Hence

$$P^n f = \sum_i \langle f, \varphi_i \rangle \varphi_i \rho_i^n. \tag{3.18}$$

Fix $y \in \mathbb{V}$, and let $f(x) = \mu_y^{-1} \mathbf{1}_y(x)$. Then $\langle f, \varphi_i \rangle = \varphi_i(y)$, and (3.18) gives

$$p_n(x, y) = P_n f(x) = \sum_i \varphi_i(y)\varphi_i(x)\rho_i^n. \qquad \square$$

The eigenfunctions φ_i and eigenvalues $\lambda_i = 1 - \rho_i$ of $-\Delta$ can be obtained by a variational procedure. Note that $0 = \lambda_1 \leq \lambda_2 \leq \ldots \leq \lambda_N \leq 2$.

Lemma 3.36 *Let $L_1 = \{0\}$, for $2 \leq k \leq N$ let L_k be the linear subspace of $C(\mathbb{V})$ spanned by $\varphi_1, \ldots, \varphi_{k-1}$, and let $L_k^\perp$ be the orthogonal complement of L_k in $L^2(\mathbb{V}, \mu)$. Then*

$$\lambda_k = \inf\left\{\mathscr{E}(f, f) : f \in L_k^\perp, ||f||_2^2 = 1\right\}, \quad 1 \leq k \leq N, \tag{3.19}$$

and the infimum in (3.19) is attained by φ_k. In particular we have $\lambda_1 = 0$, $\lambda_2 > 0$, and for all $f : \mathbb{V} \to \mathbb{R}$

$$\min_b \sum_x |f(x) - b|^2 \mu_x \leq \lambda_2^{-1} \mathscr{E}(f, f). \tag{3.20}$$

Proof For any $f \in C(\mathbb{V})$ we have using (3.17) and the orthogonality of the φ_i,

$$\mathscr{E}(f, f) = \langle -\Delta f, f \rangle = \sum_{i=1}^N \lambda_i \langle f, \varphi_i \rangle^2, \tag{3.21}$$

$$||f||_2^2 = \sum_{i=1}^N \langle f, \varphi_i \rangle^2. \tag{3.22}$$

Given (3.21) and (3.22) we obtain (3.19) by induction in k.

Taking $f = 1$ when $k = 1$ gives $\lambda_1 = 0$ and shows that φ_1 is a constant a. Then $1 = \langle \varphi_1, \varphi_1 \rangle = a^2 \mu(\mathbb{V})$, so $a = \mu(\mathbb{V})^{-1/2}$. Since the space of functions is finite dimensional, the infimum for the variational problem when $k = 2$ is attained by some function f_2; since f_2 is orthogonal to φ_1, the function f_2 must be non-constant, and so (since Γ is connected) we have $\mathscr{E}(f_2, f_2) > 0$, and thus $\lambda_2 > 0$.

The final inequality follows on setting $b = a\langle f, \varphi_1 \rangle$ and using (3.21) and (3.22) for $g = f - b$. $\square$

The size of the eigenvalues controls the speed of convergence of the transition density to a constant, in the L^2 sense.

Lemma 3.37 *Let $\mathbb{V}$ be finite, ν be a probability measure on $\mathbb{V}$, and f_n be the $\mathbb{P}^\nu$-density of X_n:*

$$f_n(y) = \sum_x \nu_x p_n(x, y).$$

Then, if $a = \mu(\mathbb{V})^{-1/2}$ and $\rho^* = \max_{i \geq 2} |\rho_i|$,

$$||f_k - a^2||_2^2 \leq (\rho^*)^k ||f_0 - a^2||_2^2.$$

Proof We have $f_0(x) = \nu_x/\mu_x$, so that

$$f_k(y) = \sum_x f_0(x) p_k(x, y) \mu_y = \langle f_0, p_k(\cdot, y) \rangle = \sum_{i=1}^N \langle f_0, \varphi_i \rangle \varphi_i(y) \rho_i^k.$$

Since $\varphi_1 = a$ and $\rho_1 = 1$, $\langle f_0, \varphi_1 \rangle \varphi_1(y) \rho_1^k = a^2$, and thus

$$f_k - a^2 = \sum_{i=2}^N \langle f_0, \varphi_i \rangle \rho_i^k \varphi_i.$$

So

$$||f_k - a^2||_2^2 = \sum_{i=2}^N \langle f_0, \varphi_i \rangle^2 \rho_i^{2k} \leq (\rho^*)^k ||f_0 - a^2||_2^2. \qquad \square$$

In the above ρ^* is either ρ_2 or ρ_N. In considering rates of convergence to equilibrium for discrete time random walks, the possibility of negative values for ρ_i is a nuisance. One way of avoiding this is to look at walks in continuous time; another is to consider the lazy walk defined in Remark 1.53. It is easy to check that the semigroup operator for the lazy walk is given by

$$P^{(L)} = \tfrac{1}{2}(I + P),$$

and so if φ_i is an eigenfunction for P then it is also one for $P^{(L)}$, with eigenvalue

$$\rho_i^{(L)} = \tfrac{1}{2}(1 + \rho_i).$$

As $\rho_i \in [-1, 1]$ we have $\rho_i^{(L)} \in [0, 1]$, and $\max_{i \geq 2} |\rho_i^{(L)}| = \rho_2^{(L)}$. By Lemma 3.37, applied to the lazy walk, we obtain

$$||(P^{(L)})^k f_0 - a^2||_2^2 \leq (\tfrac{1}{2} + \tfrac{1}{2}\rho_2)^k ||f_0 - a^2||_2^2.$$

The quantity $\lambda_2 = 1 - \rho_2$ is often called the *spectral gap*.

Lemma 3.38 (Cheeger's inequality) *Let Γ be finite and λ_2 be the smallest non-zero eigenvalue of $-\Delta$. Then*

$$\tfrac{1}{2}R_I^2 \le \lambda_2 \le 2R_I. \tag{3.23}$$

Proof By (3.20) and (3.21) we have

$$\lambda_2 = \inf\{\mathcal{E}(f,f) : ||f||_2 = 1, \overline{f} = 0\}.$$

Thus the first inequality is immediate from (3.11).

To prove the second inequality, let A be a set which attains the minimum in (3.8), set $b = \mu(A)/\mu(\mathbb{V})$, and let $f = \mathbf{1}_A - b$. Then $\overline{f} = 0$, $\mathcal{E}(f,f) = \mu_E(A; \mathbb{V} - A)$, and $||f||_2^2 = \mu(A)(1-b)$. So

$$\lambda_2 \le \frac{\mathcal{E}(f,f)}{||f||_2^2} = \frac{\mu_E(A; \mathbb{V} - A)}{\mu(A)(1-b)} \le 2R_I. \qquad \square$$

3.5 Strong Isoperimetric Inequality and Spectral Radius

By Theorem 3.7, Lemma 3.9, and Lemma 3.18 we have

$$(I_\infty) \Leftrightarrow (F_\infty^1) \Leftrightarrow (S_\infty^1) \Leftrightarrow (S_\infty^2).$$

Definition 3.39 Let $x, y \in \mathbb{V}$. The *spectral radius* ρ is defined by

$$\rho = \rho(\Gamma, \mu) = \limsup_{n\to\infty} \mathcal{P}_n(x,y)^{1/n} = \limsup_{n\to\infty} p_n(x,y)^{1/n}. \tag{3.24}$$

Proposition 3.40 *(a) The definition in* (3.24) *is independent of x, y.*
(b) $0 < \rho \le 1$.
(c) For all $x, y \in \mathbb{V}$, $n \ge 0$

$$p_n(x,y)(\mu_x\mu_y)^{1/2} \le \rho^n.$$

(d) For each $x \in \mathbb{V}$, $\rho = \lim_{n\to\infty} \mathcal{P}_{2n}(x,x)^{1/2n}$.

Proof (a) Write $\rho(x,y) = \limsup \mathcal{P}_n(x,y)^{1/n}$. Let $x, x', y \in \mathbb{V}$. Then as Γ is connected there exists k such that $p = \mathcal{P}_k(x,x') > 0$. Hence $\mathcal{P}_{n+k}(x,y) \ge \mathcal{P}_k(x,x')\mathcal{P}_n(x',y)$, so

$$\begin{aligned}\rho(x,y) &= \limsup_n (\mathcal{P}_{n+k}(x,y)^{1/(n+k)}) \\ &\ge \limsup_n p^{1/(n+k)} \mathcal{P}_n(x',y)^{1/(k+n)} = \rho(x',y).\end{aligned}$$

It follows that $\rho(x,y)$ is the same for all x, y.

(b) Since $\mathscr{P}_n(x,x) \le 1$ it is clear that $\rho \le 1$. Also, $\mathscr{P}_{2n}(x,x) \ge \mathscr{P}_2(x,x)^n$, so $\rho \ge \mathscr{P}_2(x,x)^{1/2} > 0$.
(c) Consider the case $y = x$ first. Suppose for some k and $\varepsilon > 0$ that $\mathscr{P}_k(x,x) = p_k(x,x)\mu_x \ge (1+\varepsilon)\rho^k$. Then $\mathscr{P}_{nk}(x,x) \ge \mathscr{P}_k(x,x)^n \ge (1+\varepsilon)^n \rho^{nk}$, so $\rho \ge \limsup_k \mathscr{P}_{nk}(x,x)^{1/nk} \ge (1+\varepsilon)^{1/k}\rho$, a contradiction.
If $x \neq y$ then

$$\rho^{2n} \ge p_{2n}(x,x)\mu_x \ge p_n(x,y)^2 \mu_y \mu_x.$$

(d) Let $a_n = \mathscr{P}_{2n}(x,x)$, so that $a_n \ge \mathscr{P}_2(x,x)^n > 0$. Now $a_{n+m} \ge a_n a_m$, so that the sequence (a_n) is supermultiplicative. Hence by Fekete's lemma (i.e. the convergence of a subadditive sequence) there exists κ such that

$$\lim_n a_n^{1/n} = \sup_n a_n^{1/n} = \kappa;$$

thus $\rho = \kappa^{1/2} = \lim_n p_{2n}(x,x)^{1/2n} = \sup_n p_{2n}(x,x)^{1/2n}$. □

Proposition 3.41 $||P||_{2\to 2} = \rho$.

Proof Write $||P|| = ||P||_{2\to 2}$. First, $||P^n|| \le ||P||^n$, so if $f = \mathbf{1}_{x_0}$ then $P^n f(x) = p_n(x, x_0)\mu_{x_0}$ and

$$||P||^{2n}||f||_2^2 \ge ||P^n f||_2^2 = \sum_x p_n(x,x_0)^2 \mu_x \mu_{x_0}^2 = p_{2n}(x_0,x_0)\mu_{x_0}^2.$$

Thus $||P|| \ge \lim_n p_{2n}(x_0,x_0)^{1/2n} = \rho$, using Proposition 3.40(d).
Let $f \in C_0^+(\mathbb{V})$ and $b_n = \langle P^n f, P^n f\rangle = \langle P^{2n} f, f\rangle$. Then

$$\lim_n b_n^{1/n} = \lim_n \Big(\sum_{x,y} p_{2n}(x,y) f(x) f(y) \mu_{xy}\Big)^{1/n} = \rho^2,$$

by Proposition 3.40(c) and (d). On the other hand, using Cauchy–Schwarz

$$b_{n+1}^2 = \langle P^{n+2} f, P^n f\rangle^2 \le \langle P^{n+2} f, P^{n+2} f\rangle \langle P^n f, P^n f\rangle = b_{n+2} b_n.$$

So,

$$d_n = \frac{b_{n+1}}{b_n} \le \frac{b_{n+2}}{b_{n+1}} = d_{n+1},$$

and thus $d_n \uparrow D$. Since $b_{n+1} \le D b_n$ it follows that $b_n \le D^n b_0$, and so $\rho^2 \le D$. But also $b_{k+n} \ge d_k^n b_k$, so $\rho^2 = \lim_n b_{k+n}^{1/n} \ge d_k$, and hence $D = \rho^2$. Finally,

$$||Pf||_2^2 = \langle Pf, Pf\rangle = b_1 \le D b_0 = \rho^2 ||f||_2^2,$$

so $||P|| \le \rho$. □

Definition 3.42 A weighted graph (Γ, μ) is *amenable* if there exists a sequence of sets A_n with

$$\frac{\mu_E(A_n; \mathbb{V} - A_n)}{\mu(A_n)} \to 0 \text{ as } n \to \infty.$$

Thus (Γ, μ) is amenable if and only if it fails to satisfy (I_∞).

Theorem 3.43 *(a)*

$$1 - \rho = \inf\left\{\frac{\mathscr{E}(f, f)}{\langle f, f\rangle} : f \neq 0, f \in C_0(\mathbb{V})\right\}.$$

(b) $\rho < 1$ *if and only if* (Γ, μ) *satisfies* (I_∞).
(c) $\rho = 1$ *if and only if* (Γ, μ) *is amenable.*
(d) The property that $\rho < 1$ *is stable under rough isometries.*

Proof (a) By Lemma A.32

$$\rho = ||P|| = \sup_{||f||_2=1} \langle Pf, f\rangle,$$

so

$$1 - \rho = \inf\left\{\frac{\langle f, f\rangle - \langle Pf, f\rangle}{\langle f, f\rangle} : f \neq 0\right\} = \inf\left\{\frac{\mathscr{E}(f, f)}{\langle f, f\rangle} : f \neq 0\right\}.$$

(b) and (c) From (a), if $\rho < 1$ then (S^2_∞) holds with constant $(1 - \rho)^{-1}$, while if (S^2_∞) holds with constant C_1 then (a) implies that $1 - \rho \geq C_1^{-1}$.
(d) follows from (b) and Theorem 3.4. □

It follows from this and Lemma 3.3 that the only graphs with $\rho < 1$ are those with exponential (or faster) volume growth. The converse is not true – 'lamplighter graphs' give an example.

Example (Lamplighter graphs) Let $\mathbb{V}' = \mathbb{Z} \times \{0, 1\}^{\mathbb{Z}}$. We write a typical element $x \in \mathbb{V}'$ as $x = (n, \xi)$, where $n \in \mathbb{Z}$ and $\xi = (\ldots, \xi_{-1}, \xi_0, \xi_1, \ldots)$ is a doubly infinite sequence of 0s and 1s. We take

$$\mathbb{V} = \{(n, \xi) : \sum_{-\infty}^{\infty} \xi_i < \infty\}.$$

We think of n as the location of the lamplighter, and ξ as giving the state of the lamps – of which only finitely many are in state 1. We define edges so that $(n, \xi) \sim (n + 1, \xi')$ if and only if $\xi'_k = \xi_k$ for $k \neq n, n + 1$. Thus each point x has eight neighbours. At each time step the lamplighter moves according to a SRW on $\mathbb{Z}$, and then randomises the state of the lamps at his old and new location.

Let η be the configuration with $\eta_k = 0$ for all k, and set $\mathbf{0} = (0, \eta)$. Let $n \geq 1$ and F be the event that the lamplighter returns to 0 at time $2n$, and has visited no more than $(2n)^{1/2}$ points of $\mathbb{Z}$. Then we have $\mathbb{P}^{\mathbf{0}}(F) \geq c_0 n^{-1/2}$. Let G be the event that each lamp visited at a time in $\{0, \ldots, 2n\}$ is switched off at the last visit before time $2n$. If k is the number of lamps visited, then $\mathbb{P}(G) = 2^{-k}$. Hence

$$\mathscr{P}_{2n}(\mathbf{0}, \mathbf{0}) = \mathbb{P}^{\mathbf{0}}(X_{2n} = \mathbf{0}) \geq \mathbb{P}^{\mathbf{0}}(F \cap G) \geq 2^{-(2n)^{1/2}} c_0 n^{1/2}.$$

Hence $\mathscr{P}_{2n}(\mathbf{0}, \mathbf{0})^{1/2n} \to 1$ as $n \to \infty$, so $\rho = 1$ and the lamplighter graph is amenable.

On the other hand, $B(\mathbf{0}, n)$ contains 2^{n+1} points of the form (n, ξ), where ξ is any configuration with $\xi_k = 0$ if $k \notin \{0, \ldots, n\}$, so that Γ has exponential volume growth.

This graph is vertex transitive, but has some surprising properties. For example, there are many points x with $d(x, \mathbf{0}) = n$ such that x has no neighbours at distance $n + 1$ from $\mathbf{0}$. See [Rev] for the behaviour of the heat kernel $p_n(x, y)$ on this graph.

Recall from Chapter 1 the definition of the potential operator G.

Corollary 3.44 *$||G||_{2\to 2} < \infty$ if and only if $\rho < 1$. In particular $||G||_{2\to 2} = \infty$ for $\mathbb{Z}^d$.*

Proof Since $G = \sum_n P^n$, we have $||G|| = ||G||_{2\to 2} \leq \sum_n \rho^n = (1 - \rho)^{-1}$ if $\rho < 1$.

If $\rho = 1$ then $||P^n|| = 1$ for all $n \geq 0$. Let $\varepsilon > 0$, and choose $m \geq 1$. Since $P|f| \geq |Pf|$ there exists $f \in L^2_+$ with $||f||_2 = 1$ such that $||P^{2m} f||_2 \geq (1 - \varepsilon)$. Hence we have $||P^k f||_2 \geq (1 - \varepsilon)$ for each $0 \leq k \leq 2m$. Since $P^j f \geq 0$ for each j, we have $Gf \geq \sum_{k=0}^m P^{2k} f$, and thus

$$||Gf||_2^2 \geq ||\sum_{k=0}^m P^{2k} f||_2^2 = \left\langle \left(\sum_{k=0}^m P^{2k} \right)^2 f, f \right\rangle = \sum_{j=0}^{2m} a_j \langle P^{2j} f, f \rangle,$$

where $(1 + t^2 + \cdots + t^{2m})^2 = \sum_{j=0}^{2m} a_j t^{2j}$. So

$$||Gf||_2^2 \geq (1 - \varepsilon)^2 \sum_{j=0}^{2m} a_j = (1 - \varepsilon)^2 m^2,$$

and $||G||_{2\to 2}$ must be infinite. □

Definition 3.45 We say the random walk X is *ballistic* if

$$\liminf_{n\to\infty} \frac{1}{n} d(x, X_n) > 0, \quad \mathbb{P}^x\text{-a.s. for each } x \in \mathbb{V}.$$

Theorem 3.46 *Suppose that* (Γ, μ) *satisfies (H5). If* $\rho < 1$ *then* X *is ballistic.*

Proof This follows from Proposition 3.40(d) and Lemma 1.3 by a simple 'exhaustion of mass' argument. Let $x \in \mathbb{V}$, $\lambda > 0$ and set $B_n = B(x, \lambda n)$. Set $F_n = \{X_n \in B_n\}$. Then

$$\mathbb{P}^x\big(X_n \in B_n\big) = \sum_{y \in B_n} p_n(x, y)\mu_y \le \sum_{y \in B_n} \rho^n \mu_y^{1/2} \mu_x^{-1/2}$$
$$\le \mu_x^{-1/2} \rho^n |B_n|^{1/2} \mu(B_n)^{1/2} \le c C_1^{\lambda n} \rho^n.$$

Thus if λ is chosen small enough so that $\rho C_1^{\lambda} < 1$ then $\sum_n \mathbb{P}^x(F_n) < \infty$, and by Borel–Cantelli

$$\mathbb{P}^x(F_n \text{ occurs infinitely often }) = 0.$$

Hence $\liminf_n n^{-1} d(x, X_n) \ge \lambda$. □

Examples (1) Theorem 3.46 gives a sufficient condition for ballistic behaviour, but it is far from necessary. Consider for example the graph $\mathbb{V}$ which is the join of $\mathbb{Z}^d$ with the binary tree $\mathbb{B}$. Since $\mathbb{B}$ is transient, X can cross between the two parts only a finite number of times. If $d \ge 3$ then X will ultimately escape to infinity either in $\mathbb{B}$, in which case it moves ballistically, or else in $\mathbb{Z}^d$, in which case it escapes non-ballistically.

If $d \le 2$ then X is ultimately in $\mathbb{B}$, so is ballistic. However, it is easy to see, by looking at sets in the $\mathbb{Z}^2$ part of the graph, that $\mathbb{V}$ does not satisfy (I_∞), so that $\rho(\mathbb{V}) = 1$.

(2) The random walk on the lamplighter graph is not ballistic, although this graph has exponential volume growth.

4

Discrete Time Heat Kernel

4.1 Basic Properties and Bounds on the Diagonal

We will now use the inequalities in Chapter 3 to prove 'heat kernel' bounds, that is, bounds on the transition density $p_n(x, y)$. We begin with some basic properties of the heat kernel.

Lemma 4.1 *For $x, y \in \mathbb{V}$, $n \geq 0$,*

$$p_{2n+1}(x, x) \leq p_{2n}(x, x), \tag{4.1}$$

$$p_{2n+2}(x, x) \leq p_{2n}(x, x), \tag{4.2}$$

$$p_{2n}(x, y) \leq p_{2n}(x, x)^{1/2} p_{2n}(y, y)^{1/2}, \tag{4.3}$$

$$p_{2n+1}(x, y) \leq p_{2n}(x, x)^{1/2} p_{2n+2}(y, y)^{1/2}. \tag{4.4}$$

Further, if $r_n(x, y) = p_n(x, y) + p_{n+1}(x, y)$, then

$$\mathscr{E}(r_n^x, r_m^y) = r_{n+m}(x, y) - r_{n+m+2}(x, y), \tag{4.5}$$

$$\langle r_n^x, r_m^y \rangle = r_{n+m}(x, y) + r_{n+m+1}(x, y). \tag{4.6}$$

In particular $r_{2n}(x, x)$ is decreasing in n.

Proof By (1.12)–(1.14)

$$p_{2n}(x, x) - p_{2n+1}(x, x) = \langle p_n^x - p_{n+1}^x, p_n^x \rangle = \langle -\Delta p_n^x, p_n^x \rangle = \mathscr{E}(p_n^x, p_n^x) \geq 0,$$

proving (4.1). To prove (4.2) we use Proposition 1.16:

$$p_{2n+2}(x, x) = \langle p_{n+1}^x, p_{n+1}^x \rangle = \langle P p_n^x, P p_n^x \rangle = ||P p_n^x||_2^2 \leq ||p_n^x||_2^2 = p_{2n}(x, x).$$

Equations (4.3) and (4.4) are immediate from (1.12) and Cauchy–Schwarz. (Note that we cannot expect $p_k(x, y)^2 \leq p_k(x, x) p_k(y, y)$ in general when k is odd; any bipartite graph gives a counterexample.)

For the equalities involving r^x_n, note first that

$$\Delta r^x_n = \Delta(p^x_n + p^x_{n+1}) = (p^x_{n+1} - p^x_n) + (p^x_{n+2} - p^x_{n+1}) = p^x_{n+2} - p^x_n.$$

So,

$$\begin{aligned}\mathscr{E}(r^x_n, r^y_m) &= \langle -\Delta r^x_n, r^y_m\rangle \\ &= \langle p^x_n - p^x_{n+2}, p^y_m + p^y_{m+1}\rangle \\ &= p_{n+m}(x,y) + p_{n+m+1}(x,y) - p_{n+m+2}(x,y) - p_{n+m+3}(x,y) \\ &= r_{n+m}(x,y) - r_{n+m+2}(x,y).\end{aligned}$$

A similar calculation gives (4.6). □

Lemma 4.2 *Let $f \in L^2$. Then for $n \ge 0$*

$$0 \le ||P^{n+1}f||_2^2 - ||P^{n+2}f||_2^2 \le ||P^n f||_2^2 - ||P^{n+1}f||_2^2. \tag{4.7}$$

Further, for $n \ge 1$,

$$||f||_2^2 - ||P^n f||_2^2 \le 2n\mathscr{E}(f,f). \tag{4.8}$$

Proof The first inequality in (4.7) holds since $||P||_{2\to 2} \le 1$ by Proposition 1.16. For the second, writing $g = P^n f$, we have

$$0 \le \langle P^2 g - g, P^2 g - g\rangle = \langle g, g\rangle - 2\langle g, P^2 g\rangle + \langle P^2 g, P^2 g\rangle,$$

and rearranging gives

$$\langle Pg, Pg\rangle - \langle P^2 g, P^2 g\rangle \le \langle g, g\rangle - \langle Pg, Pg\rangle,$$

which proves the second inequality. Since $P = I + \Delta$, we have

$$\langle f, f\rangle - \langle Pf, Pf\rangle = -2\langle f, \Delta f\rangle - \langle \Delta f, \Delta f\rangle \le -2\langle f, \Delta f\rangle = 2\mathscr{E}(f,f),$$

which proves (4.8) when $n = 1$. If $n \ge 2$ then, using (4.7),

$$||f||_2^2 - ||P^n f||_2^2 = \sum_{k=0}^{n-1}\left(||P^k f||_2^2 - ||P^{k+1}f||_2^2\right) \le n\left(||f||_2^2 - ||Pf||_2^2\right),$$

and (4.8) follows. □

Theorem 4.3 *Let $\alpha \ge 1$. The following are equivalent.*

(a) Γ satisfies the Nash inequality (N_α).
(b) There exists C_H such that

$$p_n(x,x) \le \frac{C_H}{(1 \vee n)^{\alpha/2}}, \quad n \ge 0, \ x \in \mathbb{V}. \tag{4.9}$$

(c) There exists C'_H such that

$$p_n(x, y) \le \frac{C'_H}{(1 \vee n)^{\alpha/2}}, \quad n \ge 0, \ x, y \in \mathbb{V}. \tag{4.10}$$

Proof (a) $\Rightarrow$ (b) Assume the Nash inequality holds with constant C_N. If $n = 0$ we have $p_0(x, x) = \mu_x^{-1} \le C_N^{-\alpha/2}$ by Remark 3.11(3). For $n \ge 1$ we work with the function $r_n^x(y) = r_n(x, y) = p_n(x, y) + p_{n+1}(x, y)$ introduced in Lemma 4.1. We have $||r_n^x||_1 = 2$, while $||r_n^x||_2^2 \ge r_{2n}(x, x)$ by (4.6). Set $\varphi_n = r_{2n}(x, x)$; then by (4.5)

$$\varphi_n - \varphi_{n+1} = \mathscr{E}(r_n^x, r_n^x) \ge C_N ||r_n^x||_1^{-4/\alpha} ||r_n^x||_2^{2+4/\alpha} \ge 2^{-4/\alpha} C_N \varphi_n^{1+2/\alpha}.$$

So we obtain the difference inequality

$$\varphi_{n+1} - \varphi_n \le -c_1 \varphi_n^{1+2/\alpha}. \tag{4.11}$$

While we could treat (4.11) directly, a simple comparison with a continuous function will give the bound we need. Let f be the continuous function on $[0, \infty)$ which is linear on each interval $(n, n+1)$ and satisfies $f(n) = \varphi_n$ for each $n \in \mathbb{Z}_+$. Since φ_n is decreasing, so is f and $f(t) \le \varphi_n$ for $t \in [n, n+1]$. Then

$$f'(t) = \varphi_{n+1} - \varphi_n \le -c_1 \varphi_n^{1+2/\alpha} \le -c_1 f(t)^{1+2/\alpha}, \quad n < t < n+1.$$

Now let $g(t) = f(t)^{-2/\alpha}$, so

$$g'(t) = -\frac{2}{\alpha} f'(t) f(t)^{-1-2/\alpha} \ge \frac{2c_1}{\alpha}.$$

Hence $g(t) \ge g(0) + c_2 t \ge c_2 t$ since $g(0) \ge 0$, and thus

$$p_{2n}(x, x) \le r_{2n}(x, x) = \varphi_n = f(n) = g(n)^{-\alpha/2} \le c_2^{-\alpha/2} n^{-\alpha/2},$$

proving (b). Inspecting the constants we can take $C_H = 1 \vee 4(\alpha/2C_N)^{\alpha/2}$.
(b) $\Leftrightarrow$ (c) Clearly (c) implies (b). If (b) holds, and $n = 2m$ is even, then (4.10) follows immediately from (4.9) and (4.3). If $n = 2m+1$ is odd, since $2m + 1 \le \frac{3}{2}(1 \vee (2m))$, using (4.3) gives

$$\begin{aligned} p_{2m+1}(x, y) &\le p_{2m}(x, x)^{1/2} p_{2m+2}(y, y)^{1/2} \\ &\le c(1 \vee 2m)^{-\alpha/4} (1 \vee (2m+2))^{-\alpha/4} \le c'(2m+1)^{-\alpha/2}, \end{aligned}$$

which is (c).
(c) $\Rightarrow$ (a) Write $\pi_n = C'_H (1 \vee n)^{-\alpha/2}$ for the upper bound in (4.10). Taking $n = 0$ we obtain $\mu_x^{-1} = p_0(x, x) \le C'_H$. For $f \in C_0(\mathbb{V})$,

$$||P_n f||_\infty = \sup_x \sum_y p_n(x, y) f(y) \mu_y \le \pi_n ||f||_1,$$

and so

$$||P_n f||_2^2 = \langle P^{2n} f, f\rangle \le ||P^{2n} f||_\infty ||f||_1 \le \pi_{2n}||f||_1^2.$$

So by Lemma 4.2

$$2\mathscr{E}(f,f) \ge \frac{\|f\|_2^2 - ||P^n f||_2^2}{n} \ge \frac{\|f\|_2^2 - \pi_{2n}\|f\|_1^2}{n}. \tag{4.12}$$

We can assume that $||f||_1 = 1$. Choose n to be the smallest integer $k \ge 1$ such that $\pi_{2k} \le \frac{1}{2}||f||_2^2$. If $n = 1$ then we have, using (1.17), Lemma 1.20, and the lower bound on μ_x,

$$c_2 2^{-\alpha/2} \le \tfrac{1}{2}||f||_2^2 \le 2\mathscr{E}(f,f) \le 2||f||_2^2 \le c'||f||_1^2 = c'.$$

Hence we obtain $\mathscr{E}(f,f) \ge c||f||_2^{2+4/\alpha}$. If $n \ge 1$ then we have $n^{-\alpha/4} \asymp ||f||_2$, and substituting in (4.12) gives (N_α). □

Remark The argument in continuous time is slightly simpler – see Remark 5.15.

Corollary 4.4 *(a) Suppose that* (Γ, μ) *satisfies* (I_α). *Then*

$$p_n(x,y) \le c(1 \vee n)^{-\alpha/2} \text{ for } n \ge 0,\ x, y \in \mathbb{V}.$$

(b) Let $\mathbb{V}$ *be infinite, and suppose that* $\mu_e \ge c_0 > 0$ *for all edges e. Then*

$$p_n(x,y) \le c(1 \vee n)^{-1/2} \text{ for } n \ge 0,\ x, y \in \mathbb{V}.$$

Proof By Corollary 3.15 we have (N_α), and hence (a) holds. For (b), Example 3.2(5) gives that I_1 holds, and the bound follows from (a). □

Remark (1) Bounds on $p_n(x,x)$ are called 'on diagonal', since they control the function $p_n : \mathbb{V} \times \mathbb{V} \to \mathbb{R}$ on the diagonal $\{(x,x), x \in \mathbb{V}\}$. Corollary 4.4 shows that (up to constants) the on-diagonal decay of p_n on $\mathbb{Z}$ is the slowest possible for an infinite graph with natural weights. Note also that (4.3) and (4.4) give global upper bounds on $p_n(x,y)$ from the on-diagonal upper bounds. (2) Corollary 4.4 is one of the main results in [V1]; in this and other papers in the early 1980s Varopoulos opened up in a very fruitful way the connections between the geometry of a graph and the properties of a random walk on it.

Examples (1) Let $\Gamma = \mathbb{Z}^d$, $\mu_{xy}^{(0)}$ be the natural weights on $\mathbb{Z}^d$, and $\mu_{xy}^{(1)}$ be weights on $\mathbb{Z}^d$ satisfying $c_0\,\mu_{xy}^{(0)} \le \mu_{xy}^{(1)} \le c_1\mu_{xy}^{(0)}$. Let $\mathscr{E}_i$ be the Dirichlet forms associated with $(\Gamma, \mu^{(i)})$, write $X^{(i)}$ for the associated SRW and $p_n^{(i)}(x,y)$ for their transition densities. By Lemma 3.13 the Nash inequality (N_d) holds for $\mathbb{Z}^d$. Since $\mathscr{E}_1(f, f) \geq c_0\mathscr{E}_0(f, f)$, we have also (N_d) with respect to $\mathscr{E}_1$, and therefore the bound $p_n^{(1)}(x, y) \leq c'n^{-d/2}$ holds.

(2) Let Γ be the join of two copies of $\mathbb{Z}^d$ at their origins. By Exercise 3.20, Γ satisfies (N_d), and so by Theorem 4.3 the heat kernel on $\mathbb{V}$ satisfies the bound

$$p_n(x, y) \leq Cn^{-d/2}.$$

(3) The graphical (pre)-Sierpinski carpet Γ_{SC} is an infinite connected graph (a subset of $\mathbb{Z}^2_+$) which is roughly isometric with the (unbounded) Sierpinski carpet (Figure 4.1). For a precise definition see [BB1].

Let $A_n = \mathbb{V} \cap [0, \frac{1}{2}3^n]^2$. Then $\mu(A_n) \approx c8^n$, and $\mu_E(A_n; \mathbb{V} - A_n) \asymp 2^n$. So if $\mathbb{V}$ satisfies (I_α) then

$$2^n \geq c(8^n)^{(\alpha-1)/\alpha},$$

which implies that $\alpha \leq 3/2$. In fact (as is intuitively plausible) the sets A_n are the optimal case as far as isoperimetric properties of $\mathbb{V}$ are concerned.

Theorem 4.5 (Osada; see [Os]) Γ_{SC} *satisfies* (I_α) *with* $\alpha = 3/2$.

Using this and Theorem 4.3 it follows that Γ_{SC} satisfies

$$p_n(x, y) \leq cn^{-3/4}. \tag{4.13}$$

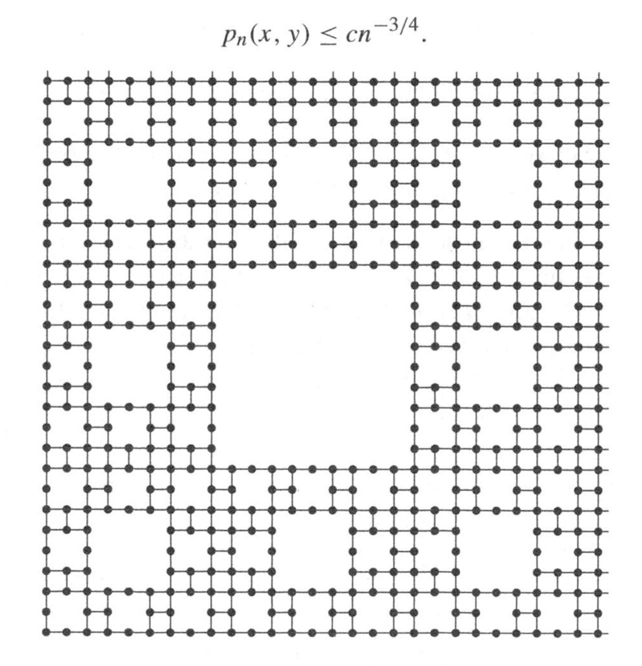

Figure 4.1. The graphical Sierpinski carpet.

However, this is not the correct rate of decay; in [BB1] it is proved that

$$p_n(x, y) \asymp n^{-\gamma},$$

where $\gamma = (\log 8)/(\log 8\rho)$. Here the exact value of ρ is unknown, but it is close to $5/4$, so that $\gamma \simeq 0.90 > 0.75$.

The reason (4.13) is not the best possible is that there is a loss of information in passing from (I_α) to (N_α). The inequality (I_α) implies that (in a certain sense) the graph Γ has 'good geometry', while (N_α) or (S^2_α) imply 'good heat flow'. Good geometry is sufficient, but not necessary, for good heat flow.

4.2 Carne–Varopoulos Bound

The bounds in Theorem 4.3 do not give any information on the behaviour of $p_n(x, y)$ for fixed n when $d(x, y)$ is large. The next few results will lead to a very simple upper bound for $p_n(x, y)$ in this regime, by a surprising argument. For the motivation and history, see Remark 4.10 later in this section. We begin with a few lemmas.

Lemma 4.6 *Let S_n be the simple symmetric random walk on $\mathbb{Z}$ with $S_0 = 0$. Then*

$$\mathbb{P}(S_n \geq D) \leq \exp(-\frac{D^2}{2n}), \tag{4.14}$$

$$\sum_{r \in \mathbb{Z}} \lambda^r \mathbb{P}(S_n = r) = 2^{-n} \sum_{r=0}^{n} \lambda^{2r-n} \binom{n}{r}. \tag{4.15}$$

Proof We first note that $\cosh \lambda \leq \exp(\frac{1}{2}\lambda^2)$ – one proof is by comparison of the power series. Then, since $\mathbb{E}(e^{\lambda S_1}) = \cosh(\lambda)$,

$$\begin{aligned}\mathbb{P}(S_n \geq D) &= \mathbb{P}(e^{\lambda S_n} \geq e^{\lambda D}) \leq e^{-\lambda D}\mathbb{E}e^{\lambda S_n} \\ &\leq e^{-\lambda D}(\mathbb{E}e^{\lambda S_1})^n = e^{-\lambda D}(\cosh \lambda)^n \leq \exp(-\lambda D + \tfrac{1}{2}n\lambda^2).\end{aligned}$$

Setting $\lambda = D/n$ gives (4.14).

For (4.15), fix n and $r = S_n$, and let u and d be the number of upward and downward steps by $(S_1, \ldots, S_n)$. Thus $u + d = n$ and $u - d = r$, and so if $n + r$ is odd then S_n cannot equal r, while if $n + r$ is even

$$\mathbb{P}(S_n = r) = 2^{-n}\binom{n}{u} = 2^{-n}\binom{n}{\frac{1}{2}(n+r)}.$$

So

$$\sum_{r\in\mathbb{Z}} \lambda^r \mathbb{P}(S_n = r) = \sum_{r=0}^{n} \lambda^{2r-n}\mathbb{P}(S_n = 2r-n) = 2^{-n}\sum_{r=0}^{n} \lambda^{2r-n}\binom{n}{r}. \quad \square$$

Lemma 4.7 *Let $q(t) = \sum_{r=0}^{k} a_r t^r$ be a polynomial with $|q(t)| \le 1$ for $t \in [-1, 1]$. If $R = q(P) = \sum_{r=0}^{k} a_r P^r$ then $||Rf||_2 \le ||f||_2$.*

Proof We first assume that Γ is finite. Using the spectral representation (3.18),

$$Rf = \sum_{r=0}^{k} a_r P^r f = \sum_{r=0}^{k} a_r \Big(\sum_{i=1}^{N} \langle f, \varphi_i\rangle \rho_i^r \varphi_i \Big) = \sum_{i=1}^{N} q(\rho_i)\langle f, \varphi_i\rangle \varphi_i.$$

Since $|q(\rho_i)| \le 1$,

$$||Rf||_2^2 = \sum_{i=1}^{N} q(\rho_i)^2 \langle f, \varphi_i\rangle^2 \le \sum_{i=1}^{N} \langle f, \varphi_i\rangle^2 = \|f\|_2^2.$$

The case of infinite Γ could be handled by spectral integrals, but there is an easier way. Let $f \in C_0(\mathbb{V})$, with (finite) support A. Let $A_k = \{x : d(x, A) \ge k\}$, and note that $Rf(x) = 0$ if $x \in A_{k+1}$. So $\|f\|_2$ and $\|Rf\|_2$ are unchanged if we replace Γ by the subgraph generated by A_{k+1}, and we deduce that $\|Rf\|_2 \le \|f\|_2$. A limiting argument now handles the case of general $f \in L^2$. $\square$

We now define the Chebyshev polynomial

$$H_k(t) = \tfrac{1}{2}(t + i(1-t^2)^{1/2})^k + \tfrac{1}{2}(t - i(1-t^2)^{1/2})^k, \quad -1 \le t \le 1.$$

Lemma 4.8 *For each k, $H_k(t)$ is a real polynomial in t of degree k, and*

$$|H_k(t)| \le 1, \quad -1 \le t \le 1.$$

Further, for each $n \ge 0$,

$$t^n = \sum_{k\in\mathbb{Z}} \mathbb{P}(S_n = k) H_{|k|}(t), \quad -1 \le t \le 1. \tag{4.16}$$

Proof Fix t and let $s = (1-t^2)^{1/2}$. Then the binomial theorem gives

$$2H_k(t) = \sum_{j=0}^{k} \binom{k}{j} t^{k-j}((is)^j + (-is)^j).$$

The terms with j odd cancel, and so $2H_k(t)$ is a real polynomial in t, which is of degree k. Let $z_1 = t + is$, $z_2 = t - is$. Then $|z_i|^2 = 1$, $z_1 z_2 = 1$, and $H_k(t) = \frac{1}{2}(z_1^k + z_2^k) = H_{-k}(t)$. So

$$|H_k(t)| \le \tfrac{1}{2}|z_1|^k + \tfrac{1}{2}|z_2|^k = 1.$$

Since $t = 2^{-1}(z_1 + z_2)$ we have

$$t^n = 2^{-n} \sum_{k=0}^{n} \binom{n}{k} z_1^k z_2^{n-k} = 2^{-n} \sum_{k=0}^{n} \binom{n}{k} z_1^{2k-n} = \sum_{r \in \mathbb{Z}} z_1^r \mathbb{P}(S_n = r).$$

Similarly $t^n = \sum_r z_2^r \mathbb{P}(S_n = r)$, and adding gives (4.16). □

Theorem 4.9 ('Carne–Varopoulos bound' [Ca, V2]) *Let (Γ, μ) be a weighted graph. Then*

$$p_n(x, y) \le \frac{2}{(\mu_x \mu_y)^{1/2}} \exp\Big(-\frac{d(x,y)^2}{2n}\Big), \quad x, y \in \mathbb{V},\ n \ge 1. \tag{4.17}$$

Proof Let $x_1, x_2 \in \mathbb{V}$ and set $f_i(x) = \mathbf{1}_{x_i}(x) \mu_{x_i}^{-1/2}$. Thus $||f_i||_2 = 1$ for $i = 1, 2$. Let $R = d(x_1, x_2)$, and note that

$$\langle f_1, P^k f_2 \rangle = (\mu_{x_1} \mu_{x_2})^{1/2} p_k(x_1, x_2). \tag{4.18}$$

If $k < R$ then this is zero, and so $\langle f_1, H_{|k|}(P) f_2 \rangle = 0$.

The polynomial identity (4.16) implies that $P^n = \sum_{k \in \mathbb{Z}} \mathbb{P}(S_n = k) H_{|k|}(P)$, and therefore that

$$\begin{aligned}
\langle f_1, P^n f_2 \rangle &= \sum_{k \in \mathbb{Z}} \mathbb{P}(S_n = k) \langle f_1, H_{|k|}(P) f_2 \rangle \\
&= \sum_{|k| \ge R} \mathbb{P}(S_n = k) \langle f_1, H_{|k|}(P) f_2 \rangle \\
&\le \sum_{|k| \ge R} \mathbb{P}(S_n = k) ||f_1||_2 ||H_{|k|}(P) f_2||_2.
\end{aligned}$$

Using Lemma 4.7 we have $||H_{|k|}(P) f_2||_2 \le 1$, so

$$\langle f_1, P^n f_2 \rangle \le \sum_{|k| \ge R} \mathbb{P}(S_n = k) = \mathbb{P}(|S_n| \ge R) \le 2e^{-R^2/2n},$$

and combining this with (4.18) completes the proof. □

Remark 4.10 (1) The proof of Theorem 4.9 does not make it very clear why the argument works. Let $k_t(x)$ be the density of the $\mathcal{N}(0,t)$ distribution, so that the Fourier transform of k_t is

$$\int_{-\infty}^{\infty} (2\pi t)^{-1/2} e^{-x^2/(2t)} e^{i\theta x} dx = \int_{-\infty}^{\infty} (2\pi t)^{-1/2} e^{-x^2/(2t)} \cos(\theta x) dx = e^{-\frac{1}{2}\theta^2 t}.$$

Setting $\lambda = \frac{1}{2}\theta^2$ gives the identity

$$\int_{-\infty}^{\infty} (2\pi t)^{-1/2} e^{-x^2/(2t)} \cos(x\sqrt{2\lambda}) dx = e^{-\lambda t}.$$

Formally, if $\lambda = -\Delta$ one obtains

$$e^{t\Delta} = \int_{-\infty}^{\infty} k_t(s) \cos(s\sqrt{-2\Delta}) ds. \tag{4.19}$$

On the left side of (4.19) is the heat semigroup, while $\cos(x\sqrt{-2\Delta})$ is the wave semigroup: this is Hadamard's transmutation formula. The wave equation in $\mathbb{R}$ has finite velocity a, so there is no contribution to the heat kernel $p_t(x,y)$ from s such that $as < d(x,y)$. Varopoulos [V2] used this idea for graphs, replacing the graph Γ by what he called its 'cable system'. (One thinks of each edge as a wire of length 1, and runs a Brownian motion on the wires, with suitable behaviour at the junctions. Many people now call these quantum graphs, perhaps because it makes them sound more important.) Carne then found the fully discrete argument given here.

(2) If f_i have support A_i and $d(A_1, A_2) = R$ then the proof of Theorem 4.9 gives the 'Davies–Gaffney bound' (see [Da])

$$\langle P_n f_1, f_2 \rangle \le 2||f_1||_2||f_1||_2 e^{-R^2/2n}. \tag{4.20}$$

These bounds lead to an upper bound on the rate of escape of X_n in graphs with polynomial volume growth.

Definition 4.11 We say (Γ, μ) has *polynomial volume growth* if there exist C_V and θ such that

$$|B(x,r)| \vee \mu(B(x,r)) \le C_V r^\theta, \quad x \in \mathbb{V}, r \ge 1.$$

This condition involves both the number of points in balls and the measure of balls; of course these are the same if, for example, μ_x are bounded away from zero and infinity.

Lemma 4.12 *Suppose* (Γ, μ) *has polynomial volume growth, with index* θ. *If* $r^2 > 4n(1 + \theta \log 2)$ *then*

$$\mathbb{P}^x(d(x, X_n) > r) \le c r^\theta \exp(-r^2/4n). \tag{4.21}$$

Hence there exists c_2 *such that*

$$d(x, X_n) \le c_2 (n \log n)^{1/2} \text{ for all large } n,\ \mathbb{P}^x\text{-a.s.}$$

Proof Let $r \ge 1$, and set $D_n = B(x, 2^n r) - B(x, 2^{n-1} r)$ for $n \ge 1$. Then

$$\mathbb{P}^x(d(X_n, x) > r) = \sum_{k=1}^{\infty} \sum_{y \in D_k} p_n(x, y)\mu_y \le 2\mu_x^{-1/2} \sum_{k=1}^{\infty} \sum_{y \in D_k} \mu_y^{1/2} e^{-(2^{k-1}r)^2/2n}.$$

By Cauchy–Schwarz and the hypotheses on (Γ, μ),

$$\sum_{y \in D_k} \mu_y^{1/2} \le |D_k|^{1/2} \mu(D_k)^{1/2} \le c(2^k r)^\theta.$$

Since $4^k \ge 4k$ for $k \ge 1$, if $r^2 > 4n(1 + \theta \log 2)$ then

$$\begin{aligned}\mathbb{P}^x(d(X_n, x) > r) &\le c r^\theta \sum_{k=1}^{\infty} 2^{k\theta} e^{-(2^k r)^2/8n} \\ &= c r^\theta e^{-r^2/4n} \sum_{k=1}^{\infty} \exp(k\theta \log 2 - k r^2/4n) \le c' r^\theta e^{-r^2/4n},\end{aligned}$$

which proves (4.21). Now set $r_n^2 = 4An \log n$, so that the bound (4.21) holds for all large n. Then

$$\mathbb{P}^x(d(X_n, x) > r_n) \le c r_n^\theta n^{-A} \le c'(\log n)^\theta n^{\theta - A}.$$

Taking A large enough, $\sum_n \mathbb{P}^x(d(X_n, x) > r_n)$ converges, and the result follows by Borel–Cantelli. □

Remark 4.13 On $\mathbb{Z}^d$ we expect Gaussian upper bounds of the form

$$p_n(x, y) \le c_1 n^{-d/2} e^{-c_2 d(x,y)^2/n}. \tag{4.22}$$

(These bounds do indeed hold – see Theorem 6.28 in Chapter 6.) Let $R = d(x, y)$. If $R \le n^{1/2}$ then Corollary 4.4 gives

$$p_n(x, y) \le c_3 n^{-d/2} \le c_3 n^{-d/2} e^{1 - R^2/2n} = c_4 n^{-d/2} e^{-R^2/2n}.$$

On the other hand if $R \ge (2dn \log n)^{1/2}$ then Theorem 4.9 gives

$$p_n(x, y) \le e^{-R^2/2n} = e^{-R^2/4n - R^2/4n} \le e^{-R^2/4n} e^{-(d/2)\log n} = n^{-d/2} e^{-R^2/4n}.$$

So for fixed n, the two bounds we have proved so far – that is the global upper bound in Theorem 4.3 and the 'long range' Carne–Varopoulos bound, prove (4.22) for all values of R except for the narrow range $cn^{1/2} \le R \le c(n \log n)^{1/2}$. Unfortunately there is no simple way to combine these bounds so as to cover this interval, and this is the range of R which determines many important properties of the random walk.

Bridging this gap, as well as obtaining matching lower bounds, will occupy much of the remainder of this book.

4.3 Gaussian and Sub-Gaussian Heat Kernel Bounds

We now introduce the main family of heat kernel bounds that we will treat in the remainder of this book. It includes Gaussian bounds, but also the type of heat kernel bound which holds on regular fractal type graphs, such as the Sierpinski gasket graph.

Definition 4.14 Define the volume growth function by

$$V(x,r) = \mu(B(x,r)), \quad x \in \mathbb{V},\ r \ge 0.$$

(1) (Γ, μ) satisfies the condition (V_α) if there exist positive constants C_L, C_U such that $C_L \le \mu_x \le C_U$ for all $x \in \mathbb{V}$, and

$$C_L r^\alpha \le V(x,r) \le C_U r^\alpha, \qquad \text{for all } x \in \mathbb{V}, r \ge 1.$$

(2) (Γ, μ) satisfies UHK(α, β) if there exist constants c_1, c_2 such that for all $n \ge 0$, $x, y \in \mathbb{V}$,

$$p_n(x,y) \le \frac{c_1}{(n \vee 1)^{\alpha/\beta}} \exp\Big(- c_2 \Big(\frac{d(x,y)^\beta}{n \vee 1} \Big)^{\frac{1}{\beta-1}} \Big). \tag{4.23}$$

(Γ, μ) satisfies LHK(α, β) if there exist constants c_3, c_4 such that, for all $n \ge 0$, $x, y \in \mathbb{V}$ with $d(x,y) \le n$,

$$p_n(x,y) + p_{n+1}(x,y) \ge \frac{c_3}{(n \vee 1)^{\alpha/\beta}} \exp\Big(-c_4 \Big(\frac{d(x,y)^\beta}{n \vee 1} \Big)^{\frac{1}{\beta-1}} \Big). \tag{4.24}$$

(Γ, μ) satisfies HK(α, β) if it satisfies both UHK(α, β) and LHK(α, β). Note that in this case (Γ, μ) is transient if and only if $\alpha > \beta$.

Remarks 4.15 (1) We need to consider $p_n + p_{n+1}$ in the lower bound because of bipartite graphs, or more generally graphs in which large regions are nearly bipartite. It is also clear that the lower bound cannot hold if $d(x,y) > n$.

(2) If $\beta = 2$ then these are the familiar Gaussian heat kernel bounds.
(3) If HK(α, β) holds with $\beta > 2$ then one says that Γ satisfies *sub-Gaussian heat kernel bounds*; the decay of the bound on $p_n(x, y)$ as $d(x, y) \to \infty$ is slower than Gaussian.
(4) We will see in Theorem 6.28 that the heat kernel on any graph roughly isometric to $\mathbb{Z}^d$ satisfies HK$(2, d)$, and in Corollary 6.11 that the Sierpinski gasket graph satisfies HK$(\log 3/\log 2, \log 5/\log 2)$.
(5) If $(\mathbb{V}, E)$ is an infinite graph with natural weights then $|B(x, r)| \ge r$, so the condition (V_α) cannot hold for any $\alpha < 1$.
(6) It is easy to check that if (V_α) holds then (H3) (bounded geometry) holds, and the two conditions (H4) and (H5) are then equivalent.

Exercise 4.16 Show that the condition (V_α) is stable under rough isometries of weighted graphs.

Lemma 4.17 *Suppose that* (Γ, μ) *satisfies HK*(α, β)*. Then* (Γ, μ) *satisfies (H3), (H4), and (H5).*

Proof Taking $n = 0$ in (4.23) we have $\mu_x^{-1} = p_0(x, x) \le c_1$ for all x, so μ_x is bounded below. Let $y \sim x$; then taking $n = 0$ in (4.24) gives

$$\frac{\mu_{xy}}{\mu_x \mu_y} = p_1(x, y) \ge c_3 e^{-c_4}. \tag{4.25}$$

Thus $\mu_x \ge \mu_{xy} \ge c\mu_x\mu_y$, and it follows that there exists c_5 such that $c_5^{-1} \le \mu_y \le c_5$ for all $y \in \mathbb{V}$. Using (4.25) again then gives that $c^{-1} \le \mu_e \le c$ for all $e \in E$, so that (Γ, μ) satisfies (H3), (H4), and (H5). □

In the remainder of this section we will explore some consequences of HK(α, β), starting with the information it gives on the exit times from balls. In the final two sections of this chapter we will introduce some conditions on Γ which are sufficient to give HK(α, β).

Definition 4.18 Define the stopping times

$$\tau(x, r) = \inf\{n \ge 0 : X_n \notin B(x, r)\}.$$

Lemma 4.19 *Let* (Γ, μ) *be any infinite connected graph. Let* $r \ge 1$, $n \ge 1$.

(a) We have

$$\mathbb{P}^x\big(X_n \notin B(x, 2r)\big) \le \mathbb{P}^x\big(\tau(x, 2r) \le n\big).$$

(b) Suppose that $\mathbb{P}^y(X_m \notin B(y,r)) \le p$ *for all* $y \in \mathbb{V}$, $0 \le m \le n$. *Then*

$$\mathbb{P}^x(\tau(x,2r) < n) \le 2p.$$

Proof (a) is clear by inclusion of events.
(b) Writing $\tau = \tau(x,2r)$, $B = B(x,r)$, and noting that $X_\tau \in \partial B(x,2r)$,

$$\begin{aligned}\mathbb{P}^x(\tau < n) &= \mathbb{P}^x(\tau < n, X_n \notin B) + \mathbb{P}^x(\tau < n, X_n \in B)\\ &\le p + \mathbb{E}^x[\mathbf{1}_{(\tau<n)}\mathbb{P}^{X_\tau}(X_{n-\tau} \in B)]\\ &\le p + \mathbb{P}^x(\tau < n) \max_{z\in\partial B} \sup_{0\le m<n} \mathbb{P}^z(X_m \notin B(z,r))\\ &\le p + p\mathbb{P}^x(\tau < n),\end{aligned}$$

and rearranging gives that $\mathbb{P}^x(\tau < n) \le p/(1-p)$. If $p \le \frac{1}{2}$ this is less than $2p$, while if not then $2p \ge 1$ and the conclusion of (b) holds automatically. □

Lemma 4.20 *Suppose that* (Γ, μ) *satisfies the conditions*

$$V(x,r) \le C_U r^\alpha, \quad x \in \mathbb{V}, r \ge 1, \tag{4.26}$$

$$p_n(x,x) \le c_1 n^{-\alpha/\beta}, \quad x \in \mathbb{V}, n \ge 1. \tag{4.27}$$

Then for $x \in \mathbb{V}$, $r \ge 1$

$$\mathbb{P}^x(\tau(x,r) \ge n) \le c_1 \exp(-c_2 n/r^\beta). \tag{4.28}$$

In particular, $\mathbb{E}^x\tau(x,r) \le cr^\beta$ *for* $x \in \mathbb{V}$, $r \ge 1$.

Proof Let $B = B(x,r)$. By Theorem 4.3(c) we have

$$\mathbb{P}^y(X_n \in B(x,r)) \le \sum_{z\in B} p_n(x,z)\mu_z \le cn^{-\alpha/\beta}V(x,r) \le c'n^{-\alpha/\beta}r^\beta. \tag{4.29}$$

Choose $n_1 = cr^\beta$ such that the right side of (4.29) is less than $\frac{1}{2}$. So, by Lemma 4.19(a),

$$\mathbb{P}^y(\tau_B > n_1) \le \tfrac{1}{2}, \quad y \in B.$$

Applying the Markov property of X it follows that for $k \ge 1$

$$\mathbb{P}^y(\tau_B > kn_1) \le 2^{-k}, \quad y \in B.$$

This proves (4.28), and also gives $\mathbb{E}^x\tau_B \le cn_1$. □

Lemma 4.21 *Suppose that* (Γ, μ) *satisfies* (4.26) *and* $UHK(\alpha, \beta)$.

(a) For $\lambda \ge 1$,

$$\mathbb{P}^x\big(X_n \notin B(x, \lambda n^{1/\beta})\big) \le c_1 \exp(-c_2\lambda^{\beta/(\beta-1)}), \tag{4.30}$$

$$\mathbb{P}^x\big(\tau(x, 2\lambda n^{1/\beta}) < n\big) \le 2c_1 \exp(-c_2\lambda^{\beta/(\beta-1)}). \tag{4.31}$$

(b) For $x \in \mathbb{V}$, $r \ge 1$,

$$c_1 r^\beta \le \mathbb{E}^x \tau(x, r) \le c_2 r^\beta.$$

Proof (a) Let $r = \lambda n^{1/\beta}$, $B_k = B(x, kr)$, and $D_k = B_{k+1} - B_k$ for $k \ge 1$. Then writing $\gamma = 1/(\beta - 1)$, and noting that $\beta\gamma > 1$,

$$\begin{aligned}
\mathbb{P}^x\big(X_n \notin B(x,r)\big) &\le \sum_{k=1}^\infty \mathbb{P}^x(X_n \in D_k) \le \sum_{k=1}^\infty \mu(D_k) \max_{z\in D_k} p_n(x,z) \\
&\le c \sum_{k=1}^\infty r^\alpha (k+1)^\alpha n^{-\alpha/\beta} \exp(-c(kr)^{\beta\gamma} n^{-\gamma}) \\
&\le c\lambda^\alpha \sum_{k=1}^\infty (k+1)^\alpha \exp(-c\lambda^{\beta\gamma} k) \le c'\lambda^\alpha \exp(-c''\lambda^{\beta\gamma}).
\end{aligned}$$

This proves (4.30), and (4.31) is then immediate by Lemma 4.19.
(b) The upper bound was proved in the preceding lemma. For the lower bound, let $y \in B(x, r/2)$ and choose λ large enough so that the left side of (4.31) is less than $\frac{1}{2}$. Then choose n_2 so that $2\lambda n_2^{1/\beta} = r/2$. We have $\tau(y, 2\lambda n_2^{1/\beta}) \le \tau_B$, and so

$$\mathbb{E}^y \tau_B \ge \mathbb{E}^y \tau(y, 2\lambda n_2^{1/\beta}) \ge n_2 \mathbb{P}^y\big(\tau(y, 2\lambda n_2^{1/\beta}) \ge n_2\big) \ge n_2/2 \ge cr^\beta. \quad \square$$

Lemma 4.22 *Let* (Γ, μ) *satisfy HK*(α, β). *Then* (V_α) *holds.*

Proof By Lemma 4.17 we have $c^{-1} \le \mu_x \le c$ for all $x \in \mathbb{V}$. Let $r \ge 1$ and $n = \lfloor r^\beta \rfloor$. Then

$$\begin{aligned}
2 &\ge \mathbb{P}^x\big(X_n \in B(x,r)\big) + \mathbb{P}^x\big(X_{n+1} \in B(x,r)\big) \\
&\ge \sum_{y \in B(x,r)} \big(p_n(x,y) + p_{n+1}(x,y)\big)\mu_y \ge cV(x,r) n^{-\alpha/\beta},
\end{aligned}$$

so we obtain $V(x,r) \le c_1 r^\alpha$.

For the lower bound we use the upper bound in Lemma 4.21: choose $\lambda \ge 1$ large enough so the right hand of (4.30) is less than $\frac{1}{2}$. Then if $r = \lambda n^{1/\beta}$

$$\begin{aligned}
1 &= \mathbb{P}^x(X_n \notin B(x,r)) + \mathbb{P}^x(X_n \in B(x,r)) \\
&\le \tfrac{1}{2} + \sum_{y \in B(x,r)} p_n(x,y)\mu_y \le \tfrac{1}{2} + cV(x,r) n^{-\alpha/\beta},
\end{aligned}$$

which gives $V(x,r) \ge c' r^\alpha$. $\square$

Exercise 4.23 Show that if Γ satisfies HK(α, β) then $\mathbb{E}^x d(x, X_n) \asymp n^{1/\beta}$.

Exercise 4.24 Suppose that Γ satisfies HK(α, β), and let

$$f(n) = n^{1/\beta}(\log\log n)^{(\beta-1)/\beta}.$$

Show that

$$0 < \liminf_{n\to\infty} \frac{d(x_0, X_n)}{f(n)} \leq \limsup_{n\to\infty} \frac{d(x_0, X_n)}{f(n)} < \infty.$$

(When $\beta = 2$ this gives the familiar law of the iterated logarithm, but without the exact constant.) To prove this let $n_k = \lfloor e^k \rfloor$, $\lambda_k = (A \log\log n_k)^{(\beta-1)/\beta}$, and use HK$(\alpha, \beta)$ and Lemma 4.21 to bound the probability of the events

$$\{X_m \in B(x_0, n_k^{1/\beta}\lambda_k) \text{ for all } m \leq n_{k\pm 1}\}.$$

Let $D \subset \mathbb{V}$, and recall from (1.24) the definition of the heat kernel for X killed on exiting D. To simplify notation write $\gamma = 1/(\beta - 1)$ and

$$\Phi(T, R) = T^{-\alpha/\beta} \exp(-(R^\beta/T)^\gamma).$$

It is clear from the definition that

$$p_n^D(x, y) \leq p_n(x, y),$$

so our interest is in obtaining lower bounds on p^D.

Theorem 4.25 *Suppose (Γ, μ) satisfies HK(α, β), and let $B = B(x_0, R)$, $B' = B(x_0, R/2)$. Then for $x, y \in B'$, and $d(x, y) \leq n \leq R^\beta$,*

$$p_n^B(x, y) + p_{n+1}^B(x, y) \geq c_1 n^{-\alpha/\beta} \exp\Big(-c_2\Big(\frac{d(x, y)^\beta}{n}\Big)^{\frac{1}{\beta-1}}\Big). \tag{4.32}$$

Proof The following observation will be useful. Let $\theta > 0$ and

$$f(t) = t^{-\theta} \exp(-s^{\gamma\beta} t^{-\gamma}).$$

Then $f'(t) = t^{-1} f(t)(\gamma s^{\beta\gamma} t^{-\gamma} - \theta)$, so if t_0 is defined by $t_0^\gamma = \gamma s^{\beta\gamma}/\theta$ then f is unimodal with a maximum at t_0.

Let $x, y \in B(x_0, 2R/3)$ with $d(x, y) \leq \delta R$, where δ will be chosen later. Then

$$\mathbb{P}^x(X_n = y) = \mathbb{P}^x(X_n = y, \tau_B > n) + \mathbb{P}^x(X_n = y, \tau_B \leq n).$$

For the second term we have

$$\mathbb{P}^x(X_n = y, \tau_B \leq n) = \mathbb{E}^x \mathbf{1}_{(\tau_B \leq n)} \mathbb{P}^{X_{\tau_B}}(X_{n-\tau_B} = y) \leq \max_{z \in \partial B} \max_{1 \leq m \leq n} \mathbb{P}^z(X_m = y).$$

Thus

$$p_n^B(x, y) \geq p_n(x, y) - \max_{z \in \partial B} \max_{1 \leq m \leq n} p_m(z, y),$$

with a similar lower bound on $p^B_{n+1}(x,y)$. Using the observation above and the upper bounds on p_m, there exists $c_3 > 0$ such that if $n \le c_3 R^\beta$ then

$$\max_{z\in\partial B}\max_{1\le m\le n} p_m(z,y) \le c_4\Phi(n, c_5 R/3).$$

Hence

$$p^B_n(x,y) + p^B_{n+1}(x,y) \ge c_6 n^{-\alpha/\beta}\Big(e^{-c_7((\delta R)^\beta/n)^\gamma} - c_8 e^{-c_9(R^\beta/3^\beta n)^\gamma}\Big).$$

Choosing δ small enough we obtain the lower bound (4.32) when $d(x,y) \le \delta R$ and $n \le c_3 R^\beta$. A chaining argument similar to that in Proposition 4.38 then concludes the proof. □

These bounds enable us to control the Green's function for X. In the case when $\alpha \le \beta$ then $\sum_n p_n(x,x) = \infty$, so X is recurrent and it only makes sense to bound g_D for proper subsets D of $\mathbb{V}$. We will do it just for balls.

Theorem 4.26 *Suppose* (Γ,μ) *satisfies HK*(α,β). *Let* $R \ge 2$, $B = B(x_0,R)$, $x,y \in B' = B(x_0,R/2)$, *and* $r = 1 \vee d(x,y)$.

(a) If $\alpha > \beta$ *then*

$$\frac{c_1}{r^{\alpha-\beta}} \le g_B(x,y) \le g(x,y) \le \frac{c_2}{r^{\alpha-\beta}}.$$

(b) If $\alpha < \beta$ *then*

$$c_1 R^{\beta-\alpha} \le g_B(x,y) \le c_2 R^{\beta-\alpha}.$$

(c) If $\alpha = \beta$ *then*

$$c_1\log(R/r) \le g_B(x,y) \le c_2\log(R/r).$$

Proof By Theorem 4.25

$$g_B(x,y) \ge \sum_{n=r^\beta}^{R^\beta} p^B_n(x,y) \ge c\sum_{n=r^\beta}^{R^\beta} n^{-\alpha/\beta} \asymp \int_{r^\beta}^{R^\beta} s^{-\alpha/\beta}\,ds,$$

and the lower bounds in (a)–(c) follow easily.

(a) The bound $g_B(x,y) \le g(x,y)$ is immediate. If $y = x$ then

$$g(x,x) = \sum_{n=0}^{\infty} p_n(x,x) \asymp \sum_{n=1}^{\infty} n^{-\alpha/\beta} \asymp c.$$

If $y \ne x$ then note first that

$$\int_0^\infty t^{-\alpha/\beta}\exp(-c_2(r^\beta/t)^\gamma)\,dt = r^{\beta-\alpha}\int_0^\infty s^{-\alpha/\beta}\exp(-c_2 s^{-\gamma})\,ds = c r^{\beta-\alpha}.$$

The upper bound on $g(x, y)$ then follows by comparison between the sum for $g(x, y)$ and the above integral.
(b) Let $m = \lfloor R^\beta \rfloor$. By Lemma 4.21 we have for $x, y \in B$

$$\begin{aligned} p^B_{m+n}(x, y) &= \sum_{z \in B} p^B_m(x, y) p^B_n(z, y) \mu_y \\ &\le c m^{-\alpha/\beta} \mathbb{P}^y(\tau_B > n) \le c R^{-\alpha} \exp(-cn/R^\beta). \end{aligned}$$

So,

$$\begin{aligned} g_B(x, y) &= \sum_{k=0}^{\infty} \sum_{n=km}^{(k+1)n} p^B_n(x, y) \\ &\le \sum_{n=0}^{m} c(1 \vee n)^{-\alpha/\beta} + \sum_{k=1}^{\infty} c R^{\beta-\alpha} e^{-c'(k-1)} \le c R^{\beta-\alpha}. \end{aligned}$$

(c) Let m be as in (b) and $m' = \lfloor r^\beta \rfloor$. Then

$$\begin{aligned} g_B(x, y) &\le c + \sum_{n=1}^{m'} p_n(x, y) + \sum_{m'}^{m} p_n(x, y) + \sum_{k=1}^{\infty} c R^{\beta-\alpha} e^{-c'(k-1)} \\ &\le \sum_{n=0}^{m'} p_n(x, y) + c \sum_{n=m'}^{m} n^{-1} + c' \le c \log(R/r) + \sum_{n=0}^{m'} p_n(x, y). \end{aligned} \tag{4.33}$$

Now

$$\int_0^{r^\beta} t^{-1} e^{-c_1 (r^\beta/t)^\gamma} dt = \int_0^1 s^{-1} e^{-c_1 s^{-\gamma}} ds = c_2,$$

and comparing the final sum in (4.33) with this integral gives the upper bound in (c). □

We conclude this section by obtaining some general bounds on the time derivative of p_n. The example of a bipartite graph shows that we cannot expect to have good control of $|p_{n+1} - p_n|$ in general, but we can bound $|p_{n+2} - p_n|$.

Lemma 4.27 *Let $1 \le k < n$. Then*

$$||p^x_n - p^x_{n+2}||_2 \le 2k^{-1} ||p^x_{n-k}||_2.$$

Proof Since X cannot visit $B(x, n+3)^c$ before time $n+3$, we can assume that $\mathbb{V} = B(x, n+3)$. Then if $N = |B(x, n+3)|$ the spectral decomposition (3.16) gives

$$p_n^x(y) - p_{n+2}^x(y) = \sum_{i=1}^N \varphi_i(x)\varphi_i(y)\rho_i^n(1-\rho_i^2).$$

Now if $t \in [0, 1]$

$$2 \geq (1+t) = (1+t)(1-t)\sum_0^\infty t^j \geq (1-t^2)kt^k.$$

So taking $t = |\rho_i|$

$$\begin{aligned}
||p_n^x - p_{n+2}^x||_2^2 &= \sum_i \varphi_i(x)^2\rho_i^{2n}(1-\rho_i^2)^2 \\
&\leq \sum_i \varphi_i(x)^2\rho_i^{2n-2k}\max_j\left(\rho_j^k(1-\rho_j^2)\right)^2 \\
&\leq 4k^{-2}\sum_i \varphi_i(x)^2\rho_i^{2n-2k} = 4k^{-2}||p_{n-k}^x||_2^2. \quad \square
\end{aligned}$$

Theorem 4.28 *Let $n = n_1 + n_2$ and $1 \leq k < n_2$. Then*

$$|p_n(x,y) - p_{n+2}(x,y)| \leq 2k^{-1}p_{2n_1}(x,x)^{1/2}p_{2n_2-2k}(y,y)^{1/2}.$$

In particular, if UHK(α, β) holds then

$$|p_n(x,y) - p_{n+2}(x,y)| \leq c_1 n^{-1-\beta/\alpha}.$$

Proof We have

$$\begin{aligned}
&|p_n(x,y) - p_{n+2}(x,y)| \\
&= \Big|\sum_z p_{n_1}(x,z)(p_{n_2}(z,y) - p_{n_2+2}(z,y))\mu_z\Big| \\
&\leq \Big(\sum_z p_{n_1}(x,z)^2\mu_z\Big)^{1/2}\Big(\sum_z (p_{n_2}(z,y) - p_{n_2+2}(z,y))^2\mu_z\Big)^{1/2} \\
&= p_{2n_1}(x,x)^{1/2}||p_{n_2}^y - p_{n_2+2}^y||_2 \\
&\leq 2k^{-1}p_{2n_1}(x,x)^{1/2}p_{2n_2-2k}(y,y)^{1/2}.
\end{aligned}$$

This proves the first part; the second then follows on choosing $n_1 = k = \lfloor n/3 \rfloor$. (If $n \leq 3$ we can adjust the constant c_1 so that the inequality holds.) $\square$

Using (4.5) we obtain

Corollary 4.29 *Let $r_n^x = p_n^x + p_{n+1}^x$. If UHK($\alpha$, β) holds then*

$$\mathscr{E}(r_n^x, r_n^x) \le c n^{-1-\alpha/\beta}.$$

4.4 Off-diagonal Upper Bounds

We now introduce some conditions which, if satisfied, will imply UHK(α, β). In Chapter 6, we will see how these conditions can be proved from geometric or resistance properties of the graph.

Definition 4.30 Let $\beta \ge 1$. Following [GT1, GT2] we say (Γ, μ) satisfies the condition (E_β) if there exist positive constants c_1, c_2 such that

$$c_1 r^\beta \le \mathbb{E}^x \tau(x,r) \le c_2 r^\beta, \qquad \text{for } x \in \mathbb{V},\ r \ge 1.$$

Lemma 4.31 *Suppose that (Γ, μ) satisfies (E_β). Then there exist constants $c. > 0$ and $p > 0$ such that for all x, r*

$$\mathbb{E}^y \tau(x,r) \le c_3 r^\beta, \quad y \in B(x,r), \tag{4.34}$$

$$\mathbb{E}^y \tau(x,r) \ge c_4 r^\beta, \quad y \in B(x,r/2), \tag{4.35}$$

$$\mathbb{P}^x\big(\tau(x,r) \ge c_5 r^\beta\big) \ge p. \tag{4.36}$$

Proof If $y \in B(x,r)$ then $B(x,r) \subset B(y,2r)$, so

$$\mathbb{E}^y \tau(x,r) \le \mathbb{E}^y \tau(y,2r) \le c_2 2^\beta r^\beta.$$

Similarly if $y \in B(x,r/2)$ then $\mathbb{E}^y \tau(x,r) \ge \mathbb{E}^y \tau(y,r/2) \ge c_1 2^{-\beta} r^\beta$.

Let $t > 0$, and write $\tau = \tau(x,r)$, $B = B(x,r)$. Using the Markov property of X we have

$$\begin{aligned} c_1 r^\beta \le \mathbb{E}^x \tau &= \mathbb{E}^x \mathbf{1}_{(\tau \le t)} \tau + \mathbb{E}^x \mathbf{1}_{(\tau > t)} \tau \\ &\le t \mathbb{P}^x(\tau \le t) + \mathbb{E}^x\Big(\mathbf{1}_{(\tau > t)}(t + \mathbb{E}^{X_t} \tau)\Big) \\ &\le t + \mathbb{P}^x(\tau > t) \sup_{y \in B} \mathbb{E}^y \tau \le t + \mathbb{P}^x(\tau > t) c_3 r^\beta. \end{aligned}$$

Hence, taking $t = \frac{1}{2} c_1 r^\beta$,

$$\mathbb{P}^x(\tau > t) \ge \frac{c_1 r^\beta - t}{c_3 r^\beta} = \frac{c_1}{2c_3}. \qquad \square$$

Lemma 4.32 *Suppose that (Γ, μ) has polynomial growth, and satisfies (E_β). Then $\beta \ge 2$.*

Proof Let c_i be as in Lemma 4.31. Then by Markov's inequality

$$\mathbb{P}^x(\tau(x,r) > 2c_3 r^\beta) \le \tfrac{1}{2}.$$

Suppose now that $\beta < 2$, and let $n = 2c_3 r^\beta$. Then $r^2/n \ge cr^{2-\beta}$, so if r is large enough we can use (4.21) to obtain

$$\begin{aligned}\tfrac{1}{2} \le \mathbb{P}^x(\tau(x,r) \le n) &\le \sum_{k=1}^n \mathbb{P}^x(X_k \notin B(x,r)) \\ &\le n \max_{0\le k\le n} cr^\theta e^{-r^2/4k} \le cr^{\beta+\theta} e^{-cr^{2-\beta}},\end{aligned}$$

a contradiction for large r. □

Proposition 4.33 *Let $\beta \ge 2$ and suppose that there exist constants $c_1 > 0$, $p > 0$ such that for all $x \in \mathbb{V}$, $r \ge 1$*

$$\mathbb{P}^x\big(\tau(x,r) \ge c_1 r^\beta\big) \ge p. \tag{4.37}$$

Then there exist constants c_2, c_3 such that for $T \ge 1$, $R \ge 1$

$$\mathbb{P}^x(\tau(x,R) < T) \le c_2 \exp(-c_3 (R^\beta/T)^{1/(\beta-1)}). \tag{4.38}$$

In particular, (4.38) *holds if Γ satisfies (E_β).*

Proof We can assume $c_1 \le 1$. Choose positive integers m and r, and let $t_0 = c_1 r^\beta$. Define the stopping times

$$S_0 = 0, \quad S_i = \inf\{n \ge S_{i-1} : d(X_{S_{i-1}}, X_n) = r\}, \quad i \ge 1.$$

Write $\mathscr{F}_n = \sigma(X_m, m \le n)$ for the filtration of X. Since $d(X_{S_{i-1}}, X_k) < r$ for $S_{i-1} \le k < S_i$, we have $d(X_0, X_k) < mr$, for $0 \le k < S_m$, and hence $\tau(x, mr) \ge S_m$. We call the parts of the path of X between times S_{i-1} and S_i 'journeys', and will call a journey 'slow' if $S_i - S_{i-1} \ge t_0$. Let

$$\xi_i = \mathbf{1}_{(S_i - S_{i-1} \ge t_0)}, \text{ and } N = \sum_{i=1}^m \xi_i.$$

Thus N is the number of slow journeys, and $\tau(x, mr) \ge S_m \ge N t_0$. The bound (4.37) implies that $\mathbb{P}^x(\xi_i = 1|\mathscr{F}_{S_{i-1}}) \ge p$, and using a martingale inequality (see Lemma A.8) to bound N we obtain

$$\mathbb{P}^x\big(\tau(x,mr) \le \tfrac{1}{2} mpt_0\big) \le \mathbb{P}^x(Nt_0 < \tfrac{1}{2} mpt_0) \le e^{-cm}. \tag{4.39}$$

This bound then leads to (4.38). The basic idea is that, given R and T, if $r = R/m$ and $mt_0 = T$ then

$$T = c_1 m r^\beta = c_1 R^\beta m^{1-\beta},$$

giving (4.38). However, since we need both m and r to be positive integers, the argument needs a little care.

We wish to choose m (and so r) such that

$$mr \le R, \quad \tfrac{1}{2}pt_0m = \tfrac{1}{2}pc_1mr^\beta \ge T.$$

If these hold, then

$$\mathbb{P}(\tau(x,R) \le T) \le \mathbb{P}(\tau(x,mr) \le T) \le \mathbb{P}(\tau(x,mr) \le \tfrac{1}{2}pmt_0) \le e^{-cm}.$$

Given $m \in \{1,\dots,R\}$, set $r = \lfloor R/m \rfloor$. Then $r \le R/m \le r+1 \le 2r$, and so, taking $c_4 = 2^{-1-\beta}c_1$,

$$\tfrac{1}{2}pc_1mr^\beta \ge c_1p2^{-1-\beta}R^\beta m^{1-\beta} = c_4R^\beta m^{1-\beta}.$$

We now choose m as large as possible, subject to the constraints

$$m \le R, \quad m^{\beta-1} \le \frac{c_4R^\beta}{T}. \tag{4.40}$$

We consider three regimes for R and T, and in each case we obtain a bound of the form (4.38), but with different constants $c_2(j), c_3(j)$ for cases $j = 1,2,3$. The bound (4.38) then follows by taking the weakest constant in each case, that is, $c_2 = \max_j c_2(j)$ and $c_3 = \min_j c_3(j)$.

Case 1: $c_4R^\beta/T \le 2^{\beta-1}$. If this holds then the right side of (4.38) is of order 1, so the bound gives no information. However, it still does hold since

$$\mathbb{P}(\tau(x,R) \le T) \le 1 \le e^2e^{-(c_4R^\beta/T)^{1/(\beta-1)}}.$$

Case 2: $T \le c_4R$. Since $c_4 < c_1 \le 1$ we have $T < R$, so the left side of (4.38) is zero, and the bound therefore holds automatically.

Case 3: $c_4R^\beta/T \le 2^{\beta-1}$, and $c_4R \le T$. (This is the only non-trivial case.) We choose $m = \lfloor (c_4R^\beta/T)^{1/(\beta-1)} \rfloor$, so that $m \ge 2$ and also $2m \ge m+1 \ge (c_3R^\beta/T)^{1/(\beta-1)}$. Then

$$m^{\beta-1} \le c_4R^\beta/T \le R^{\beta-1},$$

so we also have $m \le R$. Hence this choice of m and r satisfies both the constraints in (4.40), and (4.38) follows from (4.39). □

Remarks (1) The bound (4.38) is (up to constants) the best possible. This may seem surprising, since the preceding argument appears to give a lot away; usually it will take many more than m journeys of length r for X to leave $B(x,mr)$. (In $\mathbb{Z}^d$ one would expect that roughly m^2 journeys would be needed.) However, the bound (4.38) only gives much information when

$R^\beta \gg T$, and in this case it is very unlikely that X will exit $B(x, mr)$ in a short time; on the rare occasions when it does so, it moves more or less directly to the boundary.

(2) If $\beta = 2$ then $1/(\beta - 1) = 1$, and the term in the exponential in (4.38) is $-cR^2/T$.

The following theorem gives conditions for UHK(α, β).

Theorem 4.34 *Let $\alpha \geq 1$, $\beta \geq 2$, and suppose that (Γ, μ) satisfies (V_α). The following are equivalent.*

(a) There exist constants $c_1, c_2 > 0$, $p_1 \in (0, 1)$ such that for all $x \in \mathbb{V}$, $r \geq 1$

$$\mathbb{P}^x(\tau(x, r) \leq c_1 r^\beta) \leq p_1, \quad x \in \mathbb{V},\ r \geq 1, \tag{4.41}$$

$$p_n(x, x) \leq c_2(1 \vee n)^{-\alpha/\beta}, \ n \geq 0,\ x, y \in \mathbb{V}. \tag{4.42}$$

(b) UHK(α, β) holds; that is, there exist constants c_3, c_4 such that for $x, y \in \mathbb{V}$, $n \geq 0$

$$p_n(x, y) \leq \frac{c_3}{(n \vee 1)^{\alpha/\beta}} \exp\Big(- c_4 \Big(\frac{d(x, y)^\beta}{n \vee 1}\Big)^{\frac{1}{\beta-1}} \Big). \tag{4.43}$$

(c) (E_β) and the on-diagonal upper bound (4.42) *hold.*

Proof The implication (b) $\Rightarrow$ (c) was proved in Lemma 4.20, while (c) $\Rightarrow$ (a) is immediate from Lemma 4.31.

(a) $\Rightarrow$ (b) The condition (4.42) implies that $\mu_x = p_0(x, x) \leq c_2$ for all x. Hence it is easy to check that (4.23) holds if $n = 0$ or $n = 1$.

Let $2R = d(x, y)$ and $m = n/2$. If $R^\beta/n \leq 1$ then the term in the exponential in UHK(α, β) is of order 1, and so the bound is immediate from (4.42). Now let $n \geq 2$, and choose m such that $n/3 \leq m \leq 2n/3$. Let $A_x = \{z : d(x, z) \leq d(y, z)\}$ and $A_y = \mathbb{V} - A_x$. Then

$$\mu_x \mathbb{P}^x(X_n = y) = \mu_x \mathbb{P}^x(X_n = y, X_m \in A_y) + \mu_x \mathbb{P}^x(X_n = y, X_m \in A_x). \tag{4.44}$$

The second term in (4.44) is

$$\begin{aligned}\mu_x \sum_{z\in A_x} p_n(x, z)\mu_z p_{n-m}(z, y)\mu_y &= \mu_y \sum_{z\in A_x} p_{n-m}(y, z)\mu_z p_n(z, x)\mu_x \\ &= \mu_y \mathbb{P}^y(X_n = x, X_{n-m} \in A_x).\end{aligned}$$

So the two terms in (4.44) are of the same form, and it is sufficient to bound the first. For this we note that if $z \in A_y$ then $2d(x, z) \geq d(x, z) + d(y, z) \geq$

$d(x, y) = 2R$, so $d(x, z) \geq R$. Hence also if $X_0 = x$ and $X_m \in A_y$ then $\tau(x, R) \leq m$. So,

$$\begin{aligned}\mu_x \mathbb{P}^x(X_n = y, X_m \in A_y) &= \mu_x \mathbb{P}^x(\tau(x, R) \leq m,\ X_m \in A_y, X_n = y) \\ &\leq \mu_x \mathbb{E}^x \left(\mathbf{1}_{\{\tau(x,R) \leq m, X_m \in A_y\}} \mathbb{P}^{X_m}(X_{n-m} = y) \right) \\ &\leq \mu_x \mathbb{P}^x(\tau(x, R) \leq m) \sup_{z \in A_y} p_{m-n}(z, y) \mu_y.\end{aligned}$$

By Proposition 4.33 $\mathbb{P}^x(\tau(x, R) \leq m) \leq c\exp(-c(R^\beta/(n/3))^{1/(\beta-1)}$, while by (4.42) $p_{m-n}(z, y) \leq c(n/3)^{-\alpha/\beta}$; combining these bounds gives UHK(α, β). □

4.5 Lower Bounds

We now turn to lower bounds for $p_n(x, y)$. On-diagonal lower bounds follow immediately from the upper bounds.

Lemma 4.35 *Suppose* (Γ, μ) *satisfies* (V_α) *and UHK*(α, β). *Then*

$$p_{2n}(x, x) \geq c_1 n^{-\alpha/\beta}, \ n \geq 1.$$

Proof Choose λ so that the left side of (4.30) is less than $\frac{1}{2}$, and let $B = B(x, \lambda n^{1/\beta})$. Then

$$\begin{aligned}\tfrac{1}{2} \leq \mathbb{P}^x(X_n \in B) &= \sum_{y \in B} p_n(x, y)\mu_y \leq \Big(\sum_{y \in B} p_n(x, y)^2 \mu_y \Big)^{1/2} \mu(B)^{1/2} \\ &\leq \Big(\sum_y p_n(x, y)^2 \mu_y \Big)^{1/2} \mu(B)^{1/2} = p_{2n}(x, x)^{1/2} \mu(B)^{1/2}.\end{aligned}$$

Thus $4 p_{2n}(x, x) \geq \mu(B)^{-1} \geq c n^{-\alpha/\beta}$. □

This argument does not give any lower bound on $p_n(x, y)$ with $y \neq x$. To obtain such a bound, even if $d(x, y) \ll n$, requires more information on the graph Γ.

Definition 4.36 Let $\beta \geq 2$. We say (Γ, μ) satisfies the *near-diagonal lower bound* (NDLB(β)) if there exist constants $c_0, c_1 \in (0, 1)$ such that

$$p_n(x, y) + p_{n+1}(x, y) \geq \frac{c_0}{V(x, n^{1/\beta})} \text{ if } 1 \vee d(x, y) \leq c_1 n^{1/\beta}.$$

If (V_α) holds then we can write this as

$$p_n(x, y) + p_{n+1}(x, y) \geq c_2 n^{-\alpha/\beta} \text{ if } 1 \vee d(x, y) \leq c_1 n^{1/\beta}. \tag{4.45}$$

Even when the upper bound UHK(α, β) holds NDLB(β) may fail – an example is the join of two copies of $\mathbb{Z}^d$ (with $d \geq 2$). We will see in Chapter 6 how the near-diagonal lower bound can be proved from the Poincaré inequality or from resistance bounds; for the moment we show that it leads to full lower bounds.

Lemma 4.37 *Suppose* (V_α) *and NDLB*(β) *hold. Let* $x, y \in \mathbb{V}$, *and* $r \geq 2$. *Then if* $r + d(x, y) \leq c_2 n^{1/\beta}$,

$$\mathbb{P}^x(X_n \in B(y, r)) \geq c r^\alpha n^{-\alpha/\beta}.$$

Proof Let $B' = B(y, r/2)$. Then summing (4.45) over B' we have

$$\mathbb{P}^x(X_n \in B') + \mathbb{P}^x(X_{n+1} \in B') \geq c n^{-\alpha/\beta}(r/2)^\alpha. \tag{4.46}$$

If $X_{n+1} \in B'$ then $X_n \in B(y, r)$, so each of the probabilities on the left side of (4.46) is bounded above by $\mathbb{P}^x(X_n \in B(y, r))$. □

Proposition 4.38 *Let* (Γ, μ) *be a graph satisfying (H5) and* (V_α). *Suppose that the near-diagonal lower bound* (4.45) *holds for some* $\beta \geq 2$. *Then if* $T \geq d(x, y) \vee 1$,

$$p_T(x, y) + p_{T+1}(x, y) \geq c_3 (T \vee 1)^{-\alpha/\beta} \exp\Big(- c_4 \Big(\frac{d(x, y)^\beta}{T \vee 1}\Big)^{\frac{1}{\beta-1}}\Big). \tag{4.47}$$

Proof If $T = 0$ and $x = y$ this bound is immediate from the condition (V_α). We continue to write c_1, c_2 for the constants in (4.45). We use a classical chaining argument which works quite generally. At first sight, like the argument in the upper bound, the proof appears to give rather a lot away. As in the upper bound, we will need to treat three cases, of which the first two are easy and the third gives the main estimate. Let $R = d(x, y)$ and $c_5 = 1 \wedge 2^{-3\beta} c_1^\beta$.

Case 1: $c_5 T \leq R \leq T$. We can find a path $\gamma = (z_0, \ldots, z_m)$ between x and y, where $m \in \{T, T+1\}$; this path does not have to be self-avoiding. As (H5) holds, we have $\mathscr{P}(z_{i-1}, z_i) \geq c_6^{-1} > 0$ for each i. Then

$$\mathbb{P}^x(X_m = y) \geq \mathbb{P}^x(X_k = z_k, 1 \leq k \leq m) \geq c_6^{T+1} \geq c e^{-c' T},$$

which gives the bound (4.47). (Note that in this regime the term $T^{-\alpha/\beta}$ plays no role.)

Case 2: $R^\beta / T \leq c_1^\beta$. In this case the exponential term in (4.47) is of order 1, and since $R \leq c_2 T^{1/\beta}$ the bound follows immediately from (4.45).

Case 3: $R^\beta / T > c_1^\beta$, $R \leq c_5 T$. The basic idea is to choose m with $m^{\beta-1} \simeq R^\beta / T$, divide the journey from x to y into m steps, and use the NDLB on each step; this is possible since this choice of m means that $(T/m)^{1/\beta} \asymp R/m$. However, since both m and the duration of the time steps have to be integers, the details of the argument need a little work.

Let $1 \le m \le R$, and $r = R/m$, $s = T/m$, $r_0 = \lfloor r \rfloor$, $s_0 = \lfloor s \rfloor$. As $m \le R \le T$, we have $r/2 \le r_0 \le r$, $s/2 \le s_0 \le s$.

Since $mr_0 \le R \le m(r_0+1)$, we can find a chain $x = z_0, z_1, \dots, z_m = y$ with $d(z_{j-1}, z_j) \in \{r_0, r_0+1\}$ for each j. Set $B_j = B(z_j, r_0)$. Note that if $w_i \in B_i$, then $d(w_{i-1}, w_i) \le 3r_0 + 1 \le 4r$. Now choose $s_i \in \{s_0, s_0+1\}$ so that $\sum_{i=1}^m s_i = T$. Set

$$p_j = \min_{w \in B_{j-1}} \mathbb{P}^w(X_{s_j} \in B_j), \; j = 1, \dots, m-1.$$

Then if $t_j = s_1 + \cdots + s_j$

$$\begin{aligned}
\mathbb{P}^x(X_T = y) &\ge \mathbb{P}^x(X_{t_j} \in B_j, j = 1, \dots, m-1, X_T = y) \\
&= \sum_{w \in B_1} \mathbb{P}^x(X_{s_1} = w)\mathbb{P}^w(X_{t_j - s_1} \in B_j, j = 2, \dots, m-1, X_{T-s_1} = y) \\
&\ge \mathbb{P}^x(X_{s_1} \in B_1) \min_{w \in B_1} \mathbb{P}^w(X_{t_j - s_1} \in B_j, j = 2, \dots, m-1, X_{T-s_1} = y) \\
&\ge \cdots \ge p_1 \dots p_{m-1} \min_{w \in B_{m-1}} \mathbb{P}^w(X_{s_m} = y).
\end{aligned}$$

Suppose we can choose $m \in \{1, \dots, R\}$ so that

$$3r_0 + 1 \le c_1 s_0^{1/\beta} \le 16 r_0. \tag{4.48}$$

Then, using (4.45) and the lower bound on $\mu(B_j)$,

$$\begin{aligned}
p_j &\ge c_2 s_0^{-\alpha/\beta} \mu(B_j) \ge c s_0^{-\alpha/\beta} r_0^\alpha \ge c_7, \\
\min\nolimits_{w \in B_{m-1}} \mathbb{P}^w(X_s = y) &\ge c_2 s_0^{-\alpha/\beta} \mu_y \ge c_2 T^{-\alpha/\beta} \mu_y.
\end{aligned}$$

Combining the bounds above we obtain

$$p_n(x, y) \ge c_7^{m-1} c_2 T^{-\alpha/\beta} = c_2 c_7^{-1} T^{-\alpha/\beta} e^{-m \log c_7}. \tag{4.49}$$

It remains to check that we can choose m so that (4.48) holds, and to evaluate the left hand side of (4.49). Using the bounds on r_0, s_0 above, (4.48) holds if

$$4r = 4(R/m) \le c_1 (T/2m)^{1/\beta}, \qquad c_1 (T/m)^{1/\beta} \le 8(R/m),$$

that is, if an integer m can be found with

$$2^{1+2\beta} c_1^{-\beta} R^\beta / T \le m^{\beta-1} \le 2^{\beta-1}(2^{1+2\beta} c_1^{-\beta} R^\beta / T). \tag{4.50}$$

As $R^\beta T c_1^{-\beta} \ge 1$, we can find m satisfying (4.50). Further, since $R \le c_5 T$, we have

$$m^{\beta-1} \le c_1^{-\beta} 2^{3\beta} R^\beta / T \le c_5 c_1^{-\beta} 2^{3\beta} R^{\beta-1} \le R^{\beta-1},$$

so that $m \le R$. Substituting the bound from (4.50) into (4.49) completes the proof. □

For convenience we summarise our sufficient conditions for UHK(α, β) and LHK(α, β). Write $\Phi(n, r) = (r^\beta / n)^{1/(\beta-1)}$.

Theorem 4.39 *Let (Γ, μ) satisfy (V_α).*

(a) If (Γ, μ) satisfies (E_β) and the on-diagonal upper bound

$$p_n(x, x) \le c_0 (1 \vee n)^{-\alpha/\beta}, \quad n \ge 0,\ x, y \in \mathbb{V},$$

then there exist constants c_1, c_2 such that for $x, y \in \mathbb{V}$, $n \ge 0$

$$p_n(x, y) \le c_1 (n \vee 1)^{-\alpha/\beta} \exp\big(- c_2 \Phi(n \vee 1, d(x, y))\big).$$

(b) If (Γ, μ) satisfies (H5) and NDLB(β) then there exist constants c_3, c_4 such that for $x, y \in \mathbb{V}$, $n \ge d(x, y)$

$$p_n(x, y) + p_{n+1}(x, y) \ge c_3 (n \vee 1)^{-\alpha/\beta} \exp\big(- c_4 \Phi(n \vee 1, d(x, y))\big).$$

5

Continuous Time Random Walks

5.1 Introduction to Continuous Time

In Chapter 4 we studied the transition density $p_n(x, y)$ of the discrete time random walk X. It is also of interest to consider the related continuous time simple random walk (CTSRW) on (Γ, μ). While the essential ideas are the same in both discrete and continuous time contexts, discrete time does introduce some extra difficulties, of which the most significant are connected to parity issues. In addition, the difference equations used in Chapter 4 take a simpler form in the continuous time context. There is, however, an initial penalty in dealing in continuous time, in that one is dealing with a more complicated object, and some extra work is needed to verify that certain interchanges of derivatives and infinite sums are valid. A further (minor) source of difficulty is that, while the support of the discrete time heat kernel $p_n(x, \cdot)$ is finite, the continuous time kernel $q_t(x, \cdot)$ is strictly positive everywhere.

We denote the continuous time walk by $Y = (Y_t, t \in [0, \infty))$, and will construct it from the discrete time walk $X_n, n \in \mathbb{Z}_+$ and an independent Poisson process with rate 1 $(N_t, t \geq 0)$. (For details on the Poisson process see Appendix A.3 or [Nor].)

Given the processes X and N we set

$$Y_t = X_{N_t}, \ t \in [0, \infty). \tag{5.1}$$

The process Y waits an `exp(1)` time at each vertex x, and then jumps to some $y \sim x$ with the same jump probabilities as X, that is, with the probability $\mathcal{P}(x, y) = \mu_{xy}/\mu_x$ given by (1.2). Informally if $Y_t = x$ then the probability that Y jumps to $y \sim x$ in the time interval $(t, t+h]$ is $\mathcal{P}(x, y)h + O(h^2)$. We use $\mathbb{P}^x$ to denote the law of Y with initial position $Y_0 = x$, and write $q_t(x, y)$ for the transition density of Y_t with respect to μ, given by

$$q_t(x,y) = \frac{\mathbb{P}^x(Y_t = y)}{\mu_y}.$$

We also call q_t the (continuous time) heat kernel on (Γ, μ). As in the case of X, we write

$$T_D = \inf\{t \geq 0 : Y_t \in D\}, \quad \tau_D = T_{D^c}.$$

If it is necessary to specify the process, we write $T_D(Y)$, $\tau_D(Y)$, etc. These are stopping times for the process Y.

Lemma 5.1 summarises some basic properties of q_t, which follow easily from the definition (5.1). As before we use the notation $q_t^x(\cdot) = q_t(x, \cdot)$.

Lemma 5.1 *(a) For $t \geq 0$,*

$$q_t(x,y) = \sum_{n=0}^{\infty} \frac{e^{-t} t^n}{n!} p_n(x,y). \tag{5.2}$$

(b) $q_t(x,y) = q_t(y,x)$.
(c) For all $x, z \in \mathbb{V}$, $s, t \geq 0$,

$$q_{t+s}(x,z) = \sum_{y \in \mathbb{V}} q_t(x,y) q_s(y,z) \mu_y = \langle q_t^x, q_s^z \rangle. \tag{5.3}$$

In particular, $||q_t^x||_2^2 \leq \mu_x^{-1}$ for each $x \in \mathbb{V}$, $t \geq 0$, so that $q_t^x \in L^2$.
(d)

$$\frac{\partial}{\partial t} q_t(x,y) = \Delta q_t(x,y).$$

Proof (a) Since X and N are independent,

$$\mathbb{P}^x(Y_t = y) = \sum_{n=0}^{\infty} \mathbb{P}^x(X_n = y, N_t = n) = \sum_{n=0}^{\infty} p_n(x,y) \mu_y \mathbb{P}^x(N_t = n),$$

and, as N_t has a Poisson distribution with mean t, this gives (5.2).
(b) This is immediate from the symmetry of $p_n(x,y)$.
(c) This follows from (5.2) and the discrete Chapman–Kolmogorov equation (1.8). We have

$$\begin{aligned} q_{t+s}(x,z) &= \sum_{n=0}^{\infty} \frac{e^{-t-s}(s+t)^n}{n!} p_n(x,z) \\ &= \sum_{n=0}^{\infty} e^{-t-s} \sum_{k=0}^{n} \frac{s^k t^{n-k}}{k!(n-k)!} p_{k+n-k}(x,z) \end{aligned}$$

$$= \sum_{i=0}^{\infty} \sum_{j=0}^{\infty} e^{-t-s} \frac{s^i t^j}{i!j!} \sum_y p_i(x, y) p_j(y, z) \mu_y$$
$$= \sum_y q_t(x, y) q_s(y, z) \mu_y.$$

(d) This follows by differentiating (5.2) – note that, since $p_n(x, y) \leq 1/\mu_x$ for all n, it is easy to justify the interchange of derivative and sum. □

Let $t \geq 0$. We define the semigroup operator Q_t by setting

$$Q_t f(x) = \mathbb{E}^x f(Y_t), \quad f \in C_+(\mathbb{V}), \tag{5.4}$$

and then let $Q_t f = Q_t f_+ - Q_t f_-$ whenever this is defined. Thus we have

$$Q_t f(x) = \sum_y q_t(x, y) f(y) \mu_y \tag{5.5}$$

whenever the sum on the right hand side of (5.5) converges (absolutely). Unlike the discrete time operators P_n, which were defined for all $f \in C(\mathbb{V})$ in (1.9), the expectation in (5.4) can be infinite. We therefore need to be more careful with the domain of these operators; for most purposes $L^2(\mathbb{V})$ will be sufficient, but we include other L^p space for completeness.

Lemma 5.2 *Let $p \in [1, \infty]$.*

(a) For each $t \geq 0$,

$$Q_t f = \sum_{n=0}^{\infty} \frac{e^{-t} t^n}{n!} P^n f, \quad f \in L^p. \tag{5.6}$$

(b) For each $t \geq 0$ the operator Q_t maps L^p to L^p, and $||Q_t||_{p \to p} \leq 1$.
(c) If $f \geq 0$ then $Q_t f \geq 0$.
(d) $Q_t \mathbf{1} = \mathbf{1}$ for each $t \in [0, \infty)$.
(e) For $s, t \geq 0$ we have $Q_t Q_s = Q_{s+t}$ so that $(Q_t, t \in [0, \infty))$ is a semigroup on $L^p(\mathbb{V})$.
(f) For $f \in L^p$,

$$|Q_t f| \leq Q_t |f|.$$

(g) We have $\Delta Q_t f = Q_t \Delta f$ for all $f \in L^p$.

Proof (a) and (b) For $f \in C_0(\mathbb{V})$, (5.6) follows immediately from (5.2). By Minkowski's inequality,

$$||Q_t f||_p = \left\| \sum_{n=0}^{\infty} \frac{e^{-t} t^n}{n!} P^n f \right\|_p \leq \sum_{n=0}^{\infty} \frac{e^{-t} t^n}{n!} \left\| P^n f \right\|_p \leq ||f||_p.$$

So (5.6) extends to $f \in L^p$, and we have $||Q_t||_{p \to p} \leq 1$.
(c) and (d) These are immediate from the definition (5.4), and the semigroup property
(e) follows from (5.3).
(f) Since $-|f| \leq f \leq |f|$, this follows from (c).
(g) Since $\Delta = P - I$ commutes with P^n, this follows by (a). □

Remark The Markov property of Y at a fixed time t follows from (5.3). Later, we will also use the strong Markov property of Y, that is, the Markov property at stopping times T. As a first encounter with the precise formulation of the Markov property can produce distress, even to experienced mathematicians, the details have been put into Appendices A.2 and A.3.

Lemma 5.3 states that Δ is the infinitesimal generator of the semigroup Q_t.

Lemma 5.3 *Let $f \in L^2(\mathbb{V}, \mu)$. Then*

$$\lim_{t \downarrow 0} \frac{Q_t f - f}{t} = \Delta f, \quad \textit{pointwise and in } L^2(\mathbb{V}, \mu). \tag{5.7}$$

Proof It is sufficient to prove this for $f \in L^2_+$. By (5.6)

$$t^{-1}(Q_t f - f) = t^{-1} \sum_{n=0}^{\infty} \frac{e^{-t} t^n}{n!} (P^n f - f) = \sum_{n=1}^{\infty} \frac{e^{-t} t^{n-1}}{n!} (P^n f - f).$$

Since $P_1 f - f = \Delta f$, we obtain

$$t^{-1}(Q_t f - f) = \Delta f + t R_t f, \tag{5.8}$$

where $R_t f = \sum_{n=2}^{\infty} (e^{-t} t^{n-2}/n!)(P^n f - f)$. Then

$$\begin{aligned} ||R_t f||_2 &\leq \sum_{n=2}^{\infty} (e^{-t} t^{n-2}/n!)(||P^n f||_2 + ||f||_2) \\ &\leq 2||f||_2 \sum_{n=2}^{\infty} (e^{-t} t^{n-2}/(n-2)!) \leq 2||f||_2, \end{aligned}$$

and letting $t \downarrow 0$ in (5.8) gives the convergence in L^2. The pointwise convergence then follows. □

Corollary 5.4 *Let $f \in L^2(\mathbb{V}, \mu)$ and $t \geq 0$. Then*

$$\frac{d}{dt} Q_t f = Q_t \Delta f. \tag{5.9}$$

The limit in the derivative holds both pointwise and in $L^2(\mathbb{V}, \mu)$.

Proof The case $t = 0$ is given in Lemma 5.3. By (5.8) applied to $g = Q_t f$, if $h \in (0, 1)$ then

$$||Q_{t+h} f - Q_t f||_2 \leq h||\Delta Q_t f||_2 + h^2 ||R_h Q_t f||_2 \leq 4h||f||_2.$$

Hence if $t > 0$, and $0 < |h| < 1 \wedge t$, then by first considering separately the cases $h > 0$ and $h < 0$ we have

$$||h^{-1}(Q_{t+h} f - Q_t f) - h\Delta Q_t f||_2 \leq ||Q_t f - Q_{t-|h|} f||_2 + 2h||f||_2,$$

which proves that the limit in (5.9) holds in L^2. Again, the existence of the pointwise limit is then immediate. □

Lemma 5.5 *Let $f \in L^2$. Then*

$$\frac{d}{dt}||Q_t f||_2^2 = -2\mathscr{E}(Q_t f, Q_t f), \tag{5.10}$$

$$\frac{d^2}{dt^2}||Q_t f||_2^2 = 4||\Delta Q_t f||_2^2 \geq 0. \tag{5.11}$$

Proof Using the L^2 convergence in Corollary 5.4, and then Lemma 5.2(g),

$$\begin{aligned} \frac{d}{dt}||Q_t f||_2^2 &= \lim_{h \to 0} h^{-1} \langle Q_{2t+2h} f - Q_t f, f \rangle \\ &= 2\langle \Delta Q_{2t} f, f \rangle = -2\mathscr{E}(Q_t f, Q_t f). \end{aligned}$$

Similarly,

$$\begin{aligned} -\frac{d}{dt}\mathscr{E}(Q_t f, Q_t f) &= \lim_{h \to 0} h^{-1} \langle Q_{2t+2h} f - Q_{2t} f, \Delta f \rangle \\ &= 2\langle Q_{2t} \Delta f, \Delta f \rangle = 2||\Delta Q_t f||_2^2. \end{aligned}$$

□

Corollary 5.6 *For $f \in L^2$ and $t \geq 0$,*

$$||f||_2^2 - ||Q_t f||_2^2 \leq 2t\mathscr{E}(f, f).$$

Proof Set $\varphi(t) = ||Q_t f||_2^2$. Then φ' is increasing, so

$$\varphi(0) - \varphi(t) = - \int_0^t \varphi'(s)ds \le -t\varphi'(0) = 2t\mathscr{E}(f, f). \qquad \square$$

Remark 5.7 We now return to the relation (5.7). Let (Γ, μ) be a weighted graph, and let $\nu_x > 0$, so ν defines a measure on $\mathbb{V}$. We can associate an operator $\mathscr{L}^{(\nu)}$ to the pair $(\mathscr{E}, \nu)$ by requiring

$$\mathscr{E}(f, g) = \langle \mathscr{L}^{(\nu)} f, g\rangle_{L^2(\nu)} = \sum_x \mathscr{L}^{(\nu)} f(x) g(x)\nu_x, \quad f, g \in C_0(\mathbb{V}).$$

Using the discrete Gauss–Green formula (1.20) we also have

$$\mathscr{E}(f, g) = \sum_x \Delta f(x) g(x)\mu_x,$$

and therefore it follows that

$$\mathscr{L}^{(\nu)} f(x) = \frac{\mu_x}{\nu_x}\Delta f(x) = \frac{1}{\nu_x}\sum_y \mu_{xy}(f(y) - f(x)).$$

The Hille–Yoshida theory of semigroups of operators (see [RW]) allows us to define a semigroup $(Q_t^{(\nu)})$ on $L^2(\nu)$ such that $\mathscr{L}^{(\nu)}$ is the infinitesimal generator of $(Q_t^{(\nu)})$; that is,

$$\lim_{t\to 0} t^{-1}(Q_t^{(\nu)} f - f) = \mathscr{L}^{(\nu)} f, \quad f \in L^2(\nu).$$

Formally one can write

$$Q_t^{(\nu)} = \exp(t\mathscr{L}^{(\nu)}), \quad t \ge 0.$$

The process $Y^{(\nu)}$ with generator $\mathscr{L}^{(\nu)}$ waits at x for an exponential time with mean ν_x/μ_x, and then jumps to $y \sim x$ with probability $\mathscr{P}(x, y) = \mu_{xy}/\mu_x$. In the general terminology of Dirichlet forms (see [FOT]), one says that $Y^{(\nu)}$ is the *Markov process associated with the Dirichlet form* $\mathscr{E}$ *on* $L^2(\nu)$. Note that the quadratic form $\mathscr{E}$ alone is not enough to specify the process: one needs the measure ν also. Given two different measures ν, ν', the processes $Y^{(\nu)}$, $Y^{(\nu')}$ have the same jump probabilities as X, and it follows that they can be expressed as time changes of each other.

Unfortunately in general the semigroup $(Q_t^{(\nu)})$ is not unique. This is because of the possibly of 'explosion' of the process $Y^{(\nu)}$ – that is, it can escape to infinity in finite time. Let $\tau_n = \inf\{t \ge 0 : Y_t^{(\nu)} \notin B(x, n)\}$ and $\zeta = \lim_n \tau_n$; we call ζ the *explosion time* of $Y^{(\nu)}$. In general one may have $\mathbb{P}^x(\zeta < \infty) > 0$, and in this case the generator $\mathscr{L}^{(\nu)}$ only determines $Y^{(\nu)}$ on the interval $[0, \zeta)$.

If the explosion time ζ is always $+\infty$, (i.e. $\mathbb{P}^x(\zeta = \infty) = 1$ for all $x \in \mathbb{V}$) then the process Y is called *stochastically complete.*

For the particular case $\nu = \mu$ explosion does not occur, since (5.1) defines Y_t for all $t \in [0, \infty)$. The intuitive explanation of why Y does not explode is that it has to wait an average of one time unit at a vertex x before jumping to the next vertex.

While any ν is possible, there are two particularly natural choices of ν. The first (chosen above) is $\nu = \mu$, while the second is $\nu_x \equiv 1$. If we wish to distinguish these processes, we call them the *constant speed* and *variable speed continuous time simple random walks on* (Γ, μ), or CSRW and VSRW for short. In this book we discuss just the CSRW. Explosion in finite time is a possibility for the VSRW on graphs with unbounded weights or vertex degree; for more on the general case and some sufficient conditions for stochastic completeness, see [GHM, Fo2]. Note also that the infinitesimal generator of the VSRW is the combinatorial Laplacian Δ_{Com} given by (1.11).

Remarks (1) Taking $f = \mathbf{1}_y$ in (5.7) gives

$$\lim_{t\to 0} \frac{\mathbb{P}^x(Y_t = y)}{t} = \mathscr{P}(x, y) = \frac{\mu_{xy}}{\mu_x}, \qquad y \neq x,$$
$$\lim_{t\to 0} \frac{1 - \mathbb{P}^x(Y_t = x)}{t} = \frac{\mu_y - \mu_{yy}}{\mu_y}.$$

Thus Y has a 'Q-matrix' (see [Nor]) given by

$$Q_{xy} = \begin{cases} \mathscr{P}(x, y), & x \neq y, \\ \mathscr{P}(x, x) - 1, & x = y. \end{cases}$$

(2) If $Y_t = x$ then the probability that Y jumps to $y \sim x$ in the time interval $(t, t + h]$ is $\mathscr{P}(x, y)h + O(h^2)$.

(3) For discrete time walks, allowing $\mu_{xx} > 0$ makes a possibly important qualitative difference, since it permits one to have $X_n = X_{n+1} = x$ with positive probability. This is helpful in removing difficulties arising in bipartite graphs, and is one reason to study the 'lazy' random walk. For the continuous time walk, the jump time out of a point x has a smooth (exponential) distribution, and it is more natural to take $\mu_{xx} = 0$ for all x. However, as having $\mu_{xx} > 0$ creates no difficulties, we will continue to allow this possibility.

Corollary 5.8 *Suppose that $|\mathbb{V}| = N$, let φ_i, ρ_i be as in Corollary 3.35, and set $\lambda_i = 1 - \rho_i$. Then*

$$q_t(x, y) = \sum_{i=1}^{N} \varphi_i(x)\varphi_i(y)e^{-\lambda_i t}.$$

Proof Since

$$\sum_{n=0}^{\infty} \rho_i^n \frac{e^{-t}t^n}{n!} = e^{-t+\rho_i t} = e^{-\lambda_i t},$$

this is immediate from Corollary 3.35 and (5.2). □

The following result on speed of convergence to equilibrium is proved in the same way as Lemma 3.37.

Corollary 5.9 *Let $\mathbb{V}$ be finite, and let $0 = \lambda_1 < \lambda_2 \le \cdots \le \lambda_N$ be the eigenvalues of $-\Delta$. Let ν_x be a probability measure on $\mathbb{V}$, $b = \mu(\mathbb{V})^{-1}$, and $f_t(y) = \sum_x \nu_x q_t(x, y)$. Then*

$$||f_t - b||_2 \le e^{-\lambda_2 t}||f_0 - b||_2.$$

Let $D \subset \mathbb{V}$, and define the killed heat kernel by

$$q_t^D(x, y) = \mu_y^{-1}\mathbb{P}^x(Y_t = y, \tau_D > t). \tag{5.12}$$

We write $q_t^{D,x} = q_t^D(x, \cdot)$. It is straightforward to verify

Lemma 5.10 *For $s, t \ge 0$, $x, y, z \in \mathbb{V}$,*

$$\begin{aligned}
q_t^D(x, y) &= \sum_{n=0}^{\infty} \frac{e^{-t}t^n}{n!} p_n^D(x, y),\\
q_t^D(x, y) &= q_t^D(y, x),\\
q_{t+s}^D(x, z) &= \langle q_t^{D,x}, q_s^{D,z}\rangle,\\
\frac{\partial}{\partial t} q_t^D(x, y) &= \Delta_D q_t^D(x, y).
\end{aligned}$$

The following relation gives one key to the control of $q_t(x, x)$.

Lemma 5.11 *Let $x \in \mathbb{V}$, and $\psi(t) = \langle q_t^x, q_t^x\rangle = q_{2t}(x, x)$. Then*

$$\psi'(t) = 2\langle \Delta q_t^x, q_t^x\rangle = -2\mathscr{E}(q_t^x, q_t^x) \le 0, \tag{5.13}$$

$$\psi''(t) = 4\langle \Delta q_t^x, \Delta q_t^x\rangle \ge 0. \tag{5.14}$$

In particular, ψ is decreasing and ψ' is increasing.

Proof This follows from Lemma 5.5 by taking $f = \mu_x^{-1}\mathbf{1}_x$ so that $Q_t f(y) = q_t(y, x) = q_t^x(y)$. □

Exercise Show that if (Γ, μ) is transient then

$$g(x, y) = \int_0^\infty q_t(x, y)\, dt.$$

5.2 Heat Kernel Bounds

We will use some basic properties of the Poisson distribution to compare the discrete and continuous time heat kernels $p_n(x, y)$ and $q_t(x, y)$. For $s \in (-1, \infty)$ set

$$f(s) = (1+s)\log(1+s) - s.$$

Lemma 5.12 *For $s \in [0, 1)$,*

$$f(-s) \geq f(s) \geq \tfrac{1}{4}s^2.$$

For $s \geq 1$, $f(s) \geq c_0(1+s)\log(1+s)$, where $c_0 = 1 - (2\log 2)^{-1}$.

Proof Note that $f'(s) = \log(1+s)$. Since $-\log(1-u) \geq \log(1+u)$ for $u \in (0, 1)$, we have, for $s \in (0, 1)$,

$$f(-s) = -\int_0^s \log(1-u)du \geq \int_0^s \log(1+u)du = f(s) \geq \int_0^s \tfrac{1}{2}u du = \tfrac{1}{4}s^2.$$

Here we used the fact that $\log(1+u) \geq u/2$ for $0 < u < 1$.

For the second assertion, set $h(s) = (1+s)\log(1+s)$ and $g(s) = h(s)/s$. Then $g'(s) \geq 0$ for $s \geq 1$, so that $\inf_{s\geq 1} g(s) = g(1) = 2\log 2$. Therefore, writing $b = 1/2\log 2$, we have $f(s) = bh(s) - s + (1-b)h(s) \geq (1-b)h(s)$. □

Write

$$a_t(k) = \mathbb{P}(N_t = k) = \frac{e^{-t}t^k}{k!}. \tag{5.15}$$

Lemma 5.13 *(a) For any $t \geq 0$, $k \in \mathbb{Z}_+$,*

$$\mathbb{P}(N_t \in 2\mathbb{Z}, N_t \leq 2k) \geq \frac{1}{3}\mathbb{P}(N_t \leq 2k).$$

(b) For any $t > 0$,

$$\mathbb{P}(N_t \geq (1+s)t) \leq \exp(-tf(s)), \quad s > 0, \tag{5.16}$$

$$\mathbb{P}(N_t \leq (1-s)t) \leq \exp(-tf(-s)), \quad 0 < s < 1. \tag{5.17}$$

(c) We have

$$\mathbb{P}(N_t \geq (e-1)t) \leq e^{-t}, \quad \mathbb{P}(N_t \leq t/e) \leq e^{-t(e-2)/e}.$$

(d) For $R > t$,

$$\mathbb{P}(N_t \geq R) \leq \exp\Big(-t + R - R\log(R/t)\Big) = \exp\Big(-t - R\log(R/et)\Big).$$

(e) For $|\lambda| < 1$, $t \geq 0$,

$$\mathbb{P}(|N_t - t| \geq \lambda t) \leq 2e^{-\lambda^2 t/4}.$$

Proof (a) We have

$$t^2 - (2k+2)t + (2k+1)(2k+2) > t^2 - 2(k+1)t + (k+1)^2 \geq 0;$$

rearranging and multiplying by $e^{-t}t^{2k}/(2k+1)!$ gives

$$a_t(2k+1) \leq a_t(2k) + a_t(2k+2).$$

So

$$\begin{aligned}\mathbb{P}(N_t \in 2\mathbb{Z}, N_t \leq 2k) &= \tfrac{1}{3}\sum_{m=0}^{k} a_t(2m) + \tfrac{1}{3}\sum_{m=0}^{k} 2a_t(2m) \\ &\geq \tfrac{1}{3}\sum_{m=0}^{k} a_t(2m) + \tfrac{1}{3}\sum_{m=0}^{k-1} a_t(2m+1) = \tfrac{1}{3}\mathbb{P}(N_t \leq 2k).\end{aligned}$$

(b) This is proved by a standard exponential moment estimate. Let $\lambda > 0$. Then

$$\begin{aligned}\mathbb{P}(N_t \geq t(1+s)) &= \mathbb{P}(e^{\lambda N_t} \geq e^{\lambda(1+s)t}) \\ &\leq e^{-\lambda(1+s)t}\mathbb{E}e^{\lambda N_t} = \exp(-\lambda t(1+s) - t + te^{\lambda}).\end{aligned}$$

The minimum on the left is attained by taking $\lambda = \log(1+s)$, and this gives (5.16). Similarly for (5.17) we write

$$\mathbb{P}(N_t \leq (1-s)t) = \mathbb{P}(e^{-\lambda N_t} \geq e^{-\lambda(1-s)t}),$$

use Markov's inequality, and finally set $\lambda = -\log(1-s)$.
The remaining estimates are immediate from (b) and Lemma 5.12. □

Theorem 5.14 *Let* $\alpha \geq 1$. *The following are equivalent.*

(a) Γ *satisfies the Nash inequality* (N_α).
(b) There exists c_1 *such that*

$$p_n(x,y) \leq c_1(1 \vee n)^{-\alpha/2}, \quad n \geq 0, \; x, y \in \mathbb{V}.$$

(c) There exists c_2 such that

$$q_t(x, y) \le \frac{c_2}{t^{\alpha/2}}, \quad t > 0, \ x, y \in \mathbb{V}. \tag{5.18}$$

Proof The equivalence of (a) and (b) has already been proved in Theorem 4.3, so it remains to prove (b) $\Leftrightarrow$ (c). First assume (b). Then

$$q_t(x, x) = \mathbb{E} p_{N_t}(x, x) \le c\mathbb{E}(1 \vee N_t)^{-\alpha/2}.$$

By Lemma 5.13(c),

$$\mathbb{E}(1 \vee N_t)^{-\alpha/2} \le \mathbb{P}(N_t < t/e) + (t/e)^{-\alpha/2}\mathbb{P}(N_t \ge t/e) \le e^{-ct} + c't^{-\alpha/2},$$

which gives (c).

Now suppose (c) holds. By Lemma 4.1 it is enough to prove

$$p_{2n}(x, x) \le c(1 \vee n)^{-\alpha/2}. \tag{5.19}$$

Using the notation $a_t(k)$ given in (5.15), we have, for $t \ge 0, n \ge 0$,

$$\begin{aligned} ct^{-\alpha/2} \ge q_t(x, x) &= \sum_{k=0}^{\infty} \big(a_t(2k) p_{2k}(x, x) + a_t(2k+1) p_{2k+1}(x, x)\big) \\ &\ge \sum_{k=0}^{n} a_t(2k) p_{2k}(x, x) \\ &\ge p_{2n}(x, x)\mathbb{P}(N_t \le 2n, N_t \in 2\mathbb{N}) \\ &\ge \tfrac{1}{3} p_{2n}(x, x)\mathbb{P}(N_t \le 2n). \end{aligned}$$

If $n = 0$ then set $t = 1$; we obtain $c_2 \ge q_1(x, x) \ge \frac{1}{3} p_0(x, x)e^{-1}$, which gives (5.19). If $n \ge 1$ then set $t = 2n/(e - 1)$, so that $t \ge 1$. Lemma 5.13(c) gives $\mathbb{P}(N_t \ge 2n) \le e^{-t} \le \frac{1}{2}$, and thus $p_{2n}(x, x) \le 6ct^{-\alpha/2} = c'n^{-\alpha/2}$, proving (b). □

Remarks 5.15 (1) Note the following quick proof of (a) $\Rightarrow$ (c). Set $\psi(t) = \langle q_t^x, q_t^x \rangle = q_{2t}(x, x)$; then $||q_t^x||_1 = 1$ and $|q_t^x||_2^2 = \psi(t)$. So by Lemma 5.11 and (N_α)

$$\psi'(t) = -2\mathscr{E}(q_t^x, q_t^x) \le -c_N ||q_t^x||_2^{2+4/\alpha} ||q_t^x||_1^{-4/\alpha} = -c_N \psi(t)^{1+2/\alpha}. \tag{5.20}$$

We have $\psi(0) = q_0(x, x) = \mu_x^{-1}$. Let $\varphi(t) = \psi(t)^{-2/\alpha}$; then

$$\varphi'(t) = -\frac{2}{\alpha}\psi'(t)\psi(t)^{-1-2/\alpha} \ge c_2,$$

so $\varphi(t) \ge \varphi(0) + c_2 t \ge c_2 t$, and thus $\psi(t) \le (c_2 t)^{-\alpha/2}$.

(2) We could also prove the implication (c) $\Rightarrow$ (a) by using the same argument as in Theorem 4.3.

Corollary 5.16 *We have*

$$\mathscr{E}(q_t^x, q_t^x) \le t^{-1} q_t(x,x).$$

Proof Using the notation of Remarks 5.15, by Lemma 5.11, $\psi'' \ge 0$, so that ψ' is increasing. Thus

$$\psi(t) - \psi(t/2) = \int_{t/2}^{t} \psi'(s)ds \le \tfrac{1}{2} t \psi'(t) = -t\mathscr{E}(q_t^x, q_t^x).$$

So $\mathscr{E}(q_t^x, q_t^x) \le t^{-1}\psi(t/2) = t^{-1} q_t(x,x)$. □

The tail of the Poisson decays more rapidly than an exponential. Combining the Carne–Varopoulos bound with our bounds on the tail of the Poisson, we obtain the following.

Theorem 5.17 *Let $x, y \in \mathbb{V}$ and $R = d(x,y) \ge 1$. Then*

(a)

$$q_t(x,y) \le \begin{cases} 4(\mu_x \mu_y)^{-1/2} e^{-R^2/(e^2 t)}, & \text{if } R \le et, \\ (\mu_x \vee \mu_y)^{-1} \exp(-t - R\log \frac{R}{et}), & \text{if } R \ge et; \end{cases} \tag{5.21}$$

(b) if Γ satisfies (H5) then

$$q_t(x,y) \ge c_1 (\mu_x \wedge \mu_y)^{-1} \exp(-c_2 R - R \log \tfrac{R}{t})), \quad \text{for } R \ge t > 0. \tag{5.22}$$

Proof (a) Note that both bounds in (5.21) are of the form e^{-cR} if $t \le R \le et$. By Theorem 4.9 and Lemma 5.13(c) we have, writing $b = 2(\mu_x\mu_y)^{-1/2}$,

$$\begin{aligned} q_t(x,y) &\le b \sum_{n=0}^{\infty} a_t(n) e^{-R^2/2n} \\ &= b \sum_{n \le et} a_t(n) e^{-R^2/2n} + b \sum_{n > et} a_t(n) e^{-R^2/2n} \\ &\le b e^{-R^2/2et} \sum_{n \le et} a_t(n) + b \sum_{n > et} a_t(n) \le b e^{-R^2/2et} + b e^{-t}, \end{aligned}$$

and since $t \ge e^{-2} R^2/t$ this gives the first bound.

If $R \ge et$ then by Lemma 5.13(b)

$$q_t(x,y)\mu_y \le \mathbb{P}(N_t \ge R) \le \exp(-t - R\log(R/et)).$$

Interchanging x and y gives the same bound for $q_t(x,y)\mu_x$, which implies the second bound in (5.21).

(b) Since $d(x, y) = R \geq 1$ there exists a path $x = z_0, z_1, \dots, z_R = y$ of length R connecting x and y. Using (H5) there exists $p_0 > 0$ such that $\mathscr{P}(z_{n-1}, z_n) \geq p_0$, for $n = 1, \dots, R$. Hence $p_R(x, y)\mu_y \geq p_0^R$, and therefore

$$\begin{aligned} q_t(y, x)\mu_x = q_t(x, y)\mu_y \geq p_R(x, y)\mu_y \mathbb{P}(N_t = R) &\geq p_0^R e^{-t} t^R R^{-R} \\ &= \exp(-R \log p_0^{-1} - t - R \log(R/t)), \end{aligned}$$

and as $t \leq R$ this implies (5.22). □

Remarks 5.18 (1) The bounds in this theorem show that if $d(x, y) = k$ then $q_t(x, y) \asymp t^k$ for small t.

(2) If we do not need the best constant for the $R \log(R/t)$ term, then, writing $U(R, t) = R + R \log(R/t)$, it is easy to obtain from (5.21) and (5.22) the simpler looking bounds

$$c_1 (\mu_x \wedge \mu_y)^{-1} e^{-c_2 U(R,t)} \leq q_t(x, y) \leq 4(\mu_x \mu_y)^{-1/2} e^{-c_3 U(R,t)}.$$

We now consider heat kernel bounds of the form HK(α, β) for $q_t(x, y)$. In view of the transition between 'bulk' and 'Poisson' behaviour, and the Poisson type bounds in Theorem 5.17, we restrict to $t \geq d(x, y)$.

Definition 5.19 (Γ, μ) satisfies UHKC(α, β) if there exist constants c_1, c_2 such that for all $x, y \in \mathbb{V}$, $t \geq d(x, y)$

$$q_t(x, y) \leq \frac{c_1}{t^{\alpha/\beta}} \exp\Big(- c_2 \Big(\frac{d(x, y)^\beta}{t}\Big)^{\frac{1}{\beta-1}} \Big). \tag{5.23}$$

We say (Γ, μ) satisfies LHKC(α, β) if there exist constants c_3, c_4 such that for all $x, y \in \mathbb{V}$ and $t \geq 1 \vee d(x, y)$

$$q_t(x, y) \geq \frac{c_3}{t^{\alpha/\beta}} \exp\Big(- c_4 \Big(\frac{d(x, y)^\beta}{t}\Big)^{\frac{1}{\beta-1}} \Big). \tag{5.24}$$

We say (Γ, μ) satisfies HKC(α, β) if both UHKC(α, β) and LHKC(α, β) hold.

We now use similar arguments to those in Chapter 4 to obtain UHKC(α, β) from bounds on volume and hitting times.

Lemma 5.20 *Let $A \subset \mathbb{V}$, $x \in \mathbb{V}$. Then*

$$\mathbb{E}^x T_A(X) = \mathbb{E}^x T_A(Y).$$

In particular, the condition (E_β) is the same for both processes.

Proof Recall that $Y_t = X_{N_t}$, where N is a Poisson process independent of X. Write $S_1, S_2, \dots$ for the jump times of N. Then $S_k - S_{k-1}$ are i.i.d. exp(1) r.v., and so $\mathbb{E} S_k = k$. We have $T_A(Y) = S_{T_A(X)}$, and so

$$\mathbb{E}^x T_A(Y) = \sum_{k=0}^{\infty} \mathbb{E}^x (S_k; T_A(X) = k) = \mathbb{E}^x T_A(X). \qquad \square$$

Write $\tau_Y(x, r) = \tau_{B(x,r)}(Y)$.

Lemma 5.21 *Let $\beta \geq 2$ and suppose that (E_β) holds. There exists $c_i > 0$ such that if $t \geq r \geq 1$ then*

$$\mathbb{P}^x(\tau_Y(x, r) < t) \leq c_4 \exp(-c_5(r^\beta/t)^{1/(\beta-1)}). \tag{5.25}$$

Proof We could follow the arguments in Lemma 4.31 and Proposition 4.33, but a direct proof from the discrete time result is simpler. Write τ_X, τ_Y for the exit times for X and Y from $B(x, r)$, and set $\gamma = 1/(\beta - 1)$. Note that $\{\tau_Y \leq t, \tau_X > 2t\} \subset \{N_t > 2t\}$. So,

$$\begin{aligned}\mathbb{P}^x(\tau_Y \leq t) &= \mathbb{P}^x(\tau_Y \leq t, \tau_X \leq 2t) + \mathbb{P}^x(\tau_Y \leq t, \tau_X > 2t) \\ &\leq \mathbb{P}^x(\tau_X \leq 2t) + \mathbb{P}(N_t > 2t) \leq ce^{-c(r^\beta/2t)^\gamma} + e^{-ct}.\end{aligned}$$

Here we used Proposition 4.33 and Lemma 5.13(b) in the final line. Since $t \geq r$, we have $t \geq (r^\beta/t)^\gamma$, giving (5.25). $\square$

The proofs of Lemmas 4.19, 4.20, 4.21, and 4.22 all go through for Y with essentially no changes. In particular, we have the following lemma.

Lemma 5.22 *Let Γ satisfy HKC(α, β). Then Γ satisfies (V_α) and (E_β).*

The following two theorems give the equivalence of upper and lower heat kernel bounds in discrete and continuous time.

Theorem 5.23 *Let $\alpha \geq 1$, $\beta \geq 2$, and let (Γ, μ) satisfy (V_α). The following are equivalent.*

(a) (Γ, μ) satisfies (E_β) and the on-diagonal upper bound

$$p_n(x, x) \leq c(1 \vee n)^{-\alpha/\beta}, \quad n \geq 0.$$

(b) (Γ, μ) satisfies (E_β) and the on-diagonal upper bound

$$q_t(x, x) \leq ct^{-\alpha/\beta}, \quad t > 0.$$

(c) (Γ, μ) satisfies UHK(α, β).
(d) (Γ, μ) satisfies UHKC(α, β).

Proof The equivalence of (a) and (c) is proved in Theorem 4.34, and of (a) and (b) in Theorem 5.14 and Lemma 5.20.

Suppose (b) holds. Let $x, y \in \mathbb{V}$ with $d(x, y) = r \geq 1$, and let $t > 0$. If $t < r$ then there is nothing to prove, so it remains to consider the case $t \geq 1 \vee r$. Lemma 5.21 gives the bound (5.25) on the tail of $\tau_Y(x, r)$, and the same argument as in the proof of (a) $\Rightarrow$ (b) in Theorem 4.34 then gives (5.23).

The proof that (d) $\Rightarrow$ (b) is the same as in the discrete time case – see Lemma 4.21. □

Theorem 5.24 *Let $\alpha \geq 1$, $\beta \geq 2$. Suppose that (Γ, μ) satisfies (V_α), (H5), and UHK(α, β). The following are equivalent.*

(a) $p_n(x, y)$ satisfies the near-diagonal lower bound (Definition 4.36).
(b) (Γ, μ) satisfies HK(α, β).
(c) $q_t(x, y)$ satisfies a near-diagonal lower bound: there exist $c_1, c_2 \in (0, 1)$ such that

$$q_t(x, y) \geq c_1 t^{-\alpha/\beta} \text{ if } 1 \vee d(x, y) \leq c_2 t^{1/\beta}. \tag{5.26}$$

(d) (Γ, μ) satisfies HKC(α, β).

Proof The implications (b) $\Rightarrow$ (a) and (d) $\Rightarrow$ (c) are trivial, and Proposition 4.38 proved that (a) $\Rightarrow$ (b). The same chaining argument also works in continuous time, so that (c) $\Rightarrow$ (d). Thus it remains to prove that the discrete and continuous time near-diagonal lower bounds are equivalent.

The harder implication is that (c) $\Rightarrow$ (a), since this uses the bounds on $|p_{k+2}(x, y) - p_k(x, y)|$ obtained in Theorem 4.28. Assume that (c) holds, fix x, y, and let $c_2 n^{1/\beta} \geq 1 \vee d(x, y)$. Let $\delta \in (0, \frac{1}{2})$, and recall the notation $a_t(k)$ from (5.15). Then by (5.26)

$$\begin{aligned} c_1 n^{-\alpha/\beta} \leq q_n(x, y) &= \sum_{|k-n| \leq \delta n} a_n(k) p_k(x, y) + \sum_{|k-n| > \delta n} a_n(k) p_k(x, y) \\ &\leq \sum_{|k-n| \leq \delta n} a_n(k) p_k(x, y) + \mathbb{P}(|N_n - n| > \delta n). \end{aligned}$$

Let $J_0 = \{k : |k - n| \in 2\mathbb{Z}, |k - n| \leq \delta n\}$, $J_1 = \{k : |k - n - 1| \in 2\mathbb{Z}, |k - n| \leq \delta n\}$, and write $a_n(J) = \sum_{k \in J} a_n(k)$. Then by Theorem 4.28, for $k \in J_0$,

$$\begin{aligned} |p_k(x, y) - p_n(x, y)| &\leq \sum_{|i| \leq \delta n/2} |p_{i+2}(x, y) - p_i(x, y)| \\ &\leq n\delta c_3 (n - n\delta)^{-1-\alpha/\beta} \leq c_4 \delta n^{-\alpha/\beta}. \end{aligned}$$

Thus

$$\sum_{k\in J_0} a_n(k)(p_k(x,y) - p_n(x,y)) \le c_4\delta n^{-\alpha/\beta},$$

and a similar bound holds for $\sum_{k\in J_1} a_n(k)(p_k(x,y) - p_{n+1}(x,y))$. Hence, using Lemma 5.13(e),

$$\begin{aligned} c_1 n^{-\alpha/\beta} &\le p_n(x,y)a_n(J_0) + p_{n+1}(x,y)a_n(J_1) + c_4\delta n^{-\alpha/\beta} + \mathbb{P}(|N_n - n| > \delta n) \\ &\le p_n(x,y) + p_{n+1}(x,y) + c_4\delta n^{-\alpha/\beta} + 2e^{-\delta^2 n/4}. \end{aligned}$$

Choosing δ so that $c_4\delta \le c_1/3$, and c_5 so that if $c_2 n^{1/\beta} \ge c_5$ then $2e^{-\delta^2 n/4} \le \frac{1}{3}c_1 n^{-\alpha/\beta}$, we obtain

$$p_n(x,y) + p_{n+1}(x,y) \ge \tfrac{1}{3}c_1 n^{-\alpha/\beta},$$

provided $c_2 n^{1/\beta} \ge c_5 \vee d(x,y)$, which is the discrete time near-diagonal lower bound.

Now assume that (a) holds. If $t \ge 1$ and $|t-k| \le \frac{1}{2}t$ then $a_t(k+1)/a_t(k) = t/(k+1) \le 2$. Since $a \wedge b \ge \frac{1}{3}(a+b)$ if $b \le 2a$, we have, for $t \ge 1$, writing $J = \{k \in 2\mathbb{Z} : |t-k| \le \frac{1}{2}t\}$,

$$\begin{aligned} q_t(x,y) &\ge \sum_{k\in J} a_t(k)p_k(x,y) + a_t(k+1)p_{k+1}(x,y) \\ &\ge \tfrac{1}{3}\sum_{k\in J}(a_t(k) + a_t(k+1))(p_k(x,y) + p_{k+1}(x,y)) \\ &\ge (1 - 2e^{-t/16}) \min_{k\in J}(p_k(x,y) + p_{k+1}(x,y)). \end{aligned}$$

Hence there exist c_6, c_7 such that if $c_6 t^{1/\beta} \ge 1 \vee d(x,y)$ then $q_t(x,y) \ge c_7 t^{-\alpha/\beta}$, giving the continuous time NDLB (5.26). □

Remark See Remark A.52 for an example which shows that if (H5) does not hold then HK(α,β) and HKC(α,β) are not equivalent. (The difficulty arises with the discrete time lower bound at small times.)

The following theorem summarises our sufficient conditions for the heat kernel bounds.

Theorem 5.25 *Let $\alpha \ge 1$ and $\beta \ge 2$, and let (Γ,μ) be a weighted graph satisfying (H5). The following are equivalent.*

(a) (Γ,μ) satisfies (V_α), (N_α), (E_β), $NDLB(\beta)$.
(b) $HK(\alpha,\beta)$ holds.
(c) $HKC(\alpha,\beta)$ holds.

Proof If (Γ, μ) satisfies (V_α) then the equivalence of (a), (b), and (c) is immediate from Theorems 4.39 and 5.24. It therefore remains to prove that each of (b) and (c) implies (V_α), and this is given by Lemma 4.22, in discrete time, and by a similar argument in continuous time. □

We conclude this section by stating a lower bound on the killed heat kernel in continuous time. Let $D \subset \mathbb{V}$, and set

$$q_t^D(x, y) = \mu_y^{-1} \mathbb{P}^x(Y_t = y, \tau_D > t).$$

Then the same argument as in the discrete time case (see Theorem 4.25) gives:

Proposition 5.26 *Suppose (Γ, μ) satisfies $HK(\alpha, \beta)$, and let $B = B(x_0, R)$, $B' = B(x_0, R/2)$. Then for $x, y \in B'$, $d(x, y) \le t \le R^\beta$*

$$q_t^B(x, y) \ge c_1 t^{-\alpha/\beta} \exp\left(-c_2 \left(\frac{d(x, y)^\beta}{t} \right)^{\frac{1}{\beta-1}} \right).$$

6

Heat Kernel Bounds

6.1 Strongly Recurrent Graphs

Chapters 4 and 5 showed that one can obtain Gaussian bounds, or the more general bounds HK(α, β) given three main inputs: a Nash inequality, a bound on the exit times from balls, and the near-diagonal lower bound. Of these, we have seen in Chapter 3 how one can prove the Nash inequality from geometric data on the graph. The other two conditions were introduced in an axiomatic fashion, and so far we have seen no indication of how they can be proved.

In this chapter we will study two different situations where these conditions can be established. The easier but less familiar case, which we will do first, is a family of graphs which we call 'strongly recurrent', and where the heat kernel bounds will follow from volume and resistance estimates.

There is no generally agreed definition of 'strongly recurrent' – see [D2, Ba1, T2] for two definitions. The general idea is that if a graph (Γ, μ) satisfies a condition of the form

$$R_{\text{eff}}(x, y) \geq c d(x, y)^{\varepsilon}, \qquad \text{as } d(x, y) \to \infty,$$

then, with some additional control on the volume growth, one can use effective resistance to obtain good control of the SRW on (Γ, μ). The standard example of a graph which is recurrent but not 'strongly recurrent' is $\mathbb{Z}^2$, where $R_{\text{eff}}(x, y) \asymp \log d(x, y)$.

Definition 6.1 (Γ, μ) satisfies the condition (R_β) if there exist constants C_1, C_2 such that

$$\frac{C_1 d(x, y)^\beta}{\mu(B(x, d(x, y)))} \leq R_{\text{eff}}(x, y) \leq \frac{C_2 d(x, y)^\beta}{\mu(B(x, d(x, y)))}, \qquad \text{for all } x, y \in \mathbb{V}.$$

If (V_α) holds and $\beta > \alpha$ then we can rewrite this condition as

$$C_0 d(x,y)^{\beta-\alpha} \le R_{\text{eff}}(x,y) \le C_0 C_A d(x,y)^{\beta-\alpha}, \qquad \text{for all } x, y \in \mathbb{V}. \quad (6.1)$$

We will study graphs satisfying the two conditions (V_α) and (R_β), with $\beta > \alpha$, and will write

$$\zeta = \beta - \alpha > 0.$$

By Theorem 2.72 the graphical Sierpinski gasket satisfies these conditions with $\alpha = \log 3/\log 2$ and $\beta = \log 5/\log 2$. We will make frequent use of the following inequality, proved in Lemma 2.25:

$$|f(x) - f(y)|^2 \le \mathscr{E}(f,f) R_{\text{eff}}(x,y), \qquad f \in C_0(\mathbb{V}). \quad (6.2)$$

We begin with the following covering result; the constants C_U and C_L are as in Definition 4.14.

Lemma 6.2 *Assume (V_α). Let $r \ge 1$. Then there exists a covering of $\mathbb{V}$ by balls $B_i = B(x_i, r)$ such that*

(a) $B(x_i, r/2)$ are disjoint;
(b) Let $\theta \ge 1$. Any $y \in \mathbb{V}$ is in at most $C_U(2\theta+1)^\alpha / C_L$ of the balls $B(x_i, \theta r)$.

Proof First, we can choose a maximal set $\{x_i\}$ of points in $\mathbb{V}$ such that $B(x_i, r/2)$ are disjoint. ('Maximal' means that no extra points can be added to the set $\{x_i, i \in \mathbb{N}\}$ – there is no assumption of optimal packing.) To do this, write $\mathbb{V}$ as a sequence $\{z_1, z_2, \dots\}$, and let $x_1 = z_1$. Given $x_1, \dots, x_{n-1}$, choose x_n to be the first z_j such that $B(z_j, r/2)$ is disjoint from $\cup_{i=1}^{n-1} B(x_i, r/2)$. Write $B_i = B(x_i, r)$, $B_i^* = B(x_i, \theta r)$.

We now show that $\mathbb{V} = \cup B_i$. Let $m \ge 2$, and let k be the smallest integer such that x_k was chosen from $\{z_{m+1}, \dots\}$. Since $x_k \ne z_m$, we must have that $B(z_m, r/2) \cap B(x_i, r/2) \ne \emptyset$ for some $i \le k-1$; hence $z_m \in B(x_i, r)$.

Now let $y \in \mathbb{V}$ be in M of the balls B_i^*; without loss of generality $B_1^*, \dots, B_M^*$. So $d(y, x_i) \le \theta r$ for $1 \le i \le M$. Thus $B(x_i, r/2) \subset B(y, (\theta + \frac{1}{2})r)$ for $i \le M$. So since $B(x_i, r/2)$ are disjoint,

$$\begin{aligned} C_U(\theta + \tfrac{1}{2})^\alpha r^\alpha &\ge V(y, (\theta + \tfrac{1}{2})r) \ge \mu(\cup_{i=1}^M B(x_i, r/2)) \\ &= \sum_{i=1}^M V(x_i, r/2) \ge M C_L (r/2)^\alpha, \end{aligned}$$

which gives the upper bound on M. □

Lemma 6.3 *Let (Γ, μ) satisfy (V_α), and (R_β) with $\beta > \alpha$. Then there exist constants c_i such that*

$$c_1 r^{\beta-\alpha} \le R_{\text{eff}}(x, B(x,r)^c) \le c_2 r^{\beta-\alpha}, \qquad \textit{for all } x \in \mathbb{V},\ r \ge 1.$$

Proof Write $B = B(x,r)$. Let $y \in \partial B$. Then $\{y\} \subset B^c$ and so by Corollary 2.27 we have $R_{\text{eff}}(x, B^c) \le R_{\text{eff}}(x, y) \le C_0 C_A (1+r)^\zeta$, which gives the upper bound.

To prove the converse let $\lambda \in (0, \frac{1}{4})$. Let $B(y_i, \lambda r)$ be the cover of $\mathbb{V}$ given by Lemma 6.2, with λr replacing r. If $B(y_i, \lambda r) \cap B(x,r) \ne \emptyset$ then $x \in B(y_i, (1+\lambda)\lambda^{-1} r)$. Thus if $A_0 = \{i : B(y_i, \lambda r) \cap B(x,r) \ne \emptyset\}$ then $|A_0| \le c\lambda^{-\alpha}$.

Let $A = \{i \in A_0 : d(x, y_i) \ge \frac{1}{2} r\}$. For each $i \in A$ let h_i be the harmonic function

$$h_i(z) = \mathbb{P}^z(T_x < T_{y_i});$$

thus $h_i(x) = 1$ and $h_i(y_i) = 0$. Further h_i is the function which attains the minimum in the variational problem for $C_{\text{eff}}(x, y_i)$, so $\mathcal{E}(h_i, h_i) = R_{\text{eff}}(x, y_i)^{-1} \le c r^{-\zeta}$.

For $i \in A$ and $z \in B(y_i, \lambda r)$ we have

$$|h_i(z) - h_i(y_i)|^2 \le \mathcal{E}(h_i, h_i) R_{\text{eff}}(y_i, z) = \frac{R_{\text{eff}}(y_i, z)}{R_{\text{eff}}(y_i, x)} \le \frac{C_0 C_A (\lambda r)^\zeta}{C_0 (\frac{1}{2} r)^\zeta} = C_A (2\lambda)^\zeta.$$

So taking λ small enough we have $h_i(z) \le \frac{1}{2}$ on each $B(y_i, \lambda r)$. (If $\lambda r < 1$ then $z = y_i$ and so $h(z) = 1$.) Now let

$$g = 2 \min_{i \in A} (h_i - \tfrac{1}{2})_+.$$

Then $g(x) = 1$, and $g = 0$ on $D = \cup_{i \in A} B(y_i, \lambda r)$. Note that D contains all points z with $\frac{3}{4} r \le d(x,z) \le r$. So if $g' = g \mathbf{1}_B$ then $\mathcal{E}(g', g') \le \mathcal{E}(g, g)$. However, by Lemma 1.27(b),

$$\mathcal{E}(g, g) \le \sum_{i \in A} \mathcal{E}(h_i, h_i) \le c m r^{-\zeta}.$$

Since $g'(x) = 1$ and $g' = 0$ on B^c, we have $R_{\text{eff}}(x, B^c) \ge \mathcal{E}(g', g')^{-1} \ge c r^\zeta$. □

Remark This implies that $R_{\text{eff}}(x, \infty) = \infty$, so that (Γ, μ) is indeed recurrent.

Lemma 6.4 *Assume* (V_α)*,* (R_β) *with* $\beta > \alpha$*, and (H5). There exists* $p_S > 0$ *such that if* $x \in \mathbb{V}$ *then*

$$\mathbb{P}^y\big(T_z < \tau(x,r)\big) \geq p_S \quad \text{if } y, z \in B(x, r/2). \tag{6.3}$$

Proof Write $B = B(x,r)$ and let $m \geq 1$. If $d(y,z) \leq r/m$ then by Lemma 2.62 we have

$$\mathbb{P}^y(T_{B^c} < T_z) \leq \frac{R_{\text{eff}}(y,z)}{R_{\text{eff}}(y,B^c)} \leq \frac{C_0 C_A d(y,z)^\zeta}{R_{\text{eff}}(y, B(y,r/2)^c)} \leq c_1 m^{-\zeta}. \tag{6.4}$$

Choose m so that the left side of (6.4) is less than $\frac{1}{2}$. If $r/m \geq 1$ then, given any $y, z \in B(x, r/2)$, we can find a chain of at most $2m$ points $y = w_0, w_1, \dots, w_m = z$ such that $w_i \in B(x,r/2)$ and $d(w_{i-1}, w_i) \leq r/m$ for $i = 1, \dots, 2m$. For each i we have $\mathbb{P}^{w_{i-1}}(T_{w_i} < T_{B^c}) \geq \frac{1}{2}$, so $\mathbb{P}^y(T_z < T^{B^c}) \geq 2^{-m}$. If $r/m < 1$ then we can use (H5) directly to obtain (6.3). □

Corollary 6.5 *Let* (Γ, μ) *satisfy* (V_α)*,* (R_β) *with* $\beta > \alpha$*, and (H5). Then the elliptic Harnack inequality holds for* (Γ, μ)*.*

Proof Let $x \in \mathbb{V}$, $R \geq 2$, $B = B(x, 2R)$, $B' = B(x,R)$, and let $h : \overline{B} \to \mathbb{R}_+$ be harmonic in B. Let $y, z \in B'$ and write $\tau = \tau_B$, $M_n = h(Y_{n \wedge \tau})$. Then M is a martingale, and so, using Lemma 6.4,

$$h(y) = \mathbb{E}^y M_0 = \mathbb{E}^y M_{T_z \wedge \tau} \geq \mathbb{E}^y(M_{T_z\wedge\tau}; T_z < \tau) = h(z)\mathbb{P}^y(T_z < \tau) \geq p_S h(z).$$

An alternative proof uses the maximum principle. Let z, y be the points in B' where h attains its maximum and minimum, respectively, and suppose that $h(z) = 1$. Set $h_z = \mathbb{P}(T_z \leq \tau)$, so that h_z is harmonic in $A = B - \{z\}$ and $h_z(y) \geq p_S$ by Lemma 6.4. Set $f = h_z - h$. If $x \in \partial B$ then $h_z(x) = 0$, while $f(z) = 0$. Thus

$$\max_{\overline{A}} f = \max_{\partial A} f = 0,$$

so that $f \leq 0$ in A. Thus $p_S = h_z(y) \leq h(y)$, proving the EHI. □

Lemma 6.6 *Let* (Γ, μ) *satisfy* (V_α) *and* (R_β)*, and let* $\beta > \alpha$*. Then*

$$q_t(x,x) \leq c_0 t^{-\alpha/\beta}, \quad x \in \mathbb{V},\ t > 0. \tag{6.5}$$

Proof Recall that (V_α) includes the condition $C_L \leq \mu_x \leq C_U$ for all x. So $q_t(x,x) \leq 1/\mu_x \leq 1/C_L$ for all t, and it is enough to prove (6.5) for $t \geq 1$.

Let $r = t^{1/\beta}$, fix $x \in \mathbb{V}$, and set $D = B(x, r)$. Let $f_t(y) = q_t(x, y)$, and let $y_t \in D$ be chosen so that

$$f_t(y_t) = \min_{y \in D} f_t(y).$$

Then

$$f_t(y_t) V(x, r) \le \sum_{y \in D} f_t(y) \mu_y \le 1,$$

so that $f_t(y_t) \le V(x, r)^{-1}$. By Corollary 5.16 we have $t\mathscr{E}(f_t, f_t) \le \varphi(t/2) = f_t(x)$, so

$$\begin{aligned} f_t(x)^2 &\le 2 f_t(y_t)^2 + 2(f_t(x) - f_t(y_t))^2 \\ &\le 2V(x, r)^{-2} + 2R_{\text{eff}}(x, y_t)\mathscr{E}(f_t, f_t) \\ &\le c_1 r^{-2\alpha} + c_1 \frac{r^{\beta-\alpha} f_t(x)}{t} \le 2c_1 r^{-\alpha} \max(r^{-\alpha}, r^\beta t^{-1} f_t(x)). \end{aligned}$$

If the first term in the maximum is larger, then $f_t(x)^2 \le 2c_1 r^{-2\alpha}$, while if the second is larger then $f_t(x) \le t r^{-\alpha-\beta} = r^{-\alpha} = t^{-\alpha/\beta}$, so in either case the bound (6.5) holds. □

Lemma 6.7 *Let $\beta > \alpha$ and suppose that (Γ, μ) satisfies (V_α) and (R_β). Then (E_β) and $UHK(\alpha, \beta)$ hold.*

Proof Let $B = B(x, r)$. We have

$$\mathbb{E}^y \tau(x, r) = \sum_{z \in B} g_B(y, z) \mu_z \le V(x, r) g_B(y, y) = V(x, r) R_{\text{eff}}(y, B^c).$$

As $B \subset B(y, 2r)$ we have $R_{\text{eff}}(y, B^c) \le c(2r)^{\beta-\alpha}$, so, combining this with the upper bound on $V(x, r)$, we obtain the upper bound on $\mathbb{E}^y \tau$.

For the lower bound let $B' = B(x, r/2)$, and note that by (1.28) and Lemma 6.4 we have for $y, z \in B'$

$$g_B(y, z) = \mathbb{P}^y(T_z < \tau_B) g_B(z, z) \ge p_S g_B(z, z).$$

So,

$$\mathbb{E}^y \tau_B \ge \sum_{z \in B'} g_B(y, z) \mu_z \ge p_S g_B(z, z) V(x, r/2).$$

As $g_B(z, z) = R_{\text{eff}}(z, B^c) \ge R_{\text{eff}}(z, B(z, r/2)^c) \ge c r^\beta$ we obtain the lower bound in the condition (E_β). UHK(α, β) then follows from Lemma 6.6 and Theorem 5.23. □

In general it can be hard to obtain the near-diagonal lower bound, but in the strongly recurrent case it follows easily from (6.2).

Lemma 6.8 *Suppose $\beta > \alpha$ and (V_α) and (R_β) hold. Then there exist constants c_1, c_2 such that*

$$q_t(x, y) \geq c_1 t^{-\alpha/\beta}, \quad \text{if } d(x, y) \leq c_2 t^{1/\beta}. \tag{6.6}$$

Proof Let $f_t(y) = q_t(x, y)$ and $d(x, y) \leq \delta t^{1/\beta}$. Since UHK$(\alpha, \beta)$ holds we also have the on-diagonal lower bound $q_t(x, x) \geq c t^{-\alpha/\beta}$ by the same argument as in Lemma 4.35. Using Corollary 5.16 we have $\mathscr{E}(f_t, f_t) \leq c_2 t^{-1-\alpha/\beta}$. So by (6.2)

$$\begin{aligned} |q_t(x, y) - q_t(x, x)|^2 &\leq c d(x, y)^{\beta-\alpha} \mathscr{E}(f_t, f_t) \\ &\leq c_3 \delta^{\beta-\alpha} t^{-2\alpha/\beta} \leq c_4 \delta^{\beta-\alpha} q_t(x, x)^2. \end{aligned}$$

Choose δ small enough so that $c_4 \delta^{\beta-\alpha} \leq \frac{1}{4}$. Then $|1 - q_t(x, y)/q_t(x, x)| \leq \frac{1}{2}$, which gives (6.6). □

Theorem 6.9 *Let $\beta > \alpha$. The following are equivalent.*

(a) (H5), (V_α), and (R_β) hold
(b) HK(α, β) holds.

Proof (a) ⇒ (b) The upper bound was proved in Lemma 6.7, while the lower bound follows from Lemma 6.8 and Theorem 5.24.

(b) ⇒ (a) (H5) and (V_α) hold by Lemmas 4.17 and 4.22. To prove (R_β) let $x, y \in \mathbb{V}$, with $r = d(x, y) \geq 1$. By Theorem 2.8 and Theorem 4.26(b),

$$R_{\text{eff}}(x, B(x, r-1)^c) = g_{B(x,r-1)}(x, x) \asymp r^{\beta-\alpha}.$$

So $R_{\text{eff}}(x, y) \geq R_{\text{eff}}(x, B(x, r-1)^c) \geq c r^{\beta-\alpha}$.

Now let $B^* = B(x, 2r)$. By Theorem 4.26(b) again we have $g_{B^*}(x, y) \asymp r^{\beta-\alpha}$, and $g_{B^*}(y, y) = R_{\text{eff}}(y, (B^*)^c) \asymp r^{\beta-\alpha}$. So by Lemma 2.62

$$0 < c \leq \frac{g_{B^*}(x, y)}{g_{B^*}(x, x)} = \mathbb{P}^x(T_y < \tau_{B^*}) \leq \frac{R_{\text{eff}}(y, (B^*)^c)}{R_{\text{eff}}(y, x)} \leq \frac{c' r^{\beta-\alpha}}{R_{\text{eff}}(x, y)},$$

and rearranging we obtain the upper bound on $R_{\text{eff}}(x, y)$. □

Corollary 6.10 *Suppose that HK(α, β) holds. Then $\alpha \geq 1$ and $2 \leq \beta \leq 1 + \alpha$.*

Proof We have $\alpha \geq 1$ by Remark 4.15(2), and $\beta \geq 2$ by Lemmas 4.21 and 4.32. Also (H4) holds, so that $\mu_e \geq c > 0$ for all edges e. Thus $R_{\text{eff}}(x, y) \leq c'$ whenever $x \sim y$, and so $R_{\text{eff}}(x, B(x, r-1)^c) \leq cr$. If $\beta > \alpha$ we have from Theorem 6.9 that $R_{\text{eff}}(x, B(x, r-1)^c) \geq cr^{\beta-\alpha}$, and so $\beta - \alpha \leq 1$. □

Using the estimates (2.52) and (2.53) we have

Corollary 6.11 *The Sierpinski gasket graph Γ_{SG} satisfies HK(α, β) with $\alpha = \log 3/\log 2$ and $\beta = \log 5/\log 2$.*

Remarks 6.12 (1) The strongly recurrent graphs considered here also satisfy a Poincaré inequality, with the power R^β rather than R^2 – see Proposition 6.26 in Section 6.3.

(2) If (α, β) satisfy the conditions of Corollary 6.10 then there exists a graph Γ which satisfies HK(α, β) – see [Ba1].

(3) The techniques used in this chapter were introduced in [BCK], and handle the case $\alpha < \beta$, which is in some sense 'low dimensional'. As we have seen the proofs are fairly quick. In Section 6.2 we will see how quite different methods, based on Poincaré inequalities, enable us to prove HK$(\alpha, 2)$, that is, the case when $\beta = 2$. The final case is of 'anomalous diffusion' (i.e. $\beta > 2$) when $\alpha \geq \beta$; this turns out to be much harder than the cases already described, and will not be dealt with in this book. (For characterisations of HK(α, β) in this case, see [GT1, GT2, BB2].) However, the most interesting applications of anomalous diffusion are in critical models in statistical physics, such as the incipient infinite percolation cluster, and it turns out that for many models one has $\alpha < \beta$. (Because of random fluctuations the conditions (V_α), (R_β) only hold with high probability, but the methods described in this chapter can be modified to handle this.) For examples of random graphs of this kind see [BK, BJKS, BM, KN].

6.2 Gaussian Upper Bounds

We begin by giving (stable) conditions for Gaussian upper bounds for graphs satisfying (V_α) and (N_α). The main work is to prove the condition (E_2), and for this we will use the method of Nash [N] as improved by Bass [Bas]. The first step is an elementary but hard real variable lemma.

Lemma 6.13 (See [N], p. 938) *Let $\alpha > 0$ and $m(t)$, $q(t)$ be real functions defined on $[1, \infty)$ with $m(1) \leq 1$, and suppose that for $t \geq 1$*

$$q(t) \geq c_1 + \tfrac{1}{2}\alpha \log t, \tag{6.7}$$

$$1 + m(t) \geq c_2 e^{q(t)/\alpha}, \tag{6.8}$$

$$m'(t)^2 \leq c_3^2 q'(t). \tag{6.9}$$

Then

$$c_4 t^{1/2} \leq 1 + m(t) \leq c_5 t^{1/2} \quad \textit{for } t \geq 1.$$

Proof We first note that if $g(t) \geq 0$ then using the inequality $(u+v)^{1/2} \leq u/2v^{1/2} + v^{1/2}$, which holds for real positive u, v, we have

$$\begin{aligned}
\int_1^t (g'(s) + b/s)^{1/2} ds &\leq \int_1^t \Big(g'(s)/2(b/s)^{1/2} + (b/s)^{1/2}\Big) ds \\
&= \frac{1}{2b^{1/2}} \int_1^t g'(s) s^{1/2} ds + 2b^{1/2}(t^{1/2} - 1) \\
&= \frac{g(t)t^{1/2} - g(1)}{2b^{1/2}} - \frac{1}{2b^{1/2}} \int_1^t g(s) \tfrac{1}{2} s^{-1/2} ds + 2b^{1/2}(t^{1/2} - 1) \\
&\leq t^{1/2}\big(2b^{1/2} + g(t)/2b^{1/2}\big).
\end{aligned}$$

We used integration by parts to obtain the third line.

Now set $r(t) = \alpha^{-1}(q(t) - c_1 - \frac{1}{2}\alpha \log t)$, so that $r(t) \geq 0$ for $t \geq 1$. Then

$$\begin{aligned}
c_2 e^{c_1/\alpha} t^{1/2} e^{r(t)} &= c_2 e^{q(t)/\alpha} \leq 1 + m(t) \\
&\leq 1 + m(1) + c_3 \int_1^t q'(s)^{1/2} ds \\
&= 1 + m(1) + c_3 \alpha^{1/2} \int_1^t \left(r'(s) + \frac{1}{2s}\right)^{1/2} ds \\
&\leq 2 + c t^{1/2}(1 + r(t)).
\end{aligned}$$

Thus $r(t)$ is bounded, and the bounds on m follow. □

Fix $x_0 \in \mathbb{V}$. Following [N] define for $t \geq 1$

$$M(t) = \mathbb{E}^{x_0} d(x_0, Y_t) = \sum_y d(x_0, y) q_t(x_0, y) \mu_y,$$

$$Q(t) = -\sum_y q_t(x_0, y) \log q_t(x_0, y) \mu_y.$$

We now verify that Q and M satisfy (6.7)–(6.9).

Lemma 6.14 *Let (Γ, μ) satisfy (V_α) and (N_α). Then $M(t) \le t$ for all t and, in particular, $M(1) \le 1$. Further, there exist constants c_i such that*

$$Q(t) \ge c_1 + \tfrac{1}{2} d \log t, \quad t \ge 1, \tag{6.10}$$

$$1 + M(t) \ge c_2 e^{Q(t)/d}, \quad t \ge 1. \tag{6.11}$$

Proof The representation (5.1) gives that $d(x_0, Y_t) \le N_t$, and therefore $M(t) \le \mathbb{E} N_t = t$. The bound (6.10) follows directly from the upper bound (5.18).

The proof of (6.11) is similar to that in [N, Bas]. Let $a > 0$, set $D_0 = B(x_0, 1)$, and $D_n = B(x_0, 2^n) - B(x_0, 2^{n-1})$ for $n \ge 1$. Then using (V_α) to bound $\mu(D_n)$ we have

$$\sum_{y \in \mathbb{V}} e^{-ad(x_0,y)} \mu_y \le c + \sum_{n=1}^{\infty} \sum_{y \in D_n} \exp(-a 2^{n-1}) \mu_y \le c + \sum_{n=1}^{\infty} C_U 2^{n\alpha} \exp(-a 2^{n-1}).$$

By Lemma A.36 this sum is bounded by $c a^{-\alpha}$ if $a \in (0, 1]$.

If $\lambda > 0$ then $u(\log u + \lambda) \ge -e^{-1-\lambda}$ for $u > 0$. So, setting $\lambda = ad(x_0, y) + b$, where $a \le 1$ and $b \ge 0$,

$$\begin{aligned} -Q(t) + aM(t) + b &= \sum_y q_t(x_0, y)\big(\log q_t(x_0, y) + ad(x_0, y) + b\big)\mu_y \\ &\ge -\sum_y e^{-1-ad(x_0,y)-b} \mu_y \\ &= -e^{-1-b} \sum_y e^{-ad(x_0,y)} \mu_y \ge -c_5 e^{-b} a^{-\alpha}. \end{aligned}$$

Setting $a = 1/(1 + M(t))$ and $e^b = (1 + M(t))^\alpha = a^{-\alpha}$ we obtain

$$-Q(t) + \frac{M(t)}{1 + M(t)} + \alpha \log(1 + M(t)) \ge -c_5.$$

Since $M/(1 + M) < 1$, rearranging gives (6.11). □

If $f(\lambda) = (1 + \lambda) \log \lambda - \lambda$ then f is increasing on $[1, \infty)$. So, writing $\lambda = u/v \ge 1$, we have $(1 + u/v) \log(u/v) - u/v \ge f(1) = -1$, and rearranging gives

$$\frac{(u - v)^2}{u + v} \le (u - v)(\log u - \log v), \quad \text{for } u, v > 0. \tag{6.12}$$

Lemma 6.15 *Let (Γ, μ) be a weighted graph, with subexponential volume growth: for each $\lambda > 0$, $x \in \mathbb{V}$,*

$$\lim_{R \to \infty} V(x, R) e^{-\lambda R} = 0. \tag{6.13}$$

Then

$$Q'(t) \geq M'(t)^2, \quad t > 0.$$

Proof Set $f_t(x) = q_t(x_0, x)$, and let $b_t(x, y) = f_t(x) + f_t(y)$. Neglecting for the moment the need to justify the interchange of limits, we have

$$M'(t) = \sum_y d(x_0, y) \frac{\partial f_t(y)}{\partial t} \mu_y = \sum_y d(x_0, y) \Delta f_t(y) \mu_y \tag{6.14}$$

$$= -\tfrac{1}{2} \sum_x \sum_y \mu_{xy} (d(x_0, y) - d(x_0, x))(f_t(y) - f_t(x)) \tag{6.15}$$

$$\leq \tfrac{1}{2} \sum_x \sum_y \mu_{xy} |f_t(y) - f_t(x)|$$

$$\leq \tfrac{1}{2} \Big(\sum_x \sum_y \mu_{xy} b_t(x, y) \Big)^{1/2} \Big(\sum_x \sum_y \mu_{xy} \frac{(f_t(y) - f_t(x))^2}{b_t(x, y)} \Big)^{1/2}$$

$$= 2^{-1/2} \Big(\sum_x \sum_y \mu_{xy} \frac{(f_t(y) - f_t(x))^2}{f_t(x) + f_t(y)} \Big)^{1/2}.$$

Using (6.12) we deduce

$$M'(t)^2 \leq \tfrac{1}{2} \sum_x \sum_y \mu_{xy} (f_t(y) - f_t(x))(\log f_t(y) - \log f_t(x)).$$

On the other hand (again using discrete Gauss–Green),

$$Q'(t) = -\sum_y (1 + \log f_t(y)) \Delta f_t(y) \mu_y \tag{6.16}$$

$$= \tfrac{1}{2} \sum_x \sum_y \mu_{xy} (\log f_t(y) - \log f_t(x))(f_t(y) - f_t(x))$$

$$\geq M'(t)^2. \tag{6.17}$$

It remains to justify the interchanges of derivatives and sums in calculating M' and Q', and the use of discrete Gauss–Green to obtain (6.15) and (6.17) from (6.14) and (6.16).

Let $Q_n(t)$, $M_n(t)$ be the approximations to Q and M obtained by summing only over $y \in B(x_0, n)$. By Theorem 5.17 if $R = d(x_0, y) > 3t$ then

$$c_1 \exp(-c_2 R(1 + \log(R/t)) \leq f_t(y) \leq c_3 \exp(-c_4 R).$$

So, using the subexponential volume growth, it follows that, for any $T \geq 1$, M_n' and Q_n' converge uniformly for $t \in [1, T]$ to the right sides of (6.14) and

(6.16), and M_n and Q_n converge uniformly to M and Q. We also have that the sums

$$\sum_x \sum_y d(x_0, x)|f_t(x) - f_t(y)|\mu_{xy}, \quad \sum_x \sum_y (1 + |\log f_t(y)|)|f_t(x) - f_t(y)|\mu_{xy}$$

converge absolutely, so we can use discrete Gauss–Green to obtain (6.15) and (6.17). □

Thus Q and M satisfy the hypotheses of Lemma 6.13, and we obtain

Proposition 6.16 *Suppose that* (Γ, μ) *satisfies* (V_α) *and* (N_α)*. Let* $x_0 \in \mathbb{V}$*. Then*

$$M(t) \le c_1 t^{1/2}, \quad t \ge 1.$$

Corollary 6.17 *Let* (Γ, μ) *satisfy* (V_α) *and* (N_α)*. Then the condition* (E_2) *holds for* (Γ, μ)*.*

Proof Let $x \in \mathbb{V}$, $r \ge 1$ and write $\tau = \tau(x, r)$. The upper bound $\mathbb{E}^y \tau \le cr^2$ follows from the upper bound $q_t(x, y) \le ct^{-\alpha/2}$ and Lemma 4.20.

By the triangle inequality, $d(x, Y_{t\wedge\tau}) \le d(x, Y_{2t}) + d(Y_{t\wedge\tau}, Y_{2t})$. Now choose t so that $2^{3/2} c_1 t^{1/2} = \frac{1}{2} r$. Then

$$\begin{aligned} r\mathbb{P}^x(\tau \le t) &\le \mathbb{E}^x d(x, Y_{t\wedge\tau}) \\ &\le \mathbb{E}^x d(x, Y_{2t}) + \mathbb{E}^x d(Y_{t\wedge\tau}, Y_{2t}) \\ &\le c_1 (2t)^{1/2} + \mathbb{E}^x \left(\mathbb{E}^{Y_{t\wedge\tau}} d(Y_0, Y_{2t - t\wedge\tau})\right) \\ &\le c_1 (2t)^{1/2} + \sup_{z \in B(x,r), s \le t} \mathbb{E}^z d(z, Y_{2t-s}) \\ &\le 2c_1 (2t)^{1/2} = \tfrac{1}{2} r. \end{aligned}$$

Thus $\mathbb{E}^x \tau(x, r) \ge t\mathbb{P}^x(\tau \ge t) \ge \frac{1}{2} t = c' r^2$, which gives the lower bound. □

Theorem 6.18 *Let* (Γ, μ) *satisfy* (V_α)*. The following are equivalent.*

(a) (Γ, μ) *satisfies* (N_α)*.*
(b) UHK$(\alpha, 2)$ *holds.*
(c) UHKC$(\alpha, 2)$ *holds.*

Proof Assume that (N_α) holds. By Corollary 6.17 (Γ, μ) satisfies (E_2), and hence by Theorem 5.23 it satisfies UHK$(\alpha, 2)$ and UHKC$(\alpha, 2)$.

The implications (b) ⇒ (a) and (c) ⇒ (a) are immediate from Theorem 5.14. □

6.3 Poincaré Inequality and Gaussian Lower Bounds

In this section we will assume that (Γ, μ) satisfies the (weak) PI, and we will write λ_P for the constant by which we have to multiply the size of the balls. The main result of this section is that for graphs which satisfy (V_α) and (H5) the PI leads to the Gaussian bounds $\mathrm{HK}(\alpha, 2)$.

Theorem 6.19 *Let (Γ, μ) satisfy (H5). The following are equivalent.*

(a) (Γ, μ) satisfies (V_α) and the (weak) PI.
(b) The continuous time heat kernel $q_t(x, y)$ satisfies $HKC(\alpha, 2)$.
(c) The discrete time heat kernel $p_n(x, y)$ satisfies $HK(\alpha, 2)$.

In consequence, the property $HK(\alpha, 2)$ is stable under rough isometries.

We begin by proving that the Poincaré inequality can be used to obtain on-diagonal upper bound.

Proposition 6.20 *Let (Γ, μ) satisfy (V_α) and the PI. Then*

$$q_t(x, y) \le c_2 t^{-\alpha/2}, \quad x, y \in \mathbb{V}, \, t \ge 0.$$

Proof As in Theorem 5.14 it is enough to prove this for $x = y$. Fix $x_0 \in \mathbb{V}$, let $f_t(y) = q_t(x_0, y)$, and let $\psi(t) = \langle f_t, f_t \rangle = q_{2t}(x_0, x_0)$. Choose $r \ge 2$, and let $B_i = B(x_i, r)$ be a covering of $\mathbb{V}$ given by Lemma 6.2. Set $M = (2\lambda_P + 1)^\alpha C_U / C_L$; we have that any $x \in \mathbb{V}$ is in at most M of the balls $B_i^* = B(x_i, \lambda_P r)$. It follows that any edge $\{x, y\}$ is contained in at most M of these balls. By (5.13)

$$-\tfrac{1}{2}\psi'(t) = \mathscr{E}(f_t, f_t) \ge M^{-1} \sum_{i=1}^{\infty} \Big(\tfrac{1}{2} \sum_{x \in B_i^*} \sum_{y \in B_i^*} (f_t(x) - f_y(y))^2 \mu_{xy} \Big).$$

Set

$$\overline{f}_{t,i} = \mu(B_i^*)^{-1} \sum_{x \in B_i^*} f_t(x) \mu_x.$$

Then using the PI in the balls $B_i \subset B_i^*$ one obtains

$$\tfrac{1}{2} \sum_{x \in B_i^*} \sum_{y \in B_i^*} (f_t(x) - f_t(y))^2 \mu_{xy} \ge (C_P r^2)^{-1} \sum_{x \in B_i} (f_t(x) - \overline{f}_{t,i})^2 \mu_x$$

$$= (C_P r^2)^{-1}\Big(\sum_{x\in B_i} f_t(x)^2\mu_x - \mu(B_i)^{-1}\Big(\sum_{x\in B_i} f_t(x)\mu_x\Big)^2\Big)$$
$$\geq (C_P r^2)^{-1}\Big(\sum_{x\in B_i} f_t(x)^2\mu_x - C_L^{-1}r^{-\alpha}\Big(\sum_{x\in B_i} f_t(x)\mu_x\Big)^2\Big).$$

Now

$$\sum_{i=1}^{\infty}\Big(\sum_{x\in B_i} f_t(x)\mu_x\Big)^2 \leq \Big(\sum_{i=1}^{\infty}\Big(\sum_{x\in B_i} f_t(x)\mu_x\Big)\Big)^2 \leq M^2,$$

and thus

$$-C_P M r^2 \tfrac{1}{2}\psi'(t) \geq \Big(\sum_{i=1}^{\infty}\sum_{x\in B_i} f_t(x)^2\mu_x\Big) - C_L^{-1}r^{-\alpha}M^2$$
$$\geq \psi(t) - C_L^{-1}r^{-\alpha}M^2.$$

So,

$$-\tfrac{1}{2}\psi'(t) \geq (MC_P r^2)^{-1}(\psi(t) - MC_L^{-1}r^{-\alpha}).$$

Choosing $r = r(t)$ such that $\psi(t) = 2M^2C_L^{-1}r^{-\alpha}$ gives $\psi'(t) \leq -c\psi(t)^{1+2/\alpha}$, and the rest of the argument is as in Remark 5.15(1). □

Combining Proposition 6.20 with Theorem 5.14 we deduce:

Corollary 6.21 *(V_α) and the PI imply (N_α).*

It is also possible to give a direct proof – see [SC1].

We now turn to lower bounds, and will use a weighted Poincaré inequality and the technique of [FS] to obtain the near-diagonal lower bound (5.26). We begin by stating the weighted Poincaré inequality. Let $x_0 \in \mathbb{V}$, $R \geq 1$, $B = B(x_0, R)$, and define

$$\varphi(y) = \varphi_B(y) = \begin{cases} 1 \wedge \frac{(R+1-d(x_0,y))^2}{R^2} & \text{if } y \in B, \\ 0 & \text{if } y \in \mathbb{V} - B. \end{cases} \tag{6.18}$$

Note that φ has support B, that $R^{-2} \leq \varphi(y) \leq 1$ on B, that $\varphi(y) = R^{-2}$ if $y \in \partial_i B$, and that if $x \sim y$ with $x, y \in B$ then $\varphi(y) \leq 4\varphi(x)$ and $|\varphi(x) - \varphi(y)| \leq R^{-2}$. The following weighted Poincaré inequality is proved in Appendix A.9.

Proposition 6.22 *Let (Γ, μ) satisfy (H5), (V_α), and the PI. Then there exists a constant C_W such that for each ball $B = B(x_0, R)$ and $f : B \to \mathbb{R}$*

$$\inf_a \sum_{x \in B} (f(x) - a)^2 \varphi_B(x) \mu_x$$
$$\leq \tfrac{1}{2} C_W R^2 \sum_{x,y \in B} (f(x) - f(y))^2 (\varphi_B(x) \wedge \varphi_B(y)) \mu_{xy}. \tag{6.19}$$

We will need the following elementary inequality.

Lemma 6.23 (See [SZ]) *Let a, b, c, d be positive reals. Then*

$$-\Big(\frac{d}{b} - \frac{c}{a}\Big)(b - a) \geq \tfrac{1}{2}(c \wedge d)(\log b - \log a)^2 - \frac{(d-c)^2}{2(c \wedge d)}.$$

Proof First note that the inequality is preserved by the transformations

$$(a, b, c, d) \to (b, a, d, c) \text{ and } (a, b, c, d) \to (sa, sb, tc, td),$$

where $s, t > 0$. Using the first, we can assume that $d \geq c$, and by the second we can take $a = c = 1$. So, writing $x = b$, the inequality reduces to proving

$$f(x) = \tfrac{1}{2}(\log x)^2 + \left(\frac{d}{x} - 1\right)(x - 1) \leq \tfrac{1}{2}(d-1)^2, \tag{6.20}$$

for $x > 0$, $d \geq 1$. Since $f(x) - f(x^{-1}) = (d-1)(x - x^{-1})$, it is enough to prove (6.20) for $x \geq 1$. Now, by Cauchy–Schwarz,

$$(\log x)^2 = \Big(\int_1^x 1.t^{-1} dt\Big)^2 \leq \Big(\int_1^x dt\Big)\Big(\int_1^x t^{-2} dt\Big)$$
$$= (x-1)(1 - x^{-1}) = \frac{(x-1)^2}{x}.$$

Hence it is enough to prove that

$$\frac{(x-1)^2}{x} \leq (d-1)^2 + 2(x-1)\left(1 - \frac{d}{x}\right).$$

Finally

$$2(x-1)\left(1 - \frac{d}{x}\right) = 2\frac{(x-1)^2}{x} - 2\frac{(x-1)}{x}(d-1) \geq \frac{(x-1)^2}{x} - (d-1)^2,$$

where we used $-2AB \geq -A^2 - B^2$ in the final line. □

We now prove the near-diagonal lower bound for Y, using an argument of Fabes and Stroock [FS]. Recall from Chapter 5 the notation $q_t^B(x, y)$ for the transition density of Y killed on exiting B.

Theorem 6.24 *Suppose Γ satisfies (H5), (V_α), and the weak PI. Then for any $R \geq 1$, and $x_0 \in \mathbb{V}$,*

$$q_t^{B(x_0,R)}(x_1, x_2) \geq c_1 t^{-\alpha/2} \text{ if } x_1, x_2 \in B(x_0, \tfrac{1}{2}R) \text{ and } \tfrac{1}{4}R^2 \leq t \leq R^2. \tag{6.21}$$

In consequence, Γ satisfies HK$(\alpha, 2)$.

Proof The bounds UHK$(\alpha, 2)$ hold by Theorem 6.18 and Corollary 6.21, and Lemma 4.21 then gives that (E_2) holds. Write $T = R^2$ and $B = B(x_0, R)$. Let $x_1 \in B(x_0, R/2)$. By Lemma 5.21

$$\begin{aligned}\sum_{x \in B(x_0, 2R/3)} q_t^B(x_1, x)\mu_x &= \mathbb{P}^{x_1}(Y_t \in B(x_0, 2R/3), \tau_B > t) \\ &\geq \mathbb{P}^{x_1}(\tau(x_1, R/6) > t) \geq 1 - ce^{c' R^2/t} \geq \tfrac{1}{2},\end{aligned} \tag{6.22}$$

provided $t \leq c_2 T$, where c_2 is chosen small enough. We take $c_2 \leq 1/8$.

Let φ be defined by (6.18), and let

$$V_0 = \sum_{x \in B} \varphi(x)\mu_x.$$

Then, as $\varphi \geq \frac{1}{4}$ on $B(x_0, R/2)$,

$$c_3 R^\alpha \leq V_0 \leq \mu(B) \leq c_4 R^\alpha. \tag{6.23}$$

Write $u_t(x) = q_t^B(x_1, x)$, and let

$$w(t, y) = w_t(y) = \log V_0 u(t, y), \quad H(t) = H(x_1, t) = V_0^{-1} \sum_{y \in B} \varphi(y) w_t(y) \mu_y.$$

Note that $w_t(y)$ is well defined for $y \in B$ and $t > 0$. It is easy to see by Jensen's inequality that $H(t) < 0$, but we will not need to use this fact.

Set

$$f_t(x) = \begin{cases} \frac{\varphi(x)}{u_t(x)} & \text{if } x \in B, \\ 0 & \text{if } x \notin B. \end{cases}$$

Then

$$\begin{aligned} V_0 H'(t) &= \sum_{y \in B} \varphi(y) \frac{\partial w_t(y)}{\partial t} \mu_y = \sum_{y \in B} \frac{\varphi(y)}{u_t(y)} \frac{\partial u_t(y)}{\partial t} \mu_y = \langle f, \Delta u_t \rangle = -\mathscr{E}(f, u_t) \\ &= -\tfrac{1}{2} \sum_{x \in \mathbb{V}} \sum_{y \in \mathbb{V}} (f(x) - f(y))(u_t(x) - u_t(u))\mu_{xy}. \end{aligned}$$

Now we use Lemma 6.23. If $x, y \in B$ then $f(x) > 0$ and $f(y) > 0$, and

$$-(f(x) - f(y))(u_t(x) - u_t(y)) = -\Big(\frac{\varphi(x)}{u_t(x)} - \frac{\varphi(y)}{u_t(y)}\Big)(u_t(x) - u_t(u))$$
$$\geq \tfrac{1}{2}(\varphi(x) \wedge \varphi(y))(\log u_t(x) - \log u_t(y))^2 - \frac{(\varphi(x) - \varphi(y))^2}{2(\varphi(x) \wedge \varphi(y))}.$$

If $x, y \in B^c$ then $f(x) = f(y) = 0$, while if $x \in B$ and $y \in B^c$ then $f(y) = u_t(y) = 0$ and

$$-(f(x) - f(y))(u_t(x) - u_t(y)) = -\varphi(x).$$

We therefore have

$$V_0 H'(t) \geq \tfrac{1}{4} \sum_{x\in B}\sum_{y\in B} (\varphi(x) \wedge \varphi(y))(w_t(x) - w_t(y))^2 \mu_{xy} \tag{6.24}$$
$$- \tfrac{1}{4} \sum_{x\in B}\sum_{y\in B} \mu_{xy} \frac{(\varphi(x) - \varphi(y))^2}{\varphi(x) \wedge \varphi(y)} \tag{6.25}$$
$$- \tfrac{1}{2} \sum_{x\in B}\sum_{y\in B^c} \mu_{xy}\varphi(x). \tag{6.26}$$

Call the sums in (6.24)–(6.26) S_1, S_2, and S_3, respectively. To bound S_2 let $x \sim y$ with $x, y \in B$, and assume that $d(y, x_0) \leq d(x, x_0)$. Writing $k = R + 1 - d(x, x_0)$, we have $R^2\varphi(x) = k^2$, and $R^2\varphi(y)$ is either k^2 or $(k+1)^2$. Hence

$$\frac{(\varphi(x) - \varphi(y))^2}{\varphi(x) \wedge \varphi(y)} \leq \frac{((k+1)^2 - k^2)^2}{k^2 R^2} \leq \frac{9}{R^2}.$$

Thus

$$S_2 = -\tfrac{1}{4} \sum_{x\in B}\sum_{y\in B} \mu_{xy} \frac{(\varphi(x) - \varphi(y))^2}{\varphi(x) \wedge \varphi(y)}$$
$$\geq -\tfrac{9}{4} R^{-2} \sum_{x\in B}\sum_{y\in B} \mu_{xy} \geq -\tfrac{9}{4} R^{-2}\mu(B) \geq cR^{-2}V_0.$$

Also, if $x \in \partial_i B$ then $\varphi(x) = R^{-2}$, so that

$$S_3 = -\tfrac{1}{2} \sum_{x\in\partial_i B}\sum_{y\in B^c} \mu_{xy} R^{-2} \geq -R^{-2} \sum_{x\in B} \mu_x \geq -cR^{-2}V_0.$$

Thus the terms S_2 and S_3 are controlled by bounds of the same size, and so, using the weighted Poincaré inequality Proposition 6.22 to bound S_1, and recalling that $T = R^2$, we obtain

$$V_0 H'(t) \geq -c_5 T^{-1} V_0 + c_5' T^{-1} \sum_{x\in B} (w_t(x) - H(t))^2 \varphi(x)\mu_x. \tag{6.27}$$

It follows immediately from (6.27) that $t \to H(t) + c_5 t/T$ is increasing.

We now prove that if $H(t)$ is small enough then the final term in (6.27) is greater than $cH(t)^2$. Let $I = [\frac{1}{2}c_2T, c_2T]$. By Theorem 6.18 we have $V_0 u_t(x) \le c_6$ for $t \in I$, $x \in B$. Write $B' = B(x_0, 2R/3)$, choose $c_7 < 1$ so that $c_6 c_7 \mu(B') \le \frac{1}{4}V_0$, and let $A = \{y \in B' : V_0 u_t(y) \ge c_6 c_7\}$. Then by (6.22)

$$\begin{aligned} \tfrac{1}{2} &\le \sum_{y \in B' \cap A} u_t(y)\mu_y + \sum_{y \in B' - A} u_t(y)\mu_y \\ &\le c_6 V_0^{-1}\mu(A) + c_6 c_7 V_0^{-1}(\mu(B') - \mu(A)) \\ &\le c_6 V_0^{-1}\mu(A) + c_6 c_7 V_0^{-1}\mu(B') \le c_6 V_0^{-1}\mu(A) + \tfrac{1}{4}, \end{aligned}$$

so that $V_0\mu(A) \ge c_6/4$. For $x \in A$ and $t \in I$ we have $c_6 c_7 \le V_0 \mu_t(x) \le c_6$, so there exists c_8 such that $|w_t(x)| \le c_8$ for $x \in A, t \in I$. Now let

$$t_1 = \inf\{t : H(t) \ge -2c_8\}.$$

If $t_1 < c_2 T$ then, as $H(t) + c_5 t/T$ is increasing,

$$H(c_2T) = H(t_1) + \int_{t_1}^{c_2T} H'(s)ds \ge -2c_8 - c_5(c_2T - t_1)/T \ge -(2c_8 + c_5 c_2).$$

If $t_1 > c_2T$ then for $x \in A, t \in I$

$$w_t(x) - H(t) \ge -c_8 + c_8 + \tfrac{1}{2}|H(t)| \ge \tfrac{1}{2}|H(t)|.$$

So, for $t \in I$,

$$\begin{aligned} TH'(t) &\ge -c_5 + c_5' \sum_{x \in A} (w_t(x) - H(t))^2 \varphi(x)\mu_x \\ &\ge -c_5 + \tfrac{1}{4}c_5' \sum_{x \in A} (w_t(x) - H(t))^2 \mu_x \\ &\ge -c_5 + \tfrac{1}{16}c_5' \mu(A) V_0^{-1} H(t)^2 \\ &\ge -c_5 + \tfrac{1}{128}c_5' c_6 c_8^2 + \tfrac{1}{128}c_5' c_6 H(t)^2. \end{aligned} \tag{6.28}$$

We can assume that c_8 is large enough so that the second term in (6.28) is greater than c_5, so that there exists c_9 such that

$$TH'(t) \ge c_9 H(t)^2 \qquad \text{for } t \in I. \tag{6.29}$$

So since $H(\frac{1}{2}c_2T) < -2c_8 < 0$,

$$\frac{1}{H(c_2T)} < \frac{1}{H(c_2T)} - \frac{1}{H(\frac{1}{2}c_2T)} = -\int_{\frac{1}{2}c_2T}^{c_2T} H'(s)H(s)^{-2}ds \le -\tfrac{1}{2}c_2c_9.$$

Thus in both cases we deduce that

$$H(c_2T) \geq -c_{10}.$$

Using again the fact that $H(t) + c_5T^{-1}t$ is increasing, it follows that for $x_1 \in B(x_0, R/2)$

$$H(x_1, t) \geq -c_{11}, \quad \tfrac{1}{8}T \leq t \leq \tfrac{1}{2}T. \tag{6.30}$$

To turn this into a pointwise bound, note first that

$$V_0 q_t^B(x_1, x_2) = V_0^{-1} \sum_{y \in B} V_0 q_{t/2}^B(x_1, y) V_0 q_{t/2}^B(x_2, y) \varphi(y) \mu_y.$$

So, using Jensen's inequality and (6.30), for $x_1, x_2 \in B(x_0, R/2)$, $t \in [\tfrac{1}{4}T, T]$,

$$\begin{aligned}\log(V_0 q_t^B(x_1, x_2)) &\geq V_0^{-1} \sum_{y \in B} \Big(\log(V_0 q_{t/2}^B(x_1, y)) + \log(V_0 q_{t/2}^B(x_2, y)) \Big) \varphi(y) \mu_y \\ &= H(x_1, t/2) + H(x_2, t/2) \geq -2c_{11}.\end{aligned}$$

By (6.23) $V_0 \asymp R^\alpha$, and we obtain the near-diagonal lower bound for q_t^B. The Gaussian bound HK(α, 2) then follows immediately by Theorem 5.25. □

Remarks 6.25 (1) It is not clear how to run this argument in discrete time. However, if Γ satisfies the hypotheses of Theorem 6.24 then the discrete time NDLB follows from Theorem 5.24.

(2) In Theorem 6.24 the hypothesis (H5) is not used in the proof of the continuous time near-diagonal lower bound (6.21) from the weighted PI. However, it is used to obtain HK(α, 2) from (6.21), and it is also used in the proof of the weighted PI from the weak PI. With some more work one can use the PI to remove the need for (H5) in the continuous time case – see [BC].

We now see how these Gaussian bounds, or more generally HK(α, β), lead to a Poincaré inequality.

Proposition 6.26 *Suppose that HK(α, β) holds. Let $B = B(x_0, R)$. Then if $f : B \to \mathbb{R}$, writing $B' = B(x_0, R/2)$,*

$$\sum_{x \in B'} (f(x) - \overline{f}_{B'})^2 \mu_x \leq c_1 R^\beta \mathscr{E}_B(f, f).$$

Proof Write $\Gamma_B = (B, E_B, \mu^B)$ for the subgraph induced by B with the weights μ_{xy} restricted to B. Let $\tilde{Y}$ be the continuous time random walk on

Γ_B, and write $\tilde{q}_t(x, y)$ for the (continuous time) heat kernel associated with $\tilde{Y}$. Then, using Proposition 5.26, if $T = R^\beta$,

$$\tilde{q}_T(x, y) \geq q_T^B(x, y) \geq c_2 T^{-\alpha/\beta}, \qquad \text{for } x, y \in B'. \tag{6.31}$$

Write $\tilde{Q}_t$ for the semigroup of $\tilde{Y}$, given by

$$\tilde{Q}_t f(x) = \sum_{y \in B} \tilde{q}_t(x, y) f(y) \mu_y^B.$$

Now let $f : B \to \mathbb{R}$. For $x \in B'$, set $a_x = \tilde{Q}_T f(x)$, and let $h^{(x)} = \tilde{Q}_T (f - a_x)^2$. Then $h^{(x)}(y) = (\tilde{Q}_T f^2)(y) - 2a_x \tilde{Q}_T f(y) + a_x^2$, and therefore

$$h^{(x)}(x) = (\tilde{Q}_T f^2)(x) - a_x^2 = (\tilde{Q}_T f^2)(x) - (\tilde{Q}_T f(x))^2.$$

So, writing $\langle \cdot, \cdot \rangle_B$ for the inner product in $L^2(\mu^B)$,

$$\begin{aligned}
\sum_x h^{(x)}(x) \mu_x^B &= \sum_x (\tilde{Q}_T f^2)(x) \mu_x^B - \sum_x (\tilde{Q}_T f(x))^2 \mu_x^B \\
&= \langle \tilde{Q}_T f^2, 1 \rangle_B - \langle \tilde{Q}_T f, \tilde{Q}_T f \rangle_B \\
&= \langle f, f \rangle_B - \langle \tilde{Q}_T f, \tilde{Q}_T f \rangle_B \leq 2T \mathscr{E}_B(f, f).
\end{aligned}$$

Here we used Corollary 5.6 in the final line.

On the other hand, using the lower bound (6.31), for $x \in B'$,

$$\begin{aligned}
h^{(x)}(x) &= \sum_{y \in B} (f(y) - a_x)^2 \tilde{q}_T(x, y) \mu_y^B \\
&\geq c T^{-\alpha/\beta} \sum_{y \in B'} (f(y) - a_x)^2 \mu_y^B \geq c R^{-\alpha} \sum_{y \in B'} (f(y) - \overline{f}_{B'})^2 \mu_y^B.
\end{aligned}$$

Combining the bounds above,

$$\sum_{y \in B'} (f(y) - \overline{f}_{B'})^2 \mu_y^B \leq c R^\alpha \min_{x \in B} h^{(x)}(x) \leq c' \sum_{x \in B'} h^{(x)}(x) \mu_x \leq c R^\beta \mathscr{E}_B(f, f),$$

which proves the weak Poincaré inequality with $\lambda_P = 2$. □

Proof of Theorem 6.19 The equivalence between (b) and (c) is proved in Theorem 5.24. The implication (a) ⇒ (b) was proved in Theorem 6.24, and (b) ⇒ (a) follows by Lemma 4.22 and Proposition 6.26.

By Proposition 3.33 the weak Poincaré inequality is stable under rough isometries. Since (V_α) is also stable under rough isometries, the equivalence of (a) and (b) imply that HK$(\alpha, 2)$ is stable under rough isometries. □

Remark 6.27 Note that all that was needed to prove the Poincaré inequality in Proposition 6.26 was the near-diagonal lower bound (6.31) for the killed heat kernel, together with (V_α). Suppose that (Γ, μ) satisfies (H5) and (V_α). If (6.31) holds (with $\beta = 2$), then the PI holds, and so, by Theorem 6.19, HK$(\alpha, 2)$ holds.

For convenience we summarise the heat kernel bounds which one has for $\mathbb{Z}^d$ and for graphs which are roughly isometric to $\mathbb{Z}^d$.

Theorem 6.28 *Let (Γ, μ) be connected, locally finite, satisfy the condition (H5) of controlled weights:*

$$\frac{\mu_{xy}}{\mu_x} \geq \frac{1}{C_2} \quad \text{whenever } x \sim y, \tag{6.32}$$

and be roughly isometric to $\mathbb{Z}^d$. Let $p_n(x, y)$ and $q_t(x, y)$ be the discrete and continuous time heat kernels on Γ. There exist constants $c_i = c_i(d)$ such that the following estimates hold.

(a) For $x, y \in \mathbb{V}$, $n \geq d(x, y) \vee 1$,

$$p_n(x, y) \leq c_1 n^{-d/2} \exp(-c_2 d(x, y)^2/n),$$
$$p_n(x, y) + p_{n+1}(x, y) \geq c_3 n^{-d/2} \exp(-c_4 d(x, y)^2/n).$$

(b) If $t \geq 1 \vee d(x, y)$ then

$$c_3 t^{-d/2} \exp(-c_4 d(x, y)^2/t) \leq q_t(x, y) \leq c_1 t^{-d/2} \exp(-c_2 d(x, y)^2/t).$$

(c) If $d(x, y) \geq 1$ and $t \leq d(x, y)$ then, writing $R = d(x, y)$,

$$c_3 \exp(-c_4 R(1 + \log(R/t))) \leq q_t(x, y) \leq c_1 \exp(-c_2 R(1 + \log(R/t))).$$

(d) $\mu_x^{-1}(1 - e^{-t}) \leq q_t(x, x) \leq \mu_x^{-1}$ for $t \in (0, 1)$.

Proof It is clear that Γ satisfies (V_d) and (H5). $\mathbb{Z}^d$ satisfies the (weak) PI by Corollary 3.30, so by the stability of the PI (Proposition 3.33), Γ satisfies the PI. Theorem 6.19 then gives that HK$(d, 2)$ holds in both discrete and continuous time, proving (a) and (b). The bound (d) is elementary, and (c) follows from Theorem 5.17 and Remark 5.18(2). □

6.4 Remarks on Gaussian Bounds

We have seen that, starting with an isoperimetric inequality, it is fairly straightforward to obtain on-diagonal upper bounds, but that we had to work much harder in order to obtain full Gaussian upper bounds. There are at least seven methods in the literature for obtaining Gaussian upper bounds:

(1) The approach given in Section 6.3, using differential inequalities, originally due to Nash and improved by Bass [Bas].
(2) An analytic interpolation argument due to Hebisch and Saloff-Coste [HSC].
(3) Davies' method, using a perturbation of the semigroup Q_t – see [CKS].
(4) A 'two-point' method due to Grigor'yan – see [CGZ, Fo1, Ch] for the graph case. (This approach works well for random graphs, where global information may not be available.)
(5) A proof from a parabolic Harnack inequality, as was done by Aronsen in [Ar] for divergence form diffusions. See [D1] for such arguments in the graph context. (The parabolic Harnack inequality can be proved by Moser's iteration argument [Mo1, Mo2, Mo3].)
(6) Combining the Carne–Varopoulos bound and its extensions with a mean value inequality, as in [CG].
(7) For manifolds, there is a complex interpolation method due to Coulhon and Sikora [CS]. However, the method relies on having a full Gaussian tail and does not seem to work in the graph context.

We now state without proof two additional theorems on Gaussian bounds. The first, due to Hebisch and Saloff-Coste, gives that on-diagonal upper bounds (or equivalently a Nash inequality) imply Gaussian upper bounds, with no additional geometric hypotheses.

Theorem 6.29 (See [HSC]) *Let (Γ, μ) be a weighted graph and $\alpha \geq 1$. The following are equivalent.*

(a)

$$p_n(x, x) \leq c n^{-\alpha/2}, \quad n \geq 1.$$

(b) The Gaussian upper bounds $UHK(\alpha, 2)$ hold.

Theorem 6.31 gives a characterisation of Gaussian upper bounds and the parabolic Harnack inequality.

Definition 6.30 We say (Γ, μ) satisfies the *volume doubling* (VD) *condition* if there exists a constant C such that

$$V(x, 2r) \leq C V(x, r) \quad \text{for all } x \in \mathbb{V},\ r \geq 1.$$

Remarks (1) If (V_α) holds with constants C_L and C_U then, if $r \geq 1$,

$$V(x, 2r) \leq C_U 2^\alpha r^\alpha = \left(C_U 2^\alpha / C_L\right) C_L r^\alpha = \left(C_U 2^\alpha / C_L\right) V(x, r),$$

and thus VD holds.

(2) $\mathbb{Z}^d$ satisfies (V_d) and hence volume doubling.

(3) If (Γ, μ) satisfies VD then, iterating, one deduces that $V(x, 2^k) \leq C^k V(x, 1)$, from which it follows that $V(x, r) \leq c r^\alpha V(x, 1)$, that is, that (Γ, μ) has polynomial volume growth.

(4) The binary tree, which has exponential volume growth, therefore does not satisfy VD.

(5) Let Γ_i, $i = 1, 2$ be graphs satisfying (V_{α_i}), with $\alpha_1 \neq \alpha_2$. Then the join of Γ_1 and Γ_2 does not satisfy VD.

Theorem 6.31 *Let (Γ, μ) satisfy (H5).*

(a) VD and PI are weight stable.
(b) VD is stable under rough isometries.
(c) The (weak) PI is stable under rough isometries.
(d) VD and the (weak) PI imply the strong PI.

Proof (a) and (b) are easy, and (c) was proved in Proposition 3.33. For (d) see Appendix A.9 and Corollary A.51. □

Remark 6.32 On its own it is not clear that the strong PI is stable under rough isometries: a rough isometry φ need not map balls into balls. However, using Theorem 6.31(b), (c), and (d), one deduces that the combined condition '(VD) + (strong PI)' is stable under rough isometries.

Definition 6.33 We say (Γ, μ) satisfies the *parabolic Harnack inequality* (PHI(β)) if whenever $u(t, x) \geq 0$ is defined on the cylinder

$$\overline{Q} = [0, 4T] \times \overline{B(y, 2R)},$$

and satisfies

$$\frac{\partial}{\partial t} u(t, x) = \Delta u(t, x), \qquad (t, x) \in [0, 4T] \times B(y, 2R), \tag{6.33}$$

then, writing $T = R^\beta$, $Q_- = [T, 2T] \times B(y, R)$ and $Q_+ = [3T, 4T] \times B(y, R)$,

$$\sup_{(t,x) \in Q_-} u(t, x) \leq c_1 \inf_{(t,x) \in Q_+} u(t, x). \tag{6.34}$$

A function which satisfies (6.33) is called *caloric*. If $h(x)$ is harmonic then $u(t,x) = h(x)$ is caloric, so that PHI(β) immediately implies EHI. Somewhat surprisingly, the standard parabolic Harnack inequality (with $\beta = 2$) is easier to characterise than the EHI; this was done by Grigor'yan [Gg1] and Saloff-Coste [SC1] for manifolds. Here is the graph version.

Theorem 6.34 (See [D1]) *Let (Γ, μ) be a weighted graph satisfying (H5). The following are equivalent.*

(a) (Γ, μ) satisfies VD and PI.
(b) There exist constants C_1, C_2 such that for $t \ge d(x,y) \vee 1$,

$$\frac{C_1}{V(x,t^{1/2})} e^{-C_2 d(x,y)^2/t} \le q_t(x,y) \le \frac{C_2}{V(x,t^{1/2})} e^{-C_1 d(x,y)^2/t}. \quad (6.35)$$

(c) (Γ, μ) satisfies PHI(2).

7

Potential Theory and Harnack Inequalities

7.1 Introduction to Potential Theory

We will use Green's functions, so throughout this section we consider a domain $D \subset \mathbb{V}$, with the assumption that if (Γ, μ) is recurrent then $D \neq \mathbb{V}$.

Definition 7.1 Let $A \subset D \subset \mathbb{V}$. We define the hitting measure and escape measure (or capacitary measure) by

$$H_A(x, y) = \mathbb{P}^x(X_{\tau_A} = y; \tau_A < \infty),$$

$$\mathfrak{e}_{A,D}(x) = \begin{cases} \mathbb{P}^x(T_A^+ \geq \tau_D)\mu_x & \text{if } x \in A, \\ 0 & \text{otherwise.} \end{cases}$$

We write $\mathfrak{e}_A$ for $\mathfrak{e}_{A,\mathbb{V}}$. If ν is a finite measure on $\mathbb{V}$ we define

$$\nu G_D(y) = \sum_{x \in \mathbb{V}} \nu_x g_D(x, y). \tag{7.1}$$

Recall that we count the event $\{T_A \geq \tau_D\}$ as occurring if $T_A = \tau_D = \infty$.

In this discrete context, the distinction between measures and functions is not as pronounced as in general spaces. However, it is still helpful to distinguish between the two; note that there is no term involving μ_x in (7.1). We adopt the convention that G acts to the right of measures and to the left of functions. If $\nu_x = f(x)\mu_x$ then the symmetry of G_D gives that $\nu G_D = G_D f$.

Proposition 7.2 (Last exit decomposition) *Let $A \subset D$ with A finite. Then*

$$\mathbb{P}^x(T_A < \tau_D) = \sum_{y \in A} g_D(x, y)\mathbb{P}^y(T_A^+ \geq \tau_D)\mu_y = \mathfrak{e}_{A,D}G_D(x). \tag{7.2}$$

In particular, $\mathfrak{e}_{A,D}G_D = 1$ on A.

Proof Note that both sides of (7.2) are zero if $x \notin D$. Set $\sigma = \max\{n \geq 0 : X_n \in A, n < \tau_D\}$. As A is finite, $\mathbb{P}^y(\sigma < \infty) = 1$ for all y. Since $A \subset D$ we have $\{X_j \notin A, j = 1, \dots, \tau_D\} = \{T_A^+ > \tau_D\}$. So

$$\begin{aligned}
\mathbb{P}^x(T_A < \tau_D) &= \sum_{n=0}^{\infty} \sum_{y \in A} \mathbb{P}^x(\sigma = n, X_n = y, \tau_D > n) \\
&= \sum_{n=0}^{\infty} \sum_{y \in A} \mathbb{P}^x(X_n = y, \tau_D > n, X_j \notin A, j = n+1, \dots, \tau_D) \\
&= \sum_{n=0}^{\infty} \sum_{y \in A} \mathbb{P}^x(X_n = y, \tau_D > n) \mathbb{P}^y(X_j \notin A, j = 1, \dots, \tau_D) \\
&= \sum_{y \in A} g_D(x, y) \mu_y \mathbb{P}^y(T_A^+ \geq \tau_D).
\end{aligned}$$

□

Remark 7.3 This result can fail if $\mathbb{P}^\cdot(\sigma = \infty) > 0$. For example, consider $\mathbb{Z}$ with edge weights $\mu_{n,n+1} = \lambda^n$ for all n, where $\lambda > 1$. Let $D = \mathbb{N}$ and $A = \{k, k+1, \dots\}$, where $k \geq 2$. If $x \geq k$ then $\mathbb{P}^x(T_A < \tau_D) = 1$. If $y \geq k+1$ then $\mathbb{P}^y(T_A^+ \geq \tau_D) = 0$, so the left side of (7.2) is $g_D(x, k)\mathbb{P}^k(T_A^+ \geq \tau_D)\mu_k$. Since $g_D(x, k)/g_D(k, k) = \mathbb{P}^x(T_k < \infty) = \lambda^{k-x}$ for $x \geq k$, the left side of (7.2) is not constant, and converges to zero as $x \to \infty$, showing that (7.2) cannot hold.

Taking $D = \mathbb{V}$ we deduce:

Corollary 7.4 *Let (Γ, μ) be transient and A be finite. Then*

$$\mathbb{P}^x(T_A < \infty) = \mathfrak{e}_A G(x),$$

and $\mathfrak{e}_A G = 1$ on A.

Lemma 7.5 *Let $A \subset \mathbb{V}$ with $A \neq \mathbb{V}$. Then if $x \in A$, $y \in \partial A$*

$$H_A(x, y) = \sum_{z \in \partial_i A} g_A(x, z) p_1(z, y) \mu_z \mu_y.$$

Proof We have

$$\begin{aligned}
\mathbb{P}^x(X_{\tau_A} = y) &= \sum_{n=0}^{\infty} \sum_{z \in \partial_i A} \mathbb{P}^x(\tau_A = n+1, X_n = z, X_{n+1} = y) \\
&= \sum_{n=0}^{\infty} \sum_{z \in \partial_i A} \mathbb{P}^x(\tau_A > n, X_n = z, X_{n+1} = y)
\end{aligned}$$

$$= \sum_{n=0}^{\infty} \sum_{z \in \partial_i A} \mathbb{P}^x(\tau_A > n, X_n = z)\mathbb{P}^z(X_1 = y)$$
$$= \sum_{z \in \partial_i A} g_A(x, z)\mu_z p_1(z, y)\mu_y.$$

(This argument does not use the last exit decomposition.) □

Proposition 7.6 ('Principle of domination') *Let $A \subset D$ with A finite, and for $i = 1, 2$ let $\nu^{(i)}$ be measures with support A. Let $f_i = \nu^{(i)} G_D$, and suppose that $f_1 \geq f_2$ on A. Then $f_1 \geq f_2$ on $\mathbb{V}$, and $\nu^{(1)}(A) \geq \nu^{(2)}(A)$.*

Proof Let $D_n \uparrow\uparrow D$, with $A \subset D_1$, and write $f_{i,n} = \nu^{(i)} G_{D_n}$. Then by Theorem 1.34 $f_{i,n}$ converges to f_i pointwise. Let $\varepsilon > 0$. As A is finite there exists n such that $(1+\varepsilon) f_{1,n} \geq f_{2,n}$ on A. Since $f_{i,n}$ is harmonic in $D_n - A$ and is zero on ∂D_n, the maximum principle implies that $(1+\varepsilon) f_{1,n} \geq f_{2,n}$ on $\mathbb{V}$. As ε is arbitrary, it follows that $f_1 \geq f_2$.

By Proposition 7.2 we have $\mathfrak{e}_{A,D} G_D = 1$ on A. So for $i = 1, 2$

$$\nu^{(i)}(A) = \sum_{x \in A} \nu_x^{(i)} \sum_{y \in A} \mathfrak{e}_{A,D}(y) G_D(y, x)$$
$$= \sum_{y \in A} \mathfrak{e}_{A,D}(y)(\nu^{(i)} G_D(y)) = \sum_{x \in A} \mathfrak{e}_{A,D}(y) f_i(y).$$

Since $f_1 \geq f_2$ on A it follows that $\nu^{(1)}(A) \geq \nu^{(2)}(A)$. □

Definition 7.7 Let $A \subset D$, with A finite. Define the capacity of A by

$$\mathrm{Cap}_D(A) = \textstyle\sum_{x \in A} \mathfrak{e}_{A,D}(x) = \mathfrak{e}_{A,D}(A).$$

We write $\mathrm{Cap}(A)$ for $\mathrm{Cap}_{\mathbb{V}}(A)$.

Theorem 7.8 *Let $A \subset D$, with A finite, and set*

$$h = \mathfrak{e}_{A,D} G_D.$$

Then $h = 1$ on A, $0 < h \leq 1$ on $\mathbb{V}$, and h is harmonic on $D - A$. Further,

$$\mathrm{Cap}_D(A) = \sup\{\nu(A) : \mathrm{supp}(\nu) \subset A,\ \nu G_D \leq 1 \text{ on } D\} \tag{7.3}$$
$$= \inf\{\nu(A) : \mathrm{supp}(\nu) \subset A, \nu G_D \geq 1 \text{ on } A\}.$$

Proof The properties of h are immediate from Proposition 7.2.

To prove the final assertion let ν be a measure satisfying the conditions in (7.3).

Then $\nu G_D \le \mathfrak{e}_A G_D$ on A, so by Proposition 7.6 $\nu(A) \le \mathfrak{e}_{A,D}(A) = \mathrm{Cap}_D(A)$. Thus the supremum in (7.3) is attained by $\mathfrak{e}_A$. If ν' is a measure on A and $\nu' G_D \ge 1$ on A then $\nu' G_D \ge \mathfrak{e}_{A,D}$ on A, so by Proposition 7.6 $\nu'(A) \ge \mathfrak{e}_{A,D}(A)$. □

We can also connect capacity with the variational problems of Chapter 2.

Proposition 7.9 *(a) Suppose that Γ is transient. Let A be finite. Then $h = \mathfrak{e}_A G$ is the unique minimiser of* (VP1$'$) *with $B_1 = A$, and*

$$C_{\mathrm{eff}}(A, \infty) = \mathrm{Cap}(A).$$

(b) Suppose that D is finite, $A \subset D$. Then $h = \mathfrak{e}_{A,D} G_D$ is the unique minimiser for (VP1) *with $B_1 = A$, $B_0 = \mathbb{V} - D$, and*

$$C_{\mathrm{eff}}(A, \mathbb{V} - D) = \mathrm{Cap}_D(A).$$

Proof In case (a) we set $D = \mathbb{V}$. By Theorem 2.37 h is the unique minimiser for (VP1$'$), and $\mathscr{E}(h,h) = C_{\mathrm{eff}}(A, \infty)$. In case (b) since D is finite the minimiser is the unique solution to the Dirichlet problem $\Delta h = 0$ in $D - A$, $h = 1$ on A, and $h = 0$ on $\mathbb{V} - D$. Since h also satisfies these conditions, h is the minimiser, and $\mathscr{E}(h,h) = C_{\mathrm{eff}}(A, \mathbb{V} - D)$.

In both cases h has support D, so, writing $g_D^x = g_D(x, \cdot)$, by Theorem 1.34 we have $\mathscr{E}(g_D^x, h) = h(x) = 1$ for $x \in A$. Thus

$$\mathscr{E}(h,h) = \mathscr{E}\Big(\sum_{x \in A} \mathfrak{e}_{A,D}(x) g_D^x, h\Big) = \sum_{x \in A} \mathfrak{e}_{A,D}(x) \mathscr{E}(g_D^x, h) = \mathfrak{e}_{A,D}(A). \quad \square$$

We can now generalise the commute time identity Theorem 2.63 to sets.

Theorem 7.10 *Let Γ be finite and A_0, A_1 be disjoint subsets of $\mathbb{V}$. Then there exist probability measures π_i on A_i such that*

$$\mathbb{E}^{\pi_0} T_{A_1} + \mathbb{E}^{\pi_1} T_{A_2} = \mu(\mathbb{V}) R_{\mathrm{eff}}(A_0, A_1). \tag{7.4}$$

Further, $\pi_0 = R_{\mathrm{eff}}(A_0, A_1) \mathfrak{e}_{A_0, \mathbb{V} - A_1}$.

Proof Set $D = \mathbb{V} - A_1$ and let $h_0(x) = \mathbb{P}^x(T_{A_0} < T_{A_1})$. Then by Proposition 7.2, Theorem 7.8, and Proposition 7.9 we have $h_0 = \mathfrak{e}_{A_0,D} G_D$ and $\mathfrak{e}_{A_0,D}(A_0) = R_{\mathrm{eff}}(A_0, A_1)^{-1}$. Set $\pi_1 = \mathfrak{e}_{A_0,D} R_{\mathrm{eff}}(A_0, A_1)$. Then

$$\sum_{y \in \mathbb{V}} h_0(y) \mu_y = \sum_{y \in \mathbb{V}} \sum_{x \in A_0} \mathfrak{e}_{A_0,D}(x) g_D(x,y) \mu_y$$

$$= R_{\text{eff}}(A_0, A_1)^{-1} \sum_{x \in A_0} \pi_0(x) \sum_{y \in \mathbb{V}} g_D(x, y)\mu_y$$
$$= R_{\text{eff}}(A_0, A_1)^{-1} \sum_{x \in A_0} \pi_0(x) \mathbb{E}^x T_{A_1} = R_{\text{eff}}(A_0, A_1)^{-1} \mathbb{E}^{\pi_0} T_{A_1}.$$

Similarly, setting $h_1(x) = \mathbb{P}^x(T_{A_1} < T_{A_0})$ and $\pi_1 = R_{\text{eff}}(A_0, A_1)\mathfrak{e}_{A_1, \mathbb{V}-A_0}$, we have

$$\sum_{y \in \mathbb{V}} h_1(y)\mu_y = R_{\text{eff}}(A_0, A_1)^{-1} \mathbb{E}^{\pi_1} T_{A_0}.$$

Since $h_0 + h_1 = 1$, adding the two identities gives (7.4). □

Remark 7.11 Since hitting times and Green's functions are the same for both the discrete and continuous time random walks, this result applies to both X and Y. Now consider a more general continuous time random walk $Z = (Z_t, t \geq 0)$ with generator

$$Lf(x) = \theta_x^{-1} \sum_{y \in \mathbb{V}} \mu_{xy}(f(y) - f(x)).$$

The potential operator $G_D^{(Z)}$ and Green's functions $g_D^{(Z)}(x, y)$ for Z are defined by

$$G_D^{(Z)} f(x) = \mathbb{E}^x \int_0^{\tau_D(Z)} f(Z_s)ds = \sum_{y \in D} g_D^{(Z)}(x, y) f(y)\theta_y; \tag{7.5}$$

thus $g_D^{(Z)}(x, y)$ is the density of $G_D^{(Z)}$ with respect to the measure θ. Straightforward calculations give that $g_D^{(Z)}(x, y)$ satisfies the same equations as $g_D(x, y)$, and therefore we have $g_D^{(Z)} = g_D$; this is not so evident from (7.5).

Since Green's functions and capacitary measures are the same for Z and X, as above we obtain

$$\sum_{y \in \mathbb{V}} h_0(y)\theta_y = C_{\text{eff}}(A_0, A_1) \sum_{x \in A_0} \pi_0(x) \sum_{y \in \mathbb{V}} g_D(x, y)\theta_y.$$

Since $E^x T_{A_1}(Z) = \sum_{y \in \mathbb{V}} g_D(x, y)\theta_y$, this gives

$$\mathbb{E}^{\pi_0} T_{A_1}(Z) + \mathbb{E}^{\pi_1} T_{A_2}(Z) = \theta(\mathbb{V}) R_{\text{eff}}(A_0, A_1).$$

We now give a version of the Riesz representation theorem. Recall from Chapter 1 the definition of the operator Δ_D.

Theorem 7.12 *Let $D \subset \mathbb{V}$ with $D \neq \mathbb{V}$ if (Γ, μ) is recurrent. Let $u : \overline{D} \to \mathbb{R}$ be non-negative on $\overline{D}$ and superharmonic in D. Then there exist non-negative functions w, h with support in D, with h harmonic, such that*

$$u(x) = G_D w(x) + h(x), \quad x \in D. \tag{7.6}$$

Further, $w = \Delta_D u$, and this representation is unique.

Proof Since $u \geq 0$ on $\overline{D}$, for $x \in D$,

$$0 \leq P_D u(x) \leq P u(x) \leq u(x).$$

Consequently, $P_{n+1}^D u \leq P_n^D u$, and, since these functions are non-negative, we can define

$$h(x) = \lim_{k \to \infty} P_k^D u.$$

By monotone convergence,

$$P_D h = P_D(\lim_{k \to \infty} P_k^D u) = \lim_{k \to \infty} P_{k+1}^D u = h \text{ on } D.$$

Since $h = 0$ on D^c we have $P_D h = P h$ on D, so that h is harmonic on D. Now define w by setting

$$w = I_D u - P_D u.$$

Then

$$G_D w = \sum_{k=0}^{\infty} P_k^D w = \lim_{N \to \infty} \sum_{k=0}^{N} P_k^D w = \lim_{N \to \infty} (I_D u - P_{N+1}^D u) = I_D u - h,$$

which gives the representation (7.6). Applying Δ_D to both sides of (7.6) gives

$$\Delta_D u = \Delta_D G_D w + \Delta_D h = -w,$$

which implies that the representation is unique. □

Remark 7.13 Note that $h = 0$ on ∂D and $\Delta h = 0$ in D, so that $h = 0$ in D is finite. If Γ is a graph for which the strong Liouville property fails, and u is a non-constant positive harmonic function, then $h = u$ and $w = 0$.

We now prove a 'balayage' formula, which will be used to obtain an elliptic Harnack inequality from Green's functions bounds.

Let $D \subset \mathbb{V}$ be finite, and let $A \subset D$. Let h be non-negative in $\overline{D}$ and harmonic in D. Define the *reduite* h_A by

$$h_A(x) = \mathbb{E}^x(h(X_{T_A}); T_A < \tau_D).$$

Lemma 7.14 *(a) The function h_A satisfies $h_A \geq 0$ on $\mathbb{V}$, $h_A = 0$ on D^c, $h_A = h$ on A, and $0 \leq h_A \leq h$ on $D - A$.*
(b) Let $g = \mathbf{1}_A h$. Then

$$h_A(x) = \mathbb{E}^x g(X_{\tau_{D-A}}).$$

(c) h_A is harmonic in $D - A$, superharmonic in D, and is the smallest non-negative superharmonic function f in D such that $f \geq h\mathbf{1}_A$.

Proof (a) and (b) are immediate from the definition.
(c) The representation of h_A in (b) as the solution to a Dirichlet problem gives that h_A is harmonic in $D - A$. Further, the definition of h_A gives that $\Delta h_A(x) = 0$ if $x \in A^o \cup (D - A)$. If $x \in \partial_i A$, then

$$h_A(x) = h(x) = Ph(x) \geq Ph_A(y),$$

so h_A is superharmonic at x.

Now let $f \geq h\mathbf{1}_A$ be superharmonic in D and non-negative in $\mathbb{V}$. Then by the optional sampling theorem

$$f(x) \geq \mathbb{E}^x\big(f(X_{T_A}); T_A < \tau_D\big) \geq \mathbb{E}^x\big(h(X_{T_A}); T_A < \tau_D\big) = h_A(x).$$

Alternatively, we can use the minimum principle. Let f be superharmonic in D with $f \geq \mathbf{1}_A h$. Since h_A is harmonic in $D - A$, $f - h_A$ is superharmonic in $D - A$. Then $f - h_A \geq 0$ on $(D - A)^c$, and so $\min_{D-A}(f - h_A) = \min_{\partial(D-A)}(f - h_A) \geq 0$. □

Theorem 7.15 (Balayage formula) *Let $D \subset \mathbb{V}$ be finite, and let $A \subset D$. Let h be non-negative in $\overline{D}$ and harmonic in D. Then there exists $w \geq 0$ with* $\mathrm{supp}(w) \subset \partial_i(A)$ *such that*

$$h_A(x) = G_D w(x), \quad x \in D. \tag{7.7}$$

In particular, $h = G_D w$ on A.

Proof By Lemma 7.14, h_A is non-negative and superharmonic in D, and vanishes on ∂D. So by the Riesz representation, and the remark following, we have $h_A = G_D w$, where $w = \Delta_D h_A$. Since $h_A = 0$ on ∂D, we have $w = \Delta h_A$, and so by the calculations in Lemma 7.14 $w(x) = 0$ unless $x \in \partial_i(A)$. □

This concludes our survey of potential theory. We now apply these results to graphs with 'well-behaved' Green's functions.

7.2 Applications

Definition 7.16 We say that (Γ, μ) satisfies the condition (UG) ('uniform Green') if there exists a constant c_1 such that for all balls $B = B(x, R)$ and $x_i, y_i \in B(x, 2R/3)$ with $d(x_i, y_i) \geq R/8$ for $i = 1, 2$

$$g_B(x_1, y_1) \leq c_1 g_B(x_2, y_2).$$

Using Theorem 4.26 we have

Corollary 7.17 *If (Γ, μ) satisfies HK(α, β) then (UG) holds.*

Theorem 7.18 *Let (Γ, μ) satisfy (UG). Then (Γ, μ) satisfies the elliptic Harnack inequality.*

Proof Let $B = B(x_0, R)$, and *let* h be harmonic in B and non-negative in $\overline{B}$. Set $A = B(x_0, 2R/3)$ and $B' = B(x_0, R/2)$. Then the balayage formula (7.7) applied with $D = B$ gives that there exists $f \geq 0$ with support $\partial_i A$ such that

$$h(x) = \sum_{y \in \partial_i A} g_D(x, y) f(y) \mu_y, \quad x \in A.$$

Let $x_1, x_2 \in B'$. Then, since $d(x_i, y) \geq R/6$ for $y \in \partial_i A$, using the condition (UG),

$$h(x_1) \leq \sum_{y \in \partial_i A} g_B(x_1, y) f(y) \mu_y \leq \sum_{y \in \partial_i A} c_1 g_B(x_2, y) f(y) \mu_y = c_1 h(x_2),$$

giving the EHI. □

Theorem 7.19 *Let (Γ, μ) satisfy (H5) and be roughly isometric to $\mathbb{Z}^d$, with $d \geq 1$. Then (Γ, μ) satisfies the EHI and the strong Liouville property.*

Proof By Theorem 6.28 (Γ, μ) satisfies HK$(d, 2)$, and therefore the EHI. The strong Liouville property then follows by Theorem 1.46. □

Remark 7.20 One can also use a balayage argument to show that HK(α, β) leads to PHI(β). See [BH] for the case $\beta = 2$; the case $\beta > 2$ is very similar.

We now look at hitting probabilities in graphs which satisfy HK(α, β) with $\alpha > \beta$; note that this condition implies these graphs are transient.

Lemma 7.21 *Suppose that* (Γ, μ) *satisfies* $HK(\alpha, \beta)$ *with* $\alpha > \beta$.

(a) There exist constants c_i *such that*

$$c_1 r^{\alpha-\beta} \le \mathrm{Cap}(B(x,r)) \le c_2 r^{\alpha-\beta}.$$

(b) If $A \subset B(x, r/2)$ *then*

$$\mathrm{Cap}(A) \le \mathrm{Cap}_B(A) \le c_2\, \mathrm{Cap}(A).$$

Proof (a) By Theorem 4.26 we have $g(x, y) \asymp d(x, y)^{\beta-\alpha}$ when $x \neq y$. Let $B = B(x, r)$; then

$$1 \ge \mathfrak{e}_B G(x) \ge \mathfrak{e}_B(B) \min_{y \in B} g(x, y) \ge \mathrm{Cap}(B) c r^{\beta-\alpha},$$

giving the upper bound. For the lower bound, it is sufficient to consider a measure ν on B which puts mass λ on each point in B. For $y \in B$,

$$\nu G(y) \le \lambda \sum_{z \in B} g(z, y).$$

To bound this sum, choose k as small as possible so that $2r \le 2^k$. Then, since (Γ, μ) satisfies (V_α), writing $B_j = B(y, 2^j)$,

$$\sum_{z \in B} g(z, y) \le c + \sum_{j=1}^{k} \sum_{z \in B_{j+1} - B_j} g(y, z) \le c + c \sum_{j=1}^{k} (2^j)^{\beta-\alpha} 2^{\alpha j} \le c r^{\beta}.$$

So taking $\lambda = c r^{-\beta}$ we have $\nu G \le 1$ and hence $\mathrm{Cap}(B) \ge \nu(B) \ge c r^{\alpha-\beta}$.
(b) We have $\mathfrak{e}_A G_B \le \mathfrak{e} G \le 1$ on A, so $\mathrm{Cap}_B(A) \ge \mathfrak{e}_A(A) = \mathrm{Cap}(A)$, which proves the first inequality. For the second, by Theorem 4.26(a) there exists c_2 such that

$$g(y, z) \le c_2 g_B(y, z) \qquad \text{for } y, z \in B(x, r/2).$$

Then for $y \in A$

$$c_2^{-1} \mathfrak{e}_{A,B} G(y) \le \mathfrak{e}_{A,B} G_B(y) = 1,$$

and so $\mathrm{Cap}(A) \ge c_2^{-1} \mathfrak{e}_{A,B}(A) = c_2^{-1}\, \mathrm{Cap}_B(A)$. □

Lemma 7.22 *Let* (Γ, μ) *satisfy* $HK(\alpha, \beta)$, *and let* $B(y, r) \subset B(x_0, R/2)$. *Write* $B = B(x_0, R)$.

(a) If $\alpha < \beta$ *then there exist constants* c_i *such that*

$$\frac{c_1}{R^{\beta-\alpha}} \le \mathrm{Cap}_B(B(y, r)) \le \frac{c_2}{R^{\beta-\alpha}}.$$

(b) If $\alpha = \beta$ then there exist constants c_i such that

$$\frac{c_1}{\log(R/r)} \leq \mathrm{Cap}_B(B(y,r)) \leq \frac{c_2}{\log(R/r)}.$$

Proof (a) follows easily by the same methods as in Lemma 7.21, on noting that by Theorem 4.26 we have $g_B(x,y) \asymp R^{\beta-\alpha}$ for $x, y \in B(y,r)$.
(b) Let $A = B(y,r)$. Then

$$1 = \mathfrak{e}_{A,B} G_B \geq \mathrm{Cap}_B(A) \min_{z,z' \in A} g_B(z,z') \geq c\, \mathrm{Cap}_B(A) \log(R/r).$$

Let $z \in A$ and $A_j = \{z' \in A : 2^j \leq d(z,z') < 2^{j+1}\}$. Then

$$\sum_{z' \in A} g_B(z,z') \leq c + c' \sum_{j=1}^{\infty} |A_j| \log(R/2^j) \leq c'' r^\alpha \log(R/r).$$

So if ν places a mass $\lambda = c'' r^\alpha \log(R/r)^{-1}$ at each point in A then $\nu G_B \leq 1$ everywhere, $\nu(A) = c r^\alpha \lambda$, and hence $\mathrm{Cap}_B(A) \geq c/\log(R/r)$. □

Theorem 7.23 (Wiener's test) *Suppose that (Γ, μ) satisfies $HK(\alpha,\beta)$ with $\alpha > \beta$. Let $o \in \mathbb{V}$. Let $A \subset \mathbb{V}$, and $A_n = A \cap (B(o, 2^{n+1}) - B(o, 2^n))$. Then $\mathbb{P}^o$(X hits A infinitely often) is 0 or 1 according to whether*

$$\sum_{k=1}^{\infty} 2^{-k(\alpha-\beta)} \mathrm{Cap}(A_k) \tag{7.8}$$

is finite or infinite.

Proof As (Γ, μ) is transient, X will only hit each set A_n a finite number of times, and so, writing $F_n = \{X \text{ hits } A_n\}$,

$$\{X \text{ hits } A \text{ infinitely often}\} = \{F_n \text{ occurs for infinitely many } n\}.$$

By Theorem 4.26 we have if $x, y \in \mathbb{V}$ with $d(x,y) = r \geq 1$,

$$c_1 r^{-(\alpha-\beta)} \leq g(x,y) \leq c_2 r^{-(\alpha-\beta)}.$$

Further, if $x, y \in B(o, R/2)$ then we have $g_{B(o,R)}(x,y) \geq c_1' r^{-(\alpha-\beta)}$.
One implication is easy. We have

$$\mathbb{P}^o(F_n) = \mathfrak{e}_{A_n} G(o) \leq \mathfrak{e}_{A_n}(A_n) \max_{y \in A_n} g(o,y) \leq c\ \mathrm{Cap}(A_n) 2^{-n(\alpha-\beta)}.$$

So if the sum in (7.8) is finite then $\sum_n \mathbb{P}^o(F_n) < \infty$, and thus, $\mathbb{P}^o$-a.s., X hits A finitely often.

Now suppose that the sum in (7.8) is infinite. Set $B_k = B(o, 2^k)$, and let $G_k = \{T_{A_k} < \tau_{B_{k+2}}\}$, so that $G_k \subset F_k$. Then if $y \in \partial B_k$

$$\begin{aligned}\mathbb{P}^y(G_k) &= \mathfrak{e}_{A_k, B_{k+2}} G_{B_{k+2}}(y) \\ &\geq \mathrm{Cap}_{B_{k+2}}(A_{2k+i}) \min_{z \in A_k} g_{B_{k+2}}(y, z) \geq c \,\mathrm{Cap}(A_k) 2^{-(\alpha-\beta)k}.\end{aligned}$$

Here we used Lemma 7.21 and the fact that both $y, z \in B_{k+2}$.

Let $\mathscr{F}_k = \sigma(X_1, X_2, \dots, X_{T_k})$. Then the bound above implies that

$$\mathbb{P}^o(G_k | \mathscr{F}_k) \geq c \,\mathrm{Cap}(A_k) 2^{-(\alpha-\beta)k},$$

so that $\sum_k \mathbb{P}^o(G_k|\mathscr{F}_k) = \infty$ a.s. The conclusion that F_k occurs infinitely often, $\mathbb{P}^o$-a.s., then follows by a conditional version of the second Borel–Cantelli lemma – see [Dur], Chap. 5.3.1. $\square$

Appendix

This appendix contains a number of results which are used in this book but are not part of the main flow of the arguments.

A.1 Martingales and Tail Estimates

Let $(\Omega, \mathscr{F}, \mathbb{P})$ be a probability space. A *filtration* is an increasing sequence of sub-σ-fields $(\mathscr{F}_n, n \in \mathbb{Z}_+)$ of $\mathscr{F}$. We call $(\Omega, \mathscr{F}, \mathscr{F}_n, \mathbb{P})$ a *filtered probability space*.

Definition A.1 A *random time* is a r.v. $T : \Omega \to \mathbb{Z}_+ \cup \{+\infty\}$. A random time is a *stopping time* if for each $n \in \mathbb{Z}_+$

$$\{T \le n\} \in \mathscr{F}_n.$$

We also define the σ-field $\mathscr{F}_T$ by

$$\mathscr{F}_T = \{F \in \mathscr{F} : F \cap \{T \le n\} \in \mathscr{F}_n \text{ for all } n \in \mathbb{Z}_+\}.$$

If T is a stopping time then it is easy to check that $\{T = n\} \in \mathscr{F}_n$.

Definition A.2 Let $\mathbb{I} \subset \mathbb{Z}$. Let $X = (X_n, n \in \mathbb{I})$ be a stochastic process on a filtered probability space. We say X is a *supermartingale* if X is in L^1, so that $\mathbb{E}|X_n| < \infty$ for each n, X_n is $\mathscr{F}_n$-measurable for all n, and

$$\mathbb{E}(X_n | \mathscr{F}_m) \le X_m \text{ for each } n \ge m, \quad n, m \in \mathbb{I}. \tag{A.1}$$

(Thus a supermartingale is decreasing on average.) We say X is a submartingale if $-X$ is a supermartingale, and X is a martingale if X is both a supermartingale and a submartingale, or equivalently if X is in L^1 and equality holds in (A.1).

The two cases of interest here are $\mathbb{I} = \mathbb{Z}_+$ (a 'forwards supermartingale', or just a supermartingale) and $\mathbb{I} = \mathbb{Z}_-$ (a 'backwards supermartingale').

The following are the main results on martingales used in this book. For proofs see [Dur, W].

Theorem A.3 (Optional sampling theorem; see [W], Th. 10.10) *Let X be a forward supermartingale and $T \geq 0$ be a stopping time. Suppose one of the following conditions holds:*

(a) T is bounded (i.e. there exists K such that $T \leq K$ a.s.),
(b) X is bounded (i.e. there exists K such that $|X_n| \leq K$ a.s.)

Then

$$\mathbb{E}(X_T) \leq \mathbb{E}(X_0).$$

The convergence theorem for supermartingales is as follows.

Theorem A.4 (See [W], Th. 11.5) *Let X be a supermartingale bounded in L^1.*

(a) If $\mathbb{I} = \mathbb{Z}_+$ then there exists a r.v. X_∞ with $\mathbb{P}(|X_\infty| < \infty) = 1$ such that $X_n \to X_\infty$ a.s. as $n \to \infty$.
(b) If $\mathbb{I} = \mathbb{Z}_-$ then there exists a r.v. $X_{-\infty}$ with $\mathbb{P}(|X_{-\infty}| < \infty) = 1$ such that $X_n \to X_{-\infty}$ a.s. and in L^1 as $n \to -\infty$.

Corollary A.5 *Let X be a non-negative (forwards) supermartingale. Then X converges a.s. to a r.v. X_∞, and $\mathbb{E}X_\infty \leq \mathbb{E}X_0$. In particular, $\mathbb{P}(X_\infty < \infty) = 1$.*

Since $\mathbb{E}|X_n| = \mathbb{E}X_n \leq \mathbb{E}X_0$ this is immediate from the convergence theorem.

The hypotheses in Theorem A.4(a) are not strong enough to ensure that $\mathbb{E}X_n \to \mathbb{E}X_\infty$.

Definition A.6 $X = (X_n, n \geq 0)$ is *uniformly integrable* if for all $\varepsilon > 0$ there exists K such that

$$\mathbb{E}(|X_n|; |X_n| > K) \leq \varepsilon \text{ for all } n \geq 0.$$

This condition is slightly stronger than being bounded in L^1. Using Hölder's inequality it is easy to verify that if X is bounded in L^p for some $p > 1$ then

X is uniformly integrable. In particular, a bounded martingale is uniformly integrable.

Theorem A.7 (See [W], Th. 13.7) *Let M be a uniformly integral martingale. Then there exists M_∞ such that*

$$M_n \to M_\infty \text{ a.s. and in } L^1.$$

We conclude this subsection with an estimate on the tail of a sum of r.v.

Lemma A.8 *Let ξ_i be random variables taking values in $\{0, 1\}$, and suppose that there exists $p > 0$ such that $Y_k = \mathbb{E}(\xi_k | \mathscr{F}_{k-1}) \geq p$, where $\mathscr{F}_k = \sigma(\xi_1, \ldots, \xi_k)$. Let $b < p$. Then there exists a constant $c_1 = c_1(b, p)$ such that*

$$\mathbb{P}\left(\sum_{k=1}^{n} \xi_k < bn\right) \leq e^{-c_1 n}.$$

Proof Let

$$M_n = \sum_{i=1}^{n} (p - \xi_i),$$

so that M is a supermartingale. If $\lambda > 0$ then

$$\begin{aligned} \mathbb{E}(e^{-\lambda \xi_n} | \mathscr{F}_{n-1}) &= \mathbb{E}(e^{-\lambda} + (1 - e^{-\lambda}) \mathbf{1}_{(\xi_n = 0)} | \mathscr{F}_{n-1}) \\ &\leq e^{-\lambda} + (1 - p)(1 - e^{-\lambda}) = (1 - p) + pe^{-\lambda}. \end{aligned}$$

So,

$$\mathbb{E} e^{\lambda M_n} \leq \left((1 - p)e^{\lambda p} + pe^{-\lambda(1-p)}\right)^n.$$

Set $\psi(\lambda) = \log\left((1-p)e^{\lambda p} + pe^{-\lambda(1-p)}\right)$. Then it is straightforward to check that $\psi(0) = \psi'(0) = 0$, so that $\psi(\lambda) \leq c_2 \lambda^2$ on $[0, 1]$. So

$$\begin{aligned} \mathbb{P}\left(\sum_{i=1}^{n} \xi_k < bn\right) &= \mathbb{P}(M_n > (p - b)n) = \mathbb{P}(e^{\lambda M_n} > e^{\lambda(p-b)n}) \\ &\leq e^{-\lambda(p-b)n} \mathbb{E}(e^{\lambda M_n}) \leq \exp(-n(\lambda(p - b) - \psi(\lambda))). \end{aligned}$$

Taking $\lambda > 0$ small enough then gives the bound. □

Corollary A.9 *Let $X \sim$ `Bin(n,p)` and $a > p$. Then there exists a constant $b = b(a, p)$ such that*

$$\mathbb{P}(X > an) \leq e^{-bn}.$$

A.2 Discrete Time Markov Chains and the Strong Markov Property

While the Markov property for a process is quite intuitive ('only the value of X at time n affects the evolution of X after time n'), its precise formulation can be quite troublesome, particularly as for many applications one needs the *strong Markov property*, that is, the Markov property at stopping times. In the discrete context one does not have to worry about null sets or measurability problems, but at first sight the setup still appears unnecessarily elaborate.

Let $\mathbb{V}$ be a countable set and $(\Omega, \mathscr{F})$ be a measure space. We assume that we have random variables $X_n : \Omega \to \mathbb{V}$, and also that we have *shift* operators

$$\theta_n : \Omega \to \Omega$$

which are measurable (so $\theta_n^{-1}(F) \in \mathscr{F}$ for all $F \in \mathscr{F}$) and relate to (X_n) by

$$X_n(\theta_m \omega) = X_{n+m}(\omega) \text{ for all } n, m \in \mathbb{Z}_+, \ \omega \in \Omega. \tag{A.2}$$

We write $\xi \circ \theta_n$ for the random variable given by $\xi \circ \theta_n(\omega) = \xi(\theta_n(\omega))$.

One way of constructing these is to take Ω to be the canonical space

$$\Omega = \mathbb{V}^{\mathbb{Z}_+}.$$

Thus an element $\omega \in \Omega$ is a function $\omega : \mathbb{Z}_+ \to \mathbb{V}$. We then take X_n to be the coordinate maps and θ_n to be a shift by n, so that if $\omega = (\omega_0, \omega_1, \dots)$ then

$$X_n(\omega) = \omega_n, \ n \in \mathbb{Z}_+, \quad \theta_n(\omega) = (\omega_n, \omega_{n+1}, \dots).$$

Let $\mathscr{F}$ be the Borel σ-field on Ω, that is, the smallest σ-field which contains all the sets $X_n^{-1}(A)$, $n \in \mathbb{Z}_+$, $A \subset \mathbb{V}$. We define the filtration

$$\mathscr{F}_n = \sigma(X_0, X_1, \dots, X_n).$$

We write $\xi \in b\mathscr{F}_n$ to mean that ξ is a bounded $\mathscr{F}_n$-measurable r.v. To make our notation more compact we will sometimes write

$$\mathbb{E}^x_{\mathscr{F}_n}(\xi) = \mathbb{E}^x(\xi | \mathscr{F}_n).$$

Let $\mathscr{P}(x, y)$ be a transition matrix on $\mathbb{V} \times \mathbb{V}$, so that

$$\mathscr{P}(x, y) \ge 0 \text{ for all } x, y, \text{ and } \sum_{y \in \mathbb{V}} \mathscr{P}(x, y) = 1 \text{ for all } x.$$

Now let $\{\mathbb{P}^x, x \in \mathbb{V}\}$ be the probabilities on $(\Omega, \mathscr{F})$ so that, for all x, w, z, n,

$$\mathbb{P}^x(X_0 = x) = 1, \tag{A.3}$$

$$\mathbb{P}^x(X_{n+1} = z | \mathscr{F}_n) = \mathscr{P}(X_n, z). \tag{A.4}$$

We call the collection $(X_n, n \in \mathbb{Z}_+, \mathbb{P}^x, x \in \mathbb{V})$ which satisfies the conditions above a *Markov chain with transition probabilities* $\mathscr{P}$. As in (1.9) we define the semigroup operator P_n by $P_n f(x) = \mathbb{E}^x f(X_n)$, and write $P = P_1$.

The (simple) Markov property, that is, at a fixed time n, is formulated as follows.

Theorem A.10 *Let $n \geq 0$, $\xi \in b\mathscr{F}_n$, $\eta \in b\mathscr{F}$. Then*

$$\mathbb{E}^x(\xi(\eta \circ \theta_n)) = \mathbb{E}^x(\xi \, \mathbb{E}^{X_n} \eta). \tag{A.5}$$

Remark A.11 The notation in (A.5) is concise but can be confusing, particularly if η is written as a function of X. (For example if $\eta = \varphi(X_n)$ then the right hand side of (A.5) would be $\mathbb{E}^x(\xi E^{X_n} \varphi(X_n))$. The two X_n in this expression are not the same!) A clearer way of writing this example is to set $f(y) = \mathbb{E}^y(\eta)$. Then the right side of (A.5) is just $\mathbb{E}^x(\xi \; f(X_n))$.

An alternative way of stating (A.5) is that

$$\mathbb{E}^x(\eta \circ \theta_n | \mathscr{F}_n) = \mathbb{E}^{X_n}(\eta), \;\; \mathbb{P}^x\text{-a.s.} \tag{A.6}$$

Proof Let $g_k : \mathbb{V} \to \mathbb{R}$ be bounded and $\xi \in b\mathscr{F}_n$. Let first

$$\eta = \prod_{k=0}^{m} g_k(X_k).$$

Then if $\omega = (\omega_0, \omega_1, \dots)$, using (A.2),

$$(\eta \circ \theta_n)(\omega) = \prod_{j=0}^{m} g_k(X_k(\theta_n(\omega))) = \prod_{k=0}^{m} g_k(X_{n+k}(\omega)).$$

(This shows how θ_n shifts the process X to the left by n steps.) Now take $g_k(x) = \mathbf{1}_{(x=y_k)}$, so that

$$(\eta \circ \theta_n)(\omega) = \mathbf{1}_{(X_n=y_0,\dots,X_{n+m}=y_m)}.$$

It is then easy to check by induction that both $\mathbb{E}^{X_n}(\eta)$ and $\mathbb{E}^x(\eta \circ \theta_n | \mathscr{F}_n)$ equal

$$\mathbf{1}_{(X_n=y_0)} \prod_{i=1}^{m} \mathscr{P}(y_{i-1}, y_i),$$

which proves (A.6) in this case. As both sides of (A.6) are linear in η, (A.6) then follows for general bounded g_k. A monotone class argument, as in [Dur], Chap. 6.3, then concludes the proof. □

The Markov property extends to stopping times. We define the shift θ_T on $\{\omega : T(\omega) < \infty\}$ as follows:

$$\theta_T(\omega) = \theta_{T(\omega)}(\omega).$$

So $\theta_T = \theta_n$ on $\{\omega : T(\omega) = n\}$.

Theorem A.12 (Strong Markov property) *Let T be a stopping time, $\xi \in b\mathscr{F}_T$, and $\eta \in b\mathscr{F}$. Then*

$$\mathbb{E}^x(\xi(\eta \circ \theta_T); T < \infty) = \mathbb{E}^x(\xi\, \mathbb{E}^{X_T}(\eta); T < \infty). \tag{A.7}$$

Proof Set $A_n = \{T = n\}$, and note that $\xi \mathbf{1}_{A_n} \in b\mathscr{F}_n$. Then, since $T = n$ on A_n, using the simple Markov property,

$$\begin{aligned}
\mathbb{E}^x(\xi(\eta \circ \theta_T); T < \infty) &= \sum_{n \in \mathbb{Z}_+} \mathbb{E}^x(\xi \mathbf{1}_{A_n}(\eta \circ \theta_T)) \\
&= \sum_{n \in \mathbb{Z}_+} \mathbb{E}^x(\xi \mathbf{1}_{A_n}(\eta \circ \theta_n)) \\
&= \sum_{n \in \mathbb{Z}_+} \mathbb{E}^x(\xi \mathbf{1}_{A_n} \mathbb{E}^{X_n}(\eta)) \\
&= \sum_{n \in \mathbb{Z}_+} \mathbb{E}^x(\xi \mathbf{1}_{A_n} \mathbb{E}^{X_T}(\eta)) = \mathbb{E}^x(\xi\, \mathbb{E}^{X_T}(\eta); T < \infty).
\end{aligned}$$

□

We will now give a proof of Theorem 1.8, and, as an example of the use of the strong Markov property, begin by giving a careful proof of a preliminary result. Let

$$\begin{aligned}
p(x) &= \mathbb{P}^x(T_x^+ < \infty), \\
H(x, y) &= \mathbb{P}^x(T_y < \infty), \\
L_n^x &= \sum_{k=0}^{n-1} \mathbf{1}_{(X_k = x)}.
\end{aligned}$$

Lemma A.13 *(a) We have*

$$\mathbb{E}^x L_\infty^y = H(x, y) \mathbb{E}^y L_\infty^y. \tag{A.8}$$

(b) If $p(y) = 1$ then $\mathbb{P}^y(L_\infty^Y = \infty) = 1$, while if $p(y) < 1$ then

$$\mathbb{E}^y L_\infty^y = \frac{1}{1 - p(y)}.$$

Proof (a) If $x = y$ then $H(x, x) = 1$ and (A.8) is immediate. Now let $x \neq y$, and write $T = T_y$. Then $L^y_\infty(\theta_T \omega) = L^y_\infty(\omega)$ if $T(\omega) < \infty$, while $L^y_\infty(\omega) = 0$ if $T(\omega) = \infty$. Set $\xi = 1$, let $K > 0$, and set $\eta = L^y_\infty \wedge K$. Then

$$\begin{aligned}\mathbb{E}^x(K \wedge L^y_\infty) &= \mathbb{E}^x((K \wedge L^y_\infty) \circ \theta_T; T < \infty)\\ &= \mathbb{E}^x(\mathbb{E}^y(K \wedge L^y_\infty); T < \infty) = H(x, y)\mathbb{E}^y(K \wedge L^y_\infty).\end{aligned}$$

Here we used the strong Markov property to obtain the second equality, and also the fact that $X_T = y$. Let $K \to \infty$ to obtain (A.8).
(b) For $k \geq 0$ let $A_k = \{L^y_\infty \geq k\}$. Write $T = T^+_y$. If $T(\omega) < \infty$ then

$$L^y_\infty(\theta_T \omega) = \sum_{k=0}^{\infty} \mathbf{1}_{(X_k \circ \theta_T(\omega) = y)} = \sum_{k=0}^{\infty} \mathbf{1}_{(X_{k+T}(\omega) = y)} = L^y_\infty(\omega) - L^y_{T-1}(\omega).$$

If $X_0(\omega) = y$ then $L^y_{T-1}(\omega) = 1$, and so on $\{T < \infty\} \cap \{X_0 = y\}$ we have $\omega \in A_k$ if and only if $\theta_T \omega \in A_{k-1}$. So, for $k \geq 1$,

$$\begin{aligned}\mathbb{P}^y(A_k) &= \mathbb{E}^y \mathbf{1}_{A_k} = \mathbb{E}^y(\mathbf{1}_{A_{k-1}} \circ \theta_T; T < \infty)\\ &= \mathbb{E}^y(\mathbb{E}^y(\mathbf{1}_{A_{k-1}}); T < \infty) = p(y)\mathbb{P}^y(A_{k-1}).\end{aligned}$$

Since $\mathbb{P}^y(A_0) = 1$ we obtain

$$\mathbb{P}^y(L^y_\infty \geq k) = p(y)^k.$$

If $p(y) = 1$ this gives $\mathbb{P}^y(A_k) = 1$ for all k and so $\mathbb{P}^y(L^y_\infty = \infty) = 1$. If $p(y) < 1$ then under $\mathbb{P}^y$ the r.v. L^y_∞ has a geometric distribution, and therefore has expectation $(1 - p(y))^{-1}$. □

Proof of Theorem 1.8 Note that as Γ is connected we have $H(x, y) > 0$ for all x, y. Further, if $F_{xy} = \{T_y < T^+_x\}$ then, by considering the event that X travels from x to y along a shortest path between these points, we have $\mathbb{P}^x(F_{xy}) > 0$.

With the convention that $1/0 = \infty$, we have from Lemma A.13

$$\frac{1}{1 - p(x)} = \mathbb{E}^x L^x_\infty = \mathbb{E}^x\Big(\sum_{k=0}^{n} \mathbf{1}_{(X_k = x)}\Big) = \sum_{k=0}^{\infty} \mathscr{P}_k(x, x).$$

Thus $\sum_k \mathscr{P}_k(x, x)$ converges if and only if $p(x) < 1$.

Let $x, y \in \mathbb{V}$ and $m = d(x, y)$. Then $\mathscr{P}_m(x, y) > 0$ and we have

$$\mathscr{P}_{n+2m}(y, y) \geq \mathscr{P}_m(x, y)\mathscr{P}_m(y, x)\mathscr{P}_n(x, x),$$

so that $\sum_k \mathscr{P}_k(x, x)$ converges if and only if $\sum_k \mathscr{P}_k(y, y)$ converges.

Combining these observations, we obtain the equivalence of (a), (b), and (c) in both case (R) and case (T).

Suppose (a)–(c) hold in case (T). Then

$$1 > \mathbb{P}^x(T_x^+ < \infty) \geq \mathbb{P}^x(T_y < \infty, T_x^+ \circ \theta_{T_y} < \infty) = H(x,y)H(y,x).$$

So (d) holds. If (d) holds then choose x, y so that $H(x,y) < 1$. Then

$$\mathbb{P}^y(T_y^+ < \infty) = \mathbb{P}^y(T_y^+ < \infty; F_{yx}) + \mathbb{P}(F_{yx}^c) \leq \mathbb{P}^y(F_{yx})H(x,y) + \mathbb{P}(F_{yx}^c) < 1,$$

so (a) holds. Hence (d) is equivalent to (a)–(c) in case (T), and therefore also in case (R).

By Lemma A.13 and the above we have

$$\mathbb{P}^x(L_\infty^x = \infty) = 1 \text{ if and only if } p(x) = 1, \tag{A.9}$$

$$\mathbb{P}^x(L_\infty^x = \infty) = 0 \text{ if and only if } p(x) < 1. \tag{A.10}$$

Further, by the strong Markov property,

$$\mathbb{P}^x(L_\infty^y = \infty) = H(x,y)\mathbb{P}^y(L_\infty^y = \infty).$$

Hence (a)–(d) implies (e) in both case (R) and case (T). Taking $x = y$ and using (A.9) and (A.10) gives that (e) implies (a). □

A.3 Continuous Time Random Walk

In this section we give an outline of construction of the continuous time random walk.

Definition A.14 Let $f : \mathbb{R}_+ \to \mathscr{M}$, where $\mathscr{M}$ is a metric space. We say f is *right continuous with left limits* or *cadlag* if for all $t \in \mathbb{R}_+$

$$\lim_{s \downarrow t} f(s) = f(t), \qquad \text{and } \lim_{s \uparrow t} f(s) \text{ exists.}$$

(If $t = 0$ we do not consider the second property above.) The term cadlag is from the French abbreviation, which is easier to pronounce than the English one.

Let $D = D(\mathbb{R}_+, \mathbb{V})$ be the set of functions $f : \mathbb{R}_+ \to \mathbb{V}$ which are right continuous with left limits. We write Y_t for the coordinate functions on D defined by $Y_t(f) = f(t)$, $f \in D$. We also let $\mathscr{D}$ be the Borel σ-field on D, which is the smallest σ-field which contains all the sets $\{f : f(t) = x\}$ for $t \in \mathbb{R}_+$ and $x \in \mathbb{V}$.

Definition A.15 Let $V_i, i \in \mathbb{N}$ be independent exponential random variables with mean 1, so that $\mathbb{P}(V_i \le t) = 1 - e^{-t}$. Set

$$T_k = \sum_{i=1}^{k} V_i, \qquad N_t = \sum_{k=1}^{\infty} \mathbf{1}_{(T_k \le t)},$$

so that $N_t = k$ if and only if $T_k \le t < T_{k+1}$. The process $N = (N_t, t \in \mathbb{R}_+)$ is the *Poisson process.*

The definition gives that T_k is the time of the kth jump of N, and that for all ω the function $t \to N_t(\omega)$ is right continuous with left limits. We define the filtration of N by

$$\mathscr{F}_t^N = \sigma(N_s, 0 \le s \le t), \quad t \in \mathbb{R}_+.$$

Theorem A.16 *(a) N_t has a Poisson distribution with parameter t, so that*

$$\mathbb{P}(N_t = m) = \frac{e^{-t} t^m}{m!}, \quad m \in \mathbb{Z}_+.$$

(b) Let $s \ge 0$. The process $N^{(s)}$ defined by $N_t^{(s)} = N_{t+s} - N_s$ is independent of $\mathscr{F}_s^N$, and has the same distribution as N.
(c) The process N has stationary independent increments.

Proof (a) The r.v. T_{m+1} has the gamma distribution with parameters 1 and $m+1$, so integrating by parts we have

$$\begin{aligned}\mathbb{P}(N_t \ge m+1) = \mathbb{P}(T_{m+1} \le t) &= \int_0^t \frac{e^{-s} s^m}{m!} ds \\ &= \frac{-e^{-s} s^m}{m!}\Big|_0^t + \int_0^t \frac{e^{-s} s^{m-1}}{(m-1)!} ds = \frac{e^{-t} t^m}{m!} + \mathbb{P}(N_t \ge m),\end{aligned}$$

which proves (a).
(b) This follows by basic properties of the exponential distribution – see [Nor], Th. 2.4.1.
(c) Let $0 = t_0 < t_1 < \cdots < t_m$. Applying (b) with $s = t_{m-1}$ we have that $N_{t_m} - N_{t_{m-1}}$ is Poisson with parameter $t_m - t_{m-1}$ and is independent of $\mathscr{F}_{t_{m-1}}$. Continuing, it follows that the random variables

$$N_{t_0}, N_{t_1} - N_{t_0}, \ldots, N_{t_m} - N_{t_{m-1}}$$

are independent and that $N_{t_k} - N_{t_{k-1}}$ is Poisson with parameter $t_k - t_{k-1}$. □

Remark We have by (a) that $\sup_m T_m = \infty$, so that the random times T_m do not accumulate, and N_t is finite $\mathbb{P}$-a.s. for all t.

Definition A.17 Let (Γ, μ) be a weighted graph and $X = (X_n, n \in \mathbb{Z}_+)$ be the discrete time random walk with $X_0 = x$ defined on a probability space $(\Omega, \mathcal{F}, \widetilde{\mathbb{P}}^x)$. We assume that this space also carries a Poisson process N independent of X. Define

$$\widetilde{Y}_t = X_{N_t},\ t \in [0, \infty). \tag{A.11}$$

We call $\widetilde{Y}$ the *continuous time random walk* on (Γ, μ).

Before we prove the Markov property, it will be convenient to 'move' the process $\widetilde{Y}$ to the canonical space $(D, \mathcal{D})$ given in Definition A.14. Note that $\widetilde{Y}$ as defined by (A.11) is cadlag – this is immediate from the corresponding property for the Poisson process N. We thus have a map

$$\widetilde{Y} : \Omega \to D,$$

defined by $\widetilde{Y}(\omega) = (\widetilde{Y}_t(\omega), t \geq 0)$. To prove that this function is measurable, it is enough to check that $\widetilde{Y}^{-1}(A) \in \mathcal{F}$ for sets A given by $A = \{f \in D : f(t) = x\}$, where $t \geq 0$ and $x \in \mathbb{V}$. However,

$$\{\omega : \widetilde{Y}(\omega) \in A\} = \{\omega : \widetilde{Y}_t(\omega) = x\} = \bigcup_{n=0}^{\infty} \{N_t(\omega) = n\} \cap \{X_n(\omega) = x\}.$$

We define the law of Y started at x on the space D by setting

$$\mathbb{P}^x(F) = \widetilde{\mathbb{P}}^x(\widetilde{Y}^{-1}(F)),\ F \in \mathcal{D}.$$

It is then straightforward to verify that $\widetilde{Y}$ under $\widetilde{P}^x$ and Y under $\mathbb{P}^x$ have the same law.

As in the discrete time situation, we define the shift operators θ_t on D by setting $\theta_t(f)$ to be the function given by $\theta_t(f)(s) = f(t+s)$. We also define the filtration of Y by

$$\mathcal{F}_t = \sigma(Y_s, s \leq t).$$

Theorem A.18 *The process Y defined above satisfies the following.*

(a) The function $t \to Y_t$ is right continuous with left limits.
(b) The process Y has semigroup given by

$$Q_t f(x) = \mathbb{E}^x f(Y_t) = \sum_{n=0}^{\infty} \frac{e^{-t}}{n!} P^n f(x),$$

for $f \in C(\mathbb{V})$ with f bounded.

(c) Y satisfies the Markov property. That is, if $\xi \in b\mathscr{F}_t$ and $\eta \in b\mathscr{F}$ then

$$\mathbb{E}^x(\xi(\eta \circ \theta_t)) = \mathbb{E}^x(\xi \, \mathbb{E}^{Y_t} \eta). \tag{A.12}$$

Proof (a) and (b) are immediate from the construction of $\widetilde{Y}$ and Y.
(c) Let $t > 0, 0 \le t_1 \le \cdots \le t_m < t, 0 \le s_1 \le \cdots \le s_n, g_j, f_i$ be bounded, and

$$\xi = \prod_{j=1}^{m} g_j(Y_{t_j}), \quad \eta = \prod_{i=1}^{n} f_i(Y_{s_i}). \tag{A.13}$$

It is sufficient to prove that

$$\mathbb{E}^x\Big(\prod_{j=1}^{m} g_j(\widetilde{Y}_{t_j}) \prod_{i=1}^{n} f_i(\widetilde{Y}_{t+s_i})\Big) = \mathbb{E}^x\Big(\prod_{j=1}^{m} g_j(\widetilde{Y}_{t_j}) \mathbb{E}^{\widetilde{Y}_t}\Big(\prod_{i=1}^{n} f_i(\widetilde{Y}_{s_i})\Big)\Big). \tag{A.14}$$

Since $\widetilde{Y}$ and Y are equal in law, this implies (A.12) for ξ and η given by (A.13), and a monotone class argument then gives the general case.

To prove (A.14) it is sufficient to prove, for integers $0 \le k_1 \le \cdots \le k_m \le k$ and $z \in \mathbb{V}$, that

$$\mathbb{E}^x\Big(\prod_{j=1}^{m} g_j(\widetilde{Y}_{t_j}) \mathbf{1}_{(N_{t_j}=k_j)} \mathbf{1}_{(N_t=k)} \mathbf{1}_{(\widetilde{Y}_t=z)} f_1(\widetilde{Y}_{t+s_1})\Big)$$
$$= \mathbb{E}^x\Big(\prod_{j=1}^{m} g_j(\widetilde{Y}_{t_j}) \mathbf{1}_{(N_{t_j}=k_j)} \mathbf{1}_{(N_t=k)} \mathbf{1}_{(\widetilde{Y}_t=z)} Q_{s_1} f_1(z)\Big). \tag{A.15}$$

An induction argument on n then gives the general case.

Write $f = f_1, s = s_1$ and let

$$\xi_1 = \mathbf{1}_{(X_k=z)} \prod_{j=1}^{m} g_j(X_{k_j}), \quad \xi_2 = \mathbf{1}_{(N_t=k)} \prod_{j=1}^{m} \mathbf{1}_{(N_{t_j}=k_j)},$$

so that $\xi_1 \in \sigma(X_j, j \le k)$ and $\xi_2 \in \sigma(N_s, s \le t)$. Then

$$\mathbb{E}^x\Big(\prod_{j=1}^{m} g_j(\widetilde{Y}_{t_j}) \mathbf{1}_{(N_{t_j}=k_j)} \mathbf{1}_{(\widetilde{Y}_t=z)} f(\widetilde{Y}_{t+s})\Big) = \mathbb{E}^x\Big(\xi_1 \xi_2 f(\widetilde{Y}_{t+s})\Big)$$
$$= \mathbb{E}^x\Big(\xi_1 \xi_2 \mathbf{1}_{(X_k=z)} \sum_{i=0}^{\infty} \mathbf{1}_{(N_{t+s}-N_t=i)} f(X_{k+i})\Big)$$

$$= \sum_{i=0}^{\infty} \mathbb{E}^x \Big(\xi_1 \xi_2 \mathbf{1}_{(N_{t+s}-N_t=i)} f(X_{k+i}) \Big)$$
$$= \sum_{i=0}^{\infty} \mathbb{E}^x (\xi_2 \mathbf{1}_{(N_{t+s}-N_t=i)}) \mathbb{E}^x (\xi_1 f(X_{k+i})).$$

Here we used the independence of X and N in the last line. Using the Markov properties for N and X we then obtain

$$\mathbb{E}^x \Big(\xi_1 \xi_2 \mathbf{1}_{(X_k=z)} f(\widetilde{Y}_{t+s}) \Big) = \sum_{i=0}^{\infty} \mathbb{E}^x (\xi_2) e^{-s} \frac{s^i}{i!} \mathbb{E}^x (\xi_1 \mathbb{E}^z f(X_i))$$
$$= \mathbb{E}^x (\xi_1 \xi_2) \sum_{i=0}^{\infty} e^{-s} \frac{s^i}{i!} P^i f(z) = \mathbb{E}^z (\xi_1 \xi_2 Q_s f(z)).$$

This completes the proof of (A.15), and hence of the Markov property. □

A second method of constructing Y is from the transition densities $q_t(x, y)$ defined in (5.2). This method has the advantage of not requiring as much detailed work on the exponential distribution, but it is more abstract and needs more measure theory.

Let $\overline{\Omega} = \mathbb{V}^{\mathbb{R}_+}$, the space of *all* functions from $\mathbb{R}_+$ to $\mathbb{V}$. As before we define coordinate maps Z_t by setting $Z_t(\omega) = \omega(t)$. Let

$$\mathscr{F}^o = \sigma\big\{ Z_t^{-1}(\{x\}), x \in \mathbb{V}, t \in \mathbb{R}_+ \big\}.$$

Let $\mathscr{T}$ be the set of all finite subsets of $\mathbb{R}_+$. For $T \in \mathscr{T}$ let $\mathscr{F}^o(T) = \sigma(Z_t, t \in T)$. Let $T \in \mathscr{T}$, and write $T = \{t_1, \dots, t_m\}$ where $0 < t_1 < \cdots < t_m$. For $x_0 \in \mathbb{V}$ define the probability measure $P_T^{x_0}$ on $(\overline{\Omega}, \mathscr{F}^o(T))$ by

$$P_T^{x_0}(\{\omega : \omega(t_i) = x_i, 1 \le i \le m\}) = \prod_{i=1}^{m} q_{t_i - t_{i-1}}(x_{i-1}, x_i) \mu_{x_i}.$$

Using the Chapman–Kolmogorov equation (5.3) one has

$$\sum_z q_{t_i - t_{i-1}}(x_{i-1}, z) q_{t_{i+1} - t_i}(z, x_{i+1}) \mu_z = q_{t_{i+1} - t_{i-1}}(x_{i-1}, x_{i+1}),$$

and it follows that the family $(P_T^x, T \in \mathscr{T})$ is *consistent* – that is, if $T_1 \subset T_2$ then

$$P_{T_2}^x \big|_{\mathscr{F}^o(T_1)} = P_{T_1}^x.$$

The Kolmogorov extension theorem (see [Dur], App. A.3) then implies that there exists a probability P^x on $(\overline{\Omega}, \mathscr{F}^o)$ such that $P\big|_{\mathscr{F}^o(T)} = P_T^x$ for each

$T \in \mathscr{T}$. Under the probability measure P^x the process Z is then a Markov process with semigroup Q_t.

A reader who sees this construction here for the first time may feel a sense of unease. On one hand, the space $\overline{\Omega}$ is very big – much larger than the spaces usually studied in probability, which have the same cardinality as $\mathbb{R}$. Further, the construction appears to be too quick and general, and does not use any specific properties of the process. This unease is justified: while what has been done above is correct, it does not give us a useful Markov process. In Lemma 5.21, for example, we look at the first exit time $\tau_Y(x, r)$ of Y from a ball. However, the corresponding time $\tau_Z(x, r) = \inf\{t > 0 : Z_t \notin B(x, r)\}$ is not measurable with respect to $\mathscr{F}^o$. This is easy to prove: the event $\{\tau_Z(x, r) < 1\}$ depends on Z at uncountably many time coordinates, while it is not hard to verify that any event in $\mathscr{F}^o$ can only depend on countably many time coordinates.

The solution is to regularise the process Z. For simplicity we just do this on the interval $[0, 1]$, but the same argument works on any compact interval, and so (by an easy argument) on $\mathbb{R}_+$. Let $\mathbb{D} = \{k2^{-n}, k \in \mathbb{Z}\}$, and set $\mathbb{D} = \cup_n \mathbb{D}_n$. We write Z^D for the process Z restricted to the time set $\mathbb{D}_+ = \mathbb{D} \cap [0, \infty)$. Let $J(a, b)$ be the number of jumps of Z^D in the interval $[a, b]$, defined by

$$\begin{aligned} J(a, b, n) = \max\{k :\ & \text{there exist } t_1 < t_2 < \cdots < t_k \text{ with} \\ & t_i \in \mathbb{D}_n \cap [a, b] \text{ and } Z_{t_i+1} \neq Z_{t_i} \text{ for } i = 1, \ldots, n-1\}, \\ J(a, b) = \sup_n J(a, b, n).& \end{aligned}$$

Lemma A.19 *(a) For any $n \geq 1$ we have $\mathbb{P}(J(t, t+h, n) \geq 1) \leq h$.*
(b) For $t \geq 0$, $h > 0$, $n \geq 0$, $\mathbb{P}(J(t, t+h, n) \geq 2) \leq h^2$.
(c) For $h \leq \frac{1}{2}$ we have

$$\mathbb{P}^x(\sup_{0 \leq t \leq 1} J(t, t+h) \geq 2) \leq ch.$$

(d) There exists a r.v. H on $(\overline{\Omega}, \mathscr{F}^o, \mathbb{P}^x)$ with $\mathbb{P}^x(H > 0) = 1$ such that $J(t, t+H) \in \{0, 1\}$ for all $t \in [0, 1]$.

Proof (a) An easy calculation using the transition density of Z gives that $\mathbb{P}^x(Z_{t+h} \neq Z_t) \leq h$. Write $\mathbb{D}_n \cap [t, t+h] = \{t_1, t_2, \ldots, t_m\}$ with $t_{i+1} = t_i + 2^{-n}$ and set $F_j = \{Z_{t_{j+1}} \neq Z_{t_j}\}$. Then

$$\mathbb{P}^x(J(t, t+h, n) \geq 1) = \mathbb{P}^x(\cup_{j=1}^{m-1} F_j) \leq m2^{-n} \leq h.$$

(b) If $J(t, t+h, n) \geq 2$ then two distinct events F_i must occur. The Markov property gives that $\mathbb{P}^x(F_i \cap F_j) \leq 2^{-2n}$, so

$$\mathbb{P}^x(J(t, t+h, n) \geq 2) \leq \sum_i \sum_{j>i} \mathbb{P}^x(F_i \cap F_j) \leq m^2 2^{-2n} \leq h^2.$$

(c) As $J(t, t+h, n)$ is increasing in n, it follows from (b) that $\mathbb{P}^x(J(t, t+h) \geq 2) \leq h^2$. Let $G_k = \{J(a+kh, a+(k+2)h) \geq 2\}$, for $k = 0, 1, \ldots, \lceil h^{-1} \rceil$. Then

$$\mathbb{P}^x(\sup_{0 \leq t \leq 1} J(t, t+h) \geq 2) \leq \sum_k \mathbb{P}^x(G_k) \leq (2h)^2 \lceil h^{-1} \rceil \leq ch.$$

(d) This follows from (c) by a straightforward argument using Borel–Cantelli by considering the events $\{\sup_{0 \leq t \leq 1} J(t, t+2^{-k}) \geq 2\}$. □

Lemma A.19 proves that on the event $\{H > 0\}$, which has probability one, the jumps of $Z|_{\mathbb{D} \cap [0,1]}$ do not accumulate. Therefore we can define

$$\overline{Z}_t = \lim_{p \in \mathbb{D}, p \downarrow t} Z_p, \; t \in [0, 1).$$

Lemma A.20 *(a) The process $\overline{Z}$ is right continuous with left limits on* $[0, 1)$, *$\mathbb{P}^x$-a.s.*

(b) For $t \in [0, 1)$, $\mathbb{P}^x(\overline{Z}_t = Z_t) = 1$.

We call the process $\overline{Z}$ a *right-continuous modification* of Z. We can now move the process $(\overline{Z}, \mathbb{P}^x)$ onto the canonical space $D(\mathbb{R}_+, \mathbb{V})$ to obtain the continuous time simple random walk. We note that with this construction the Markov property at a fixed time t is straightforward to verify.

We now give the strong Markov property for Y.

Definition A.21 A random variable $T : \Omega \to [0, \infty]$ is a *stopping time* if

$$\{T \leq t\} \in \mathscr{F}_t \text{ for all } t \geq 0.$$

We define the σ-field $\mathscr{F}_T$ by

$$\mathscr{F}_T = \{F \in \mathscr{F} : F \cap \{T \leq t\} \in \mathscr{F}_t \text{ for all } t \geq 0\}.$$

One can check that if $S \leq T$ then $\mathscr{F}_S \subset \mathscr{F}_T$, and that T is $\mathscr{F}_T$ measurable. As in the discrete time case we define

$$\theta_T(\omega) = \theta_{T(\omega)}(\omega) \text{ when } T(\omega) < \infty.$$

Theorem A.22 (Strong Markov property in continuous time) *Let Y be as above, T be a stopping time, and $\xi \in b\mathscr{F}_T$ and $\eta \in b\mathscr{F}$. Then*

$$\mathbb{E}^x(\xi(\eta \circ \theta_T); T < \infty) = \mathbb{E}^x(\xi\ \mathbb{E}^{Y_T}(\eta); T < \infty). \tag{A.16}$$

Proof Replacing T by $M \wedge T$ we can assume that $\mathbb{P}^x(T < \infty) = 1$. Let

$$T_n = \min\{p \in \mathbb{D}_n : p \geq T\}.$$

Then T_n is a stopping time, $T_n \geq T$, and $\lim_n T_n = T$. By a monotone class argument, as in the case of the Markov property at a fixed time, it is enough to prove (A.16) when

$$\eta = \prod_{i=1}^{n} f_i(Y_{s_i}),$$

where f_i are bounded.

We first prove (A.16) for T_n. Let $U_k = \mathbf{1}_{(T_n = k2^{-n})}$. Then $U_k \in \mathscr{F}_{k2^{-n}}$ and, since $\xi \in \mathscr{T}_{T_n}$, we have $\xi U_k \in \mathscr{F}_{k2^{-n}}$. So, by the Markov property for Y at the times $k2^{-n}$,

$$\begin{aligned}\mathbb{E}^x(\xi(\eta \circ \theta_{T_n})) &= \sum_k \mathbb{E}^x(\xi U_k(\eta \circ \theta_{k2^{-n}})) \\ &= \sum_k \mathbb{E}^x(\xi U_k \mathbb{E}^{Y_{k2^{-n}}}(\eta)) = \mathbb{E}^x(\xi \mathbb{E}^{Y_{T_n}}(\eta)).\end{aligned}$$

Since Y is right continuous, $Y_{T_n} \to Y_T$, and therefore we have

$$\lim_n \eta \circ \theta_{T_n} = \eta \circ \theta_T, \quad \lim_n \mathbb{E}^{Y_{T_n}}(\eta) = \mathbb{E}^{Y_T}(\eta).$$

Using dominated convergence, we then obtain (A.16). □

A.4 Invariant and Tail σ-fields

Given probability measures $\mathbb{P}$ and $\mathbb{Q}$ on a measure space $(\mathbb{M}, \mathscr{M})$, we define the *total variation norm* by

$$||\mathbb{P} - \mathbb{Q}||_{\mathrm{TV}(\mathscr{M})} = \sup_{A \in \mathscr{M}} (\mathbb{P}(A) - \mathbb{Q}(A)). \tag{A.17}$$

When the σ-field is clear from the context we will sometimes write TV rather than TV$(\mathscr{M})$.

It is straightforward to verify that there exists a set $A \in \mathscr{M}$ which attains the supremum in (A.17). If $\mathbb{M} = \mathbb{V}$ and $\mathscr{M}$ is the σ-field of all subsets of $\mathbb{V}$, and

ν and ν' are measures on $\mathbb{V}$, then we can take $A = \{x : \nu_x > \nu'_x\}$ and deduce that

$$||\nu - \nu'||_{\mathrm{TV}} = \sum_{x \in \mathbb{V}} (\nu_x - \nu'_x)_+ = \tfrac{1}{2} \sum_{x \in \mathbb{V}} |\nu_x - \nu'_x|.$$

Remark A.23 There are different normalisations of the total variation distance. For example [De] defines it by

$$\sup\{\mathbb{E}_{\mathbb{P}}(Z) - \mathbb{E}_{\mathbb{Q}}(Z) : |Z| \le 1, \ Z \text{ is } \mathscr{M} \text{ measurable}\}, \tag{A.18}$$

which gives a value twice that in (A.17).

Lemma A.24 *Let $(\Omega, \mathscr{G}_0)$ be a measure space and $\mathbb{P}$, $\mathbb{Q}$ be probability measures on $(\Omega, \mathscr{G}_0)$. Let $(\mathscr{G}_n, n \in \mathbb{Z}_+)$ be a decreasing filtration, and let $\mathscr{G}_\infty = \cap_n \mathscr{G}_n$. Then*

$$\lim_n ||P - Q||_{\mathrm{TV}(\mathscr{G}_n)} = ||P - Q||_{\mathrm{TV}(\mathscr{G}_\infty)}.$$

Proof Suppose initially that $\mathbb{Q} \ll \mathbb{P}$ on $\mathscr{G}_0$. Let Z_n be the Radon–Nikodym derivative $d\mathbb{Q}/d\mathbb{P}$ on $\mathscr{G}_n$. Since for $G \in \mathscr{G}_n$ we have $\mathbb{Q}(G) = \mathbb{E}_{\mathbb{P}}(\mathbf{1}_G Z_n)$, it is easy to check that a set A which attains the supremum in (A.17) is given by $A = \{Z_n < 1\}$. So

$$||P - Q||_{\mathrm{TV}(\mathscr{G}_n)} = \mathbb{P}(Z_n < 1) - \mathbb{E}_{\mathbb{P}} Z_n \mathbf{1}_{(Z_n < 1)} = \mathbb{E}_{\mathbb{P}}(1 - Z_n)_+.$$

Now Z_n is a non-negative (backwards) martingale with respect to $\mathbb{P}$, and so by Theorem A.4 converges a.s. and in $L^1(\mathbb{P})$ to a r.v. Z_∞. Further, $Z_\infty = d\mathbb{Q}/d\mathbb{P}$ on $\mathscr{G}_\infty$. Dominated convergence then gives

$$\lim_n \mathbb{E}_{\mathbb{P}}(1 - Z_n)_+ = \mathbb{E}_{\mathbb{P}}(1 - Z_\infty)_+ = ||P - Q||_{\mathrm{TV}(\mathscr{G}_\infty)},$$

proving the lemma in this case.

For the general case set $\mathbb{P}_\varepsilon = (1 - \varepsilon)\mathbb{P} + \varepsilon\mathbb{Q}$. Then $\mathbb{Q} \ll \mathbb{P}_\varepsilon$, and it is easy to check that for $0 \le n \le \infty$

$$||P_\varepsilon - Q||_{\mathrm{TV}(\mathscr{G}_n)} = (1 - \varepsilon)||P - Q||_{\mathrm{TV}(\mathscr{G}_n)}.$$

Hence

$$\begin{aligned} \lim_n ||P - Q||_{\mathrm{TV}(\mathscr{G}_n)} &= (1 - \varepsilon)^{-1} \lim_n ||P_\varepsilon - Q||_{\mathrm{TV}(\mathscr{G}_n)} \\ &= (1 - \varepsilon)^{-1} ||P_\varepsilon - Q||_{\mathrm{TV}(\mathscr{G}_\infty)} = ||P - Q||_{\mathrm{TV}(\mathscr{G}_\infty)}. \end{aligned}$$

□

Now let X be a random walk on a graph $\Gamma = (\mathbb{V}, E)$, defined on the canonical space as in Section A.2. We recall the definition of the shifts θ_n, write $\theta = \theta_1$, and note that $\theta_n = \theta^n$. Let $\mathscr{F}_n = \sigma(X_0, \dots, X_n)$, $\mathscr{G}_n = \sigma(X_{n+k}, k \geq 0)$, and recall that the *tail σ-field* is defined by

$$\mathscr{T} = \bigcap_{n=0}^{\infty} \mathscr{G}_n.$$

Recall from Section 1.9 that $F \in \mathscr{I}$, the invariant σ-field, if and only if $\theta^{-1}(F) = F$.

Given a probability measure ν on $\mathbb{V}$ we write νP^n for the law of X_n under $\mathbb{P}^\nu$, so that

$$\nu P^n(y) = \sum_x \nu_x \mathscr{P}_n(x, y).$$

We write $\mathscr{M}_>$ for the set of probability measures ν on $\mathbb{V}$ with $\nu_x > 0$ for all x. If $\nu \in \mathscr{M}_>$ then we have $\nu P \ll \nu$.

Lemma A.25 *Let $\nu \in \mathscr{M}_>$.*

(a) $F \in \mathscr{G}_n$ if and only if there exists $F_n \in \mathscr{G}_0$ such that $F = \theta^{-n}(F_n)$.
(b) If $F \in \mathscr{T}$ and $\mathbb{P}^\nu(F) = 0$ then $\mathbb{P}^\nu(\theta^{-n}(F)) = 0$ for all $n \geq 0$.
(c) If $F \in \mathscr{T}$ and $\mathbb{P}^\nu(F \,\triangle\, \theta^{-1}(F)) = 0$ then

$$\tilde{F} = \limsup \theta^{-n}(F) = \bigcup_{n=0}^{\infty} \bigcap_{m=n}^{\infty} \theta^{-m}(F) \in \mathscr{I},$$

and $\mathbb{P}^\nu(F \,\triangle\, \tilde{F}) = 0$.

Proof (a) If $F = \{X_{n+k} \in A_k, 1 \leq k \leq m\}$ then $F = \theta^{-n}(F_n)$, where $F_n = \{X_k \in A_k, 1 \leq k \leq m\}$. The general case then follows by a monotone class argument.
(b) Let $F \in \mathscr{T}$ with $\mathbb{P}^\nu(F) = 0$. Since $\nu \in \mathscr{M}_>$ we have $\mathbb{P}^x(F) = 0$ for all x. So by the Markov property

$$\mathbb{E}^\nu(1_F \circ \theta_n) = \mathbb{E}^{X_n}(1_F) = 0,$$

and thus, using (1.39), $\mathbb{P}^\nu(\theta^{-n}(F)) = 0$.
(c) If $\omega \in \tilde{F}$ then there exists n such that $\omega \in \theta^{-m}(F)$ for all $m \geq n$. So $\theta(\omega) \in \theta^{-(m-1)}(F)$ for all $m - 1 \geq n - 1$, and thus $\theta(\omega) \in \tilde{F}$. Similarly if $\theta(\omega) \in \tilde{F}$ then $\omega \in \tilde{F}$. So $\theta^{-1}(\tilde{F}) = \tilde{F}$, and $\tilde{F} \in \mathscr{I}$. We have, using (b),

$$\begin{aligned}\mathbb{P}^\nu(F \,\triangle\, \theta^{-2}(F)) &= \mathbb{E}^\nu|1_F - 1_{\theta^{-2}(F)}| \\ &\leq \mathbb{E}^\nu|1_F - 1_{\theta^{-1}(F)}| + \mathbb{E}^\nu|1_{\theta^{-1}(F)} - 1_{\theta^{-2}(F)}| = 0,\end{aligned}$$

and similarly $\mathbb{P}^\nu(\theta^{-i}(F) \triangle \theta^{-j}(F)) = 0$ if $i \neq j$. Now let

$$G_{n,m} = \bigcap_{k=n}^{m} \theta^{-k}(F).$$

Then

$$\mathbb{E}^\nu |1_F - 1_{G_{n,m}}| = \mathbb{E}^\nu |1_F - \prod_{k=n}^{m} 1_{\theta^{-k}(F)}| = 0.$$

Taking the limits as $m \to \infty$ and then $n \to \infty$ gives that $\mathbb{P}^\nu(F \triangle \tilde{F}) = 0$. □

Lemma A.26 (See [Kai], Lem. 2.1) *Let ν, ν' be probability measures on $\mathbb{V}$. Then*

$$||\mathbb{P}^\nu - \mathbb{P}^{\nu'}||_{\mathrm{TV}(\mathscr{G}_n)} = ||\nu P^n - \nu' P^n||_{\mathrm{TV}}. \tag{A.19}$$

Further,

$$\lim_n ||\nu P^n - \nu' P^n||_{\mathrm{TV}} = ||\mathbb{P}^\nu - \mathbb{P}^{\nu'}||_{\mathrm{TV}(\mathscr{T})}.$$

Proof Let $n \geq 0$. Then, since $\sigma(X_n) \subset \mathscr{G}_n$, we have

$$||\mathbb{P}^\nu - \mathbb{P}^{\nu'}||_{\mathrm{TV}(\mathscr{G}_n)} \geq ||\mathbb{P}^\nu - \mathbb{P}^{\nu'}||_{\mathrm{TV}(\sigma(X_n))} = ||\nu P^n - \nu' P^n||_{\mathrm{TV}}.$$

Let $G \in \mathscr{G}_n$. Then $G = \theta^{-n}(G_n)$ for some $\mathscr{G}_n \in \mathscr{G}_0$, and

$$\begin{aligned}
\mathbb{P}^\nu(G) - \mathbb{P}^{\nu'}(G) &= \sum_{y \in \mathbb{V}} (\nu P^n_y - \nu' P^n_y) \mathbb{P}^y(\theta^{-n}(G_n)) \\
&\leq \sum_{y \in \mathbb{V}} (\nu P^n_y - \nu' P^n_y)_+ = ||\nu P^n - \nu' P^n||_{\mathrm{TV}},
\end{aligned}$$

which proves (A.19). Using Lemma A.24 completes the proof. □

Definition A.27 Let ν be a measure on $\mathbb{V}$. We say that $\mathscr{I} = \mathscr{T}$ $\mathbb{P}^\nu$-a.s. if for any $F \in \mathscr{T}$ there exists $F' \in \mathscr{I}$ such that $\mathbb{E}^\nu|\mathbf{1}_F - \mathbf{1}_{F'}| = 0$.

Write δ^x for the probability measure on $\mathbb{V}$ which assigns mass 1 to $\{x\}$. Set

$$\alpha(x) = \sup_n ||\delta^x P^n - \delta^x P^{n+1}||_{\mathrm{TV}}, \qquad \alpha = \sup_{x \in \mathbb{V}} \alpha(x).$$

The following result is often called the '02 law'; see [De]. We have '1' rather than '2' since we defined the total variation distance by (A.17) rather than (A.18).

Theorem A.28 *We have*

$$\alpha = \sup_\nu \lim_n ||\nu P^n - \nu P^{n+1}||_{\mathrm{TV}}, \text{ and is either 0 or 1 .}$$

$\alpha = 0$ *if and only if* $\mathscr{I} = \mathscr{T}$ $\mathbb{P}^\nu$*-a.s. for all probability measures* ν *on* $\mathbb{V}$.

Proof Suppose first that $\mathscr{I} = \mathscr{T}$ $\mathbb{P}^\nu$-a.s. for all ν. Let ν be a probability measure on $\mathbb{V}$, and suppose initially that $\nu P \ll \nu$. Then by Lemma A.26

$$\lim_n ||\nu P^n - \nu P^{n+1}||_{\mathrm{TV}} = ||\mathbb{P}^\nu - \mathbb{P}^{\nu P}||_{\mathrm{TV}(\mathscr{T})} = \sup_{F\in\mathscr{T}} \Big(\mathbb{E}^\nu(\mathbf{1}_F) - \mathbb{E}^{\nu P}(\mathbf{1}_F)\Big). \tag{A.20}$$

Let $F \in \mathscr{T}$. Then by hypothesis there exists $F' \in \mathscr{I}$ such that $\mathbb{P}^\nu(F \triangle F') = 0$. Let $Z = \mathbf{1}_F$, $Z' = \mathbf{1}_{F'}$. We have $\mathbb{P}^\nu(Z = Z') = 1$, and so since $\nu P \ll \nu$ it follows that $\mathbb{P}^{\nu P}(Z = Z') = 1$. Then by the Markov property

$$\begin{aligned}\mathbb{E}^{\nu P}(Z) = \mathbb{E}^{\nu P}(Z') &= \sum_x \nu_x \mathscr{P}(x, y)\mathbb{E}^y(Z')\\ &= \mathbb{E}^\nu \mathbb{E}^{X_1} Z' = \mathbb{E}^\nu(Z' \circ \theta_1) = \mathbb{E}^\nu(Z') = \mathbb{E}^\nu(Z).\end{aligned}$$

Thus the left side of (A.20) is zero. For the general case we can replace ν by $\sum_n a_n \nu P^n$, where $\sum a_n = 1$ and $a_n > 0$ for all n.

Now suppose that there exists ν such that $\mathscr{I} \neq \mathscr{T}$ $\mathbb{P}^\nu$-a.s. Replacing ν by $\sum_n a_n \nu P^n$ if necessary, we can assume that $\nu \in \mathscr{M}_>$ and so $\nu P \ll \nu$. Let $F \in \mathscr{T}$ be such that $\mathbb{P}^\nu(F \triangle F') > 0$ for all $F' \in \mathscr{I}$. Then by Lemma A.25(c) we must have $\mathbb{P}^\nu(F \triangle \theta^{-1}(F)) > 0$. Let $G = F - \theta^{-1}(F)$, so that $0 < \mathbb{P}(G) < 1$ and $G \cap \theta^{-1}(G) = \varnothing$.

Let $n \geq 0$. Since $G \in \mathscr{G}_n$ there exists $G_n \in \mathscr{G}_0$ such that $\mathbf{1}_G = \mathbf{1}_{G_n} \circ \theta_n$. Set

$$h_n(x) = \mathbb{E}^x(\mathbf{1}_{G_n}).$$

Let $M_n = \mathbb{E}^\nu(\mathbf{1}_G|\mathscr{F}_n)$. Then M is a bounded martingale and so $M_n \to \mathbf{1}_G$ a.s. and in L^1. By the Markov property

$$M_n = \mathbb{E}^\nu(\mathbf{1}_G|\mathscr{F}_n) = \mathbb{E}^\nu(\mathbf{1}_{G_n} \circ \theta_n|\mathscr{F}_n) = \mathbb{E}^{X_n}(\mathbf{1}_{G_n}) = h_n(X_n).$$

Similarly,

$$\mathbb{E}^\nu(\mathbf{1}_{\theta^{-1}(G)}|\mathscr{F}_n) = \mathbb{E}^\nu(\mathbf{1}_{G_{n-1}} \circ \theta_n|\mathscr{F}_n) = \mathbb{E}^{X_n}(\mathbf{1}_{G_{n-1}}) = h_{n-1}(X_n).$$

So we have

$$(h_n(X_n), h_{n-1}(X_n)) \to (\mathbf{1}_G, \mathbf{1}_{\theta^{-1}(G)}) \text{ a.s. and in } L^1(\mathbb{P}^\nu).$$

Hence, given $\varepsilon > 0$ there exists $n \geq 1$ and $x \in \mathbb{V}$ such that $h_n(x) > 1 - \varepsilon$ and $h_{n-1}(x) < \varepsilon$. So for $k \geq 0$

$$1 - 2\varepsilon < h_n(x) - h_{n-1}(x) = \mathbb{E}^x h_{n+k}(X_k) - \mathbb{E}^x h_{n+k}(X_{k+1})$$

$$= \mathbb{P}^{\delta^x P^k}(G_{n+k}) - \mathbb{P}^{\delta^x P^{k+1}}(G_{n+k}) \le ||\mathbb{P}^{\delta^x P^k} - \mathbb{P}^{\delta^x P^{k+1}}||_{\mathrm{TV}(\mathscr{G}_k)}$$
$$= ||\delta^x P^k - \delta^x P^{k+1}||_{\mathrm{TV}}.$$

This implies that $\alpha > 1 - 2\varepsilon$. □

Corollary A.29 *Let (Γ, μ) be a weighted graph. Then $\mathscr{T} = \mathscr{I}$ for the lazy random walk on (Γ, μ).*

Proof Let $x \in \mathbb{V}$. Since $\mathscr{P}(x, x) \ge \frac{1}{2}$, we have $\delta \ge \frac{1}{2}\delta^x\mathbb{P}$, and therefore $\mathbb{P}^x \ge \frac{1}{2}\mathbb{P}^{\delta^x P}$. So $\alpha(x) \le \frac{1}{2}$, and thus $\alpha \le \frac{1}{2}$. By the 02 law therefore $\alpha = 0$ and $\mathscr{T} = \mathscr{I}$. □

Corollary A.30 *Let (Γ, μ) be recurrent, and not bipartite. Then for any ν both $\mathscr{T}$ and $\mathscr{I}$ are trivial.*

Proof $\mathscr{I}$ is trivial by Theorem 1.48.

Since (Γ, μ) is not bipartite there exists $k \ge 1$ and a cycle C of length $2k+1$. Let z be a point in C and $w \sim z$. Let $x \in \mathbb{V}$, X_n be the SRW started at x, and X'_n be the SRW started with distribution $\delta^x P$. We can couple X and X' so that $X'_k = X_{k+1}$ for $k + 1 \le T_z(X)$. As (Γ, μ) is recurrent, X will hit x with probability 1. Let $S = T_z(X) + 2k$. After the hit on x, we run the processes X and X' independently. With positive probability, X will move back and forth k times between z and w, while X' will go round the cycle C. So there exists $p > 0$ such that $\mathbb{P}(X_S = X'_S) \ge 2p$. On the event $\{X_S = X'_S\}$ we then run X and X' together. We can find n large enough so that $\mathbb{P}(S > n) \le p$, and so we have

$$\mathbb{P}(X_n = X'_n) \ge p.$$

This implies that $\alpha(x) \le 1 - p$ for all x, and so by the 02 law we have that $\mathscr{T}$ is trivial. □

A.5 Hilbert Space Results

We begin with some basic facts about operators in Hilbert spaces.

Lemma A.31 *Let T be a self-adjoint operator in a Hilbert space $\mathscr{H}$. Then for all $n \ge 0$*

$$\|T^n\| = \|T\|^n. \tag{A.21}$$

Proof Since $||AB|| \le ||A|| \cdot ||B||$ for any linear operators on a Banach space, we have $\|T^n\| \le \|T\|^n$. Let $\mathscr{B} = \{f \in \mathscr{H} : ||f|| = 1\}$. If $n = 2m$ is even then

$$||T^{2m}|| = \sup\{\langle T^{2m} f, g\rangle : f, g \in \mathscr{B}\} \geq \sup\{\langle T^m f, T^m f\rangle : f \in \mathscr{B}\} = ||T^m||^2.$$

Hence (A.21) holds if $n = 2^k$ for any $k \geq 0$. If for some n we have $\|T^n\| < \|T\|^n$, then, taking k so that $n < 2^k$,

$$||T^{2^k}|| \leq ||T^n||||T^{2^k-n}|| < ||T||^n||T^{2^k-n}|| \leq ||T||^{2^k},$$

a contradiction. □

Lemma A.32 *For the operator P defined by* (1.9),

$$\|P\|_{2\to 2} = \sup\{\langle Pf, f\rangle : \|f\|_2 = 1\}. \tag{A.22}$$

Proof Let $B = \{f : \|f\|_2 = 1\}$, $B^+ = B \cap C_+(\mathbb{V})$, and write $\rho = \|P\|_{2\to 2}$. Note that $\langle Pf, g\rangle \leq \langle P|f|, |g|\rangle$, so that using (1.10) we have $\rho = \sup\{\langle Pf, g\rangle : f, g \in B^+\}$. Let $\varepsilon > 0$ and $f, g \in B^+$ with $\langle Pf, g\rangle \geq \rho - \varepsilon^2$. Then (as P is self-adjoint)

$$\begin{aligned}||Pf - \rho g||_2^2 &= \langle Pf - \rho g, Pf - \rho g\rangle = \|Pf\|_2^2 + \rho^2\|g\|_2^2 - 2\rho\langle Pf, g\rangle \\ &\leq 2\rho^2 - 2\rho(\rho - \varepsilon^2) = 2\rho\varepsilon^2 \leq 2\varepsilon^2.\end{aligned}$$

Similarly $\|Pg - \rho f\|_2^2 \leq 2\varepsilon^2$. So $Pf = \rho g + h_1$ and $Pg = \rho f + h_2$ with $\|h_j\|_2 \leq 2^{1/2}\varepsilon$ for $j = 1, 2$. Let $u = (f + g)$; as $f, g \geq 0$ we have $1 \leq \|u\|_2 \leq 2$. So $Pu = \rho u + h$, where $\|h\|_2 \leq 2^{3/2}\varepsilon$. Therefore

$$\langle Pu, u\rangle = \rho\|u\|_2^2 + \langle h, u\rangle \geq \rho\|u\|_2^2 - 2(2^{3/2}\varepsilon)\|u\|_2.$$

So if $v = u/\|u\|_2$ we have $\langle Pv, v\rangle \geq \rho - 2^{5/2}\varepsilon$, proving (A.22). □

The following is a discrete version of the representation theorem of Beurling, Deny, and LeYan – see [FOT], Th. 3.2.1. Let $\mathbb{V}$ be a countable set and μ be a measure on $\mathbb{V}$. Let $\mathscr{A}$ be a symmetric (real) bilinear form on $L^2(\mathbb{V})$ with the following properties:

(D1) $\mathscr{A}(f, f) \leq C_0||f||_2^2$, $f \in L^2$,
(D2) $\mathscr{A}(f, f) \geq 0$ for all $f \in L^2$,
(D3) $\mathscr{A}(1 \wedge f_+, 1 \wedge f_+) \leq \mathscr{A}(f_+, f_+) \leq A(f, f)$, $f \in L^2$.

Property (D3) is called the *Markov property* of the quadratic form $\mathscr{A}$.

Theorem A.33 *There exist J_{xy}, k_x with $J_{xx} = 0$, $J_{xy} = J_{yx} \geq 0$, $x \neq y$, $k_x \geq 0$ such that for $f, g \in L^2(\mathbb{V}, \mu)$*

$$\mathscr{A}(f, g) = \sum_x \sum_y (f(x) - f(y))(g(x) - g(y))J_{xy} + \sum_x f(x)g(x)k_x. \tag{A.23}$$

Proof Set $A_{xy} = \mathscr{A}(\mathbf{1}_x, \mathbf{1}_y)$, so that for $f, g \in C_0(\mathbb{V})$

$$\mathscr{A}(f, g) = \sum_x \sum_y f(x) A_{xy} g(y).$$

Let $x, y \in \mathbb{V}$ with $x \neq y$, and let $f = \mathbf{1}_x - \lambda \mathbf{1}_y$, where $\lambda > 0$. Then $f_+ = \mathbf{1}_x$, so by (D3)

$$A_{xx} = \mathscr{A}(\mathbf{1}_x, \mathbf{1}_x) \leq \mathscr{A}(f, f) = A_{xx} - 2\lambda A_{xy} + \lambda^2 A_{yy}.$$

Hence $A_{xy} \leq 0$, so define

$$J_{xy} = -A_{xy}, \quad x \neq y,$$
$$J_{xx} = 0.$$

Note that $J_{xy} = J_{yx}$ by the symmetry of $\mathscr{A}$. Now let $B \subset \mathbb{V}$ be finite and $g \geq 0$ with $\text{supp}(g) \subset B$. If $f = \mathbf{1}_B + \lambda g$, where $\lambda > 0$, then $f \wedge 1 = \mathbf{1}_B$. Using (D3) again we have

$$\mathscr{A}(\mathbf{1}_B, \mathbf{1}_B) \leq \mathscr{A}(\mathbf{1}_B, \mathbf{1}_B) + 2\lambda \mathscr{A}(g, \mathbf{1}_B) + \lambda^2 \mathscr{A}(g, g),$$

and therefore

$$0 \leq \mathscr{A}(\mathbf{1}_B, g) = \sum_{x' \in B} g(x') \sum_{y \in B} A_{x'y}.$$

Taking $g = \mathbf{1}_x$ with $x \in B$ it follows that $\sum_{y \in B} A_{xy} \geq 0$ for any finite B, and therefore that

$$\sum_{y \neq x} J_{xy} \leq A_{xx}.$$

Now set

$$k_x = \sum_y A_{xy} = A_{xx} - \sum_{y \neq x} J_{xy} \geq 0.$$

Then if $f, g \in C_0(\mathbb{V})$

$$\begin{aligned}
\mathscr{A}(f, g) &= \sum_x f(x) \Big(A_{xx} g(x) + \sum_{y \neq x} A_{xy} g(y) \Big) \\
&= \sum_x f(x) g(x) k_x + \sum_x f(x) \sum_{y \neq x} J_{xy}(g(x) - g(y)) \\
&= \sum_x f(x) g(x) k_x + \tfrac{1}{2} \sum_{x,y} f(x) J_{xy}(g(x) - g(y)) \\
&\quad + \tfrac{1}{2} \sum_{y,x} f(y) J_{yx}(g(y) - g(x)) \\
&= \sum_x f(x) g(x) k_x + \tfrac{1}{2} \sum_{x,y} J_{xy}(g(x) - g(y))(f(x) - f(y)).
\end{aligned}$$

This gives (A.23) for $f, g \in C_0(\mathbb{V})$.

We now extend this to $L^2(\mathbb{V}, \mu)$. For $B \subset \mathbb{V}$ write

$$S(f, f; B) = \sum_{x \in B} f(x)g(x)k_x + \tfrac{1}{2} \sum_{x \in B, y \in B} J_{xy}(f(x) - f(y))^2.$$

Now let $B_n \uparrow\uparrow \mathbb{V}$. Let $f \in L^2$ and $f_n = f \mathbf{1}_{B_n}$, so that $f_n \to f$ in $L^2(\mathbb{V}, \mu)$. Hence, by (D1) and Fatou,

$$\mathscr{A}(f, f) = \lim_n \mathscr{A}(f_n, f_n) = \lim_n S(f_n, f_n; \mathbb{V}) \leq S(f, f; \mathbb{V}).$$

On the other hand, if $m \geq n$ then $S(f_m, f_m; B_n) \leq \mathscr{A}(f_m, f_m)$, and so

$$S(f, f; B_n) = \lim_{m \to \infty} S(f_m, f_m; B_n) \leq \lim_{m \to \infty} \mathscr{A}(f_m, f_m) = \mathscr{A}(f, f).$$

Taking the limit as $n \to \infty$ we deduce that (A.23) holds when $f = g$. Finally by polarization (A.23) extends to $f, g \in L^2$. □

Since (A.23) gives $\mathscr{A}(1_{B(x,2)}, \mathbf{1}_{\{x\}}) = k_x$ we have:

Corollary A.34 *Let $\mathscr{A}$ satisfy (D1)–(D3). If $\mathscr{A}(1_{B(x,m)}, \mathbf{1}_{\{x\}}) = 0$ for some $m \geq 2$ then $\mathscr{A}$ satisfies the representation (A.23) with $k_x = 0$.*

A.6 Miscellaneous Estimates

Lemma A.35 *Let*

$$I(\gamma, x) = \int_1^\infty e^{-xt^\gamma} dt.$$

Then

$$I(\gamma, x) \asymp x^{-1/\gamma} \quad \text{for } 0 < x \leq 1, \tag{A.24}$$

$$I(\gamma, x) \asymp x^{-1} e^{-x} \quad \text{for } x \geq 1. \tag{A.25}$$

Proof Let $x \in (0, 1]$. Then

$$x^{1/\gamma} I(\gamma, x) = \int_{x^{1/\gamma}}^\infty e^{-s^\gamma} ds,$$

and (A.24) follows. Now let $x \geq 1$. Then, making the substitution $v = x(t^\gamma - 1)$,

$$I(\gamma, x) = \gamma^{-1} x^{-1} e^{-x} \int_0^\infty e^{-v}(1 + v/x)^{-1+1/\gamma} dv,$$

giving (A.25). □

Lemma A.36 *Let $\lambda > 1$, $\gamma > 0$, $x > 0$, and*

$$S(\lambda, \gamma, x) = \sum_{n=0}^{\infty} \lambda^n \exp(-x\lambda^{n\gamma}).$$

Then there exist $c_i = c_i(\lambda, \alpha)$ such that

$$\begin{aligned} c_1 x^{-1/\gamma} &\le S(\lambda, \gamma, x) \le c_2 x^{-1/\gamma}, \qquad && x \in (0, 1], \\ c_1 x^{-1} e^{-x} &\le S(\lambda, \gamma, x) \le c_2 x^{-1} e^{-x/\lambda^\gamma}, && x \ge 1. \end{aligned}$$

Proof We have

$$I(\gamma, x) = \sum_{n=0}^{\infty} \int_{\lambda^n}^{\lambda^{n+1}} e^{-xt^\gamma} dt,$$

and estimating each term in the sum gives

$$(\lambda - 1)^{-1} I(\gamma, x) \le S(\lambda, \gamma, x) \le (\lambda - 1)^{-1} I(\gamma, x\lambda^{-\gamma}).$$

The bounds on S then follow easily from Lemma A.35. □

A.7 Whitney Type Coverings of a Ball

In this section, we construct a covering of a ball with some quite specific properties, which will be used to prove a weighted Poincaré inequality. The construction is similar to that in [SC2], but has some significant differences close to the boundary of the ball.

We assume that (Γ, μ) is a weighted graph which satisfies volume doubling – see Definition 6.30. If C_V is the doubling constant, set

$$\theta = \frac{\log C_V}{\log 2}. \tag{A.26}$$

Lemma A.37 *Suppose that (Γ, μ) satisfies volume doubling. Then if $x, y \in \mathbb{V}$, $1 \le r \le R$*

$$\frac{V(y, R)}{V(x, r)} \le C_V^2 \Big(\frac{R + d(x, y)}{r} \Big)^\theta.$$

Proof This is a straightforward and informative exercise in the use of the volume doubling condition. □

Lemma A.38 *Suppose that (Γ, μ) satisfies volume doubling. Let*

$$\mathcal{B} = \{B(x_i, r_i), i \in I\}$$

be a family of disjoint balls in (Γ, μ), *with* $R_1 \le r_i \le R_2$ *for all* i. *Let* $\lambda \ge 1$. *Then*

$$|\{i : x \in B(x_i, \lambda r_i)\}| \le C_V^2 (1 + 2\lambda)^\theta \Big(\frac{R_2}{R_1}\Big)^\theta .$$

Proof Let $I' = \{i : x \in B(x_i, \lambda r_i)\}$ and $N = |I'|$. If $i \in I'$ then $B(x_i, r_i) \subset B(x, (1+\lambda) r_i)$. So

$$\mu(B(x, (1+\lambda)R_2)) \ge \mu\Big(\bigcup_{i\in I} B(x_i, r_i)\Big) = \sum_{i \in I} \mu(B(x_i, r_i)) \ge N \min_{i\in I} V(x_i, r_i).$$

Using Lemma A.37 to bound $V(x, (1+\lambda)R_2)/V(x_i, r_i)$ we obtain

$$N \le C_V^2 \max_i \Big(\frac{\lambda r_i + (1+\lambda)R_2}{r_i}\Big)^\theta ,$$

from which the result follows easily. □

For the rest of this section we fix a ball $B = B(o, R)$. Set

$$\rho(x) = R - d(o, x), \tag{A.27}$$

so that $d(x, \partial_i B) \ge \rho(x)$. We now fix $\lambda_1 \ge 10$ and a constant $R_0 \ge 10$, and assume that $\lambda_1 R_0 \le R$. We also assume that both R and $\lambda_1 R_0$ are integers.

We now choose balls $B_i = B(x_i, r_i)$, $i = 1, \dots, N$ as follows. We take $x_1 = o$ and $r_1 = R/\lambda_1$. Given B_i for $i = 1, \dots, k-1$ set $A_k = B(o, R - \lambda_1 R_0) - \cup_{i=1}^{k-1} B_i$. If A_k is empty then we let $N = k - 1$ and stop; otherwise we choose $x_k \in A_k$ to minimise $d(o, x_k)$, and set

$$r_k = \frac{\rho(x_k)}{\lambda_1} = \frac{R - d(o, x_k)}{\lambda_1}.$$

We also define

$$B_i' = B(x_i, \tfrac{1}{2} r_i).$$

Lemma A.39 *(a) The balls* B_i *cover* $B(o, R - \lambda_1 R_0)$.
(b) We have $r_1 \ge r_2 \ge r_3 \ge \cdots \ge r_N \ge R_0$.
(c) The balls B_i' *are disjoint.*
(d) If $y \in B(x_i, \theta r_i)$ *then*

$$(\lambda_1 - \theta) r_i \le \rho(y) \le (\lambda_1 + \theta) r_i . \tag{A.28}$$

Proof (a) is immediate from the construction.
(b) That r_i are decreasing is clear from the construction. Since $x_N \in B(0, R - \lambda_1 R_0)$ we have $\rho(X_N) \ge \lambda_1 R_0$ and hence $r_N \ge R_0$.

(c) Suppose that $y \in B'_i \cap B'_j$ with $i < j$. Then $d(x_i, x_j) \le d(x_i, y) + d(y, x_j) \le \frac{1}{2}(r_i + r_j) \le r_i$, a contradiction since then $x_j \in B_i$.
(d) If $y \in B_i(x_i, \theta r_i)$ then $|\rho(y) - \rho(x_i)| \le \theta r_i$, and (A.28) follows. □

We call a ball B_i a *boundary ball* if

$$r_i \le \frac{\lambda_1 R_0}{\lambda_1 - 1}.$$

It is clear from the construction that there exists M such that B_i is a boundary ball if and only if $M + 1 \le i \le N$.

For $x, y \in \mathbb{V}$ we write $\gamma(x, y)$ for a shortest path connecting x to y. (If this path is not unique, then we make some choice among the set of shortest paths.) For each $i \ge M + 1$ we define

$$\widetilde{B}_i = B_i \cup \{z \in B : \text{ there exists } y \in \gamma(o, z) \cap B_i \text{ such that } \rho(y) = \lambda_1 R_0\}.$$

Lemma A.40 *(a) If B_i is not a boundary ball then $\rho(x) > \lambda_1 R_0$ for all $x \in B_i$.*
(b) If $z \in B \cap \{x : \rho(x) \le \lambda_1 R_0\}$ then there exists a boundary ball B_i such that $z \in \widetilde{B}_i$.
(c) If B_i is a boundary ball and $z \in \widetilde{B}_i$ then there exists a path inside $\widetilde{B}_i$ of length at most $(\lambda_1 + 2)R_0$ connecting x_i and z. In particular, $\widetilde{B}_i \subset B(x_i, (\lambda_1 + 2)R_0)$.

Proof (a) This is immediate from (A.28).
(b) Let $z \in B$. Let y be the unique point on the path $\gamma(z, o)$ such that $\rho(y) = \lambda_1 R_0$. By Lemma A.39(a) there exists i such that $y \in B_i$. Then $d(y, x_i) \le r_i$, so

$$\lambda_1 r_i = \rho(x_i) \le \rho(y_i) + r_i \le \lambda_1 R_0 + r_i,$$

and thus B_i is a boundary ball and $z \in \widetilde{B}_i$.
(c) Let $z \in \widetilde{B}_i - B_i$ and let y be as above. Since B_i is a boundary ball,

$$d(z, x_i) \le d(z, y) + d(y, x_i) \le \lambda_1 R_0 + r_i \le \frac{\lambda_1^2}{\lambda_1 - 1} R_0 \le (\lambda_1 + 2)R_0.$$

Further, the whole path $\gamma(z, y)$ is contained in $\widetilde{B}_i$, and so z is connected to x_i by a path in $\widetilde{B}_i$ of length at most $(2 + \lambda_1)R_0$. □

We now define an ancestor function $\mathfrak{a} : \{2, \ldots, N\} \to \{1, \ldots, N\}$ which will give a tree structure on the set of indices $\{1, \ldots, N\}$. Let $i \ge 2$. We need to consider two cases.

Case 1: If $d(o, x_i) \geq 2r_i$ then let z_i be the first point on the path $\gamma(x_i, o)$ with $d(z_i, x_i) \geq 2r_i$. So we have $2r_i \leq d(z_i, x_i) < 2r_i + 1$. Since $\rho(z_i) > \rho(x_i) \geq \lambda_1 R_0$, there exists j such that $z_i \in B_j$. (If there is more than one such j we choose the smallest.) We have $|\rho(z_i) - \rho(x_j)| \leq r_j$, and therefore

$$2r_i + \lambda_1 r_i \leq d(z_i, x_i) + \rho(x_i) = \rho(z_i) \leq 2r_i + \lambda_1 r_i + 1,$$
$$2r_i + \lambda_1 r_i - r_j \leq \lambda_1 r_j = \rho(x_j) \leq 2r_i + \lambda_1 r_i + 1 + r_j.$$

Thus

$$\frac{2+\lambda_1}{\lambda_1+1} r_i \leq r_j \leq \frac{2+\lambda_1}{\lambda_1-1} r_i + \frac{1}{(\lambda_1-1)}.$$

In particular, we have $r_j > r_i$, so $j < i$. We set $\mathfrak{a}(i) = j$.
Case 2: If $d(o, x_i) \leq 2r_i$ then we set $\mathfrak{a}(i) = 1$. Since $2r_i \geq d(x_i, o) > r_1$, we obtain

$$\lambda_1 r_1 - 2r_i = R - 2r_i \leq \lambda_1 r_i \leq \lambda_1 r_1 - r_1,$$

and hence

$$\frac{\lambda_1}{\lambda_1 - 1} r_i \leq r_1 \leq \frac{\lambda_1 + 2}{\lambda_1} r_i.$$

Thus in both cases there exist δ_1 and δ_2 depending only on λ_1 so that if $j = \mathfrak{a}(i)$ then

$$(1 + \delta_1) r_i \leq r_j \leq (1 + \delta_2) r_i; \tag{A.29}$$

further, we can take $\delta_1 = (1 + \lambda_1)^{-1}$ and $\delta_2 = 4/9$.

We write $\mathfrak{a}_k$ for the kth iterate of $\mathfrak{a}$. Using this ancestor function, starting in B_i we obtain a sequence $B_i, B_{\mathfrak{a}(i)}, B_{\mathfrak{a}_2(i)}, \ldots, B_{\mathfrak{a}_m(i)}$ of successively larger balls, which ends with the ball B_1.

Lemma A.41 *(a) If $j = \mathfrak{a}(i)$ then $d(x_i, x_j) \leq 1 + 2r_i + r_j$, and*

$$\mu(B(x_i, 2r_i) \cap B(x_j, 2r_j)) \geq \frac{\mu(B_i') \vee \mu(B_j')}{C_V^3}.$$

(b) If $j = \mathfrak{a}_m(i)$ for some $m \geq 1$ then

$$d(x_i, x_j) \leq \frac{4r_j}{\delta_1}.$$

Consequently $B_i \subset B(x_j, K_1 r_j)$, where

$$K_1 = 5 + 4\lambda_1.$$

Proof (a) The first assertion is immediate from the triangle inequality and (A.29). In Case 1 above let w be the first point on $\gamma(x_i, o)$ with $d(x_i, w) \geq 3r_i/2$, and in Case 2 let $w = o$. Then

$$B(w, r_i/3) \subset B(x_i, 2r_i) \cap B(x_j, 2r_j),$$

while $B_i' \cup B_j' \subset B(w, 8r_i/3)$. Hence

$$\begin{aligned}\mu\big(B(x_i, 2r_i) \cap B(x_j, 2r_j)\big) &\geq \mu(B(w, r_i/3)) \\ &\geq C_V^{-3}\mu(B(w, 8r_i/3)) \geq C_V^{-3}\mu(B_i') \vee \mu(B_j').\end{aligned}$$

(b) Set $s_k = r_{\mathfrak{a}_k(i)}$ for $k = 0, \dots, m$. Then since $s_k \leq s_{k+1}(1+\delta_1)^{-1}$, using (a),

$$\begin{aligned}d(x_i, x_j) &\leq \sum_{k=0}^{m-1} d(x_{\mathfrak{a}_k(i)}, x_{\mathfrak{a}_{k+1}(i)}) \\ &\leq \sum_{k=0}^{m-1}(2s_k + s_{k+1} + 1) \leq \tfrac{7}{2}\sum_{k=0}^{m} s_k \leq \tfrac{7}{2}s_m\sum_{k=0}^{m}(1+\delta_1)^{k-m} \leq \frac{4s_m}{\delta_1}.\end{aligned}$$

The final assertion now follows easily. □

Definition A.42 Given the construction above, set

$$\begin{aligned}&B_i^* = B(x_i, 2r_i), \quad D_i = B(x_i, \tfrac{1}{2}\lambda_1 r_i), \quad D_i^* = B(x_i, Kr_i), \quad \text{if } 1 \leq i \leq M, \\ &B_i^* = D_i = \widetilde{B}_i, \quad D_i^* = B(x_i, Kr_i), \quad \text{if } M+1 \leq i \leq N.\end{aligned}$$

Note that we have for each i

$$B_i' \subset B_i^* \subset D_i \subset B \cap D_i^*.$$

Lemma A.43 *There exists a constant K_2 such that any $x \in B$ is contained in at most K_2 of the sets D_i.*

Proof It is enough to consider the families $\{D_i, i \leq M\}$ and $\{D_i, i > M\}$ separately. For the first family, the result follows from Lemma A.38 on using Lemma A.39 to control r_i. For the second, since each D_i is contained in a ball of radius $(1+\lambda_1)R_0$, this is immediate by Lemma A.38. □

A.8 A Maximal Inequality

We begin with a result, known as the 5B Theorem, on ball coverings. See [Zie], Th. 1.3.1 for the case of a general metric space; in the discrete context here we can replace 5 by 3.

Notation In this section and Section A.9 we will use the notation λB to denote the ball with the same centre as B and λ times the radius.

Theorem A.44 *Let $\Gamma = (\mathbb{V}, E)$ be a graph, and let $\mathcal{B} = \{B(x_i, r_i), i \in I\}$ be a family of balls, with $R = \sup_i r_i < \infty$. Then there exists $J \subset I$ such that $B(x_j, r_j)$, $j \in J$ are disjoint, and*

$$\cup_{i \in I} B(x_i, r_i) \subset \cup_{j \in J} B(x_j, 3r_j).$$

Indeed, for any i there exists $j \in J$ such that $B_i \subset B(x_j, 3r_j)$.

Proof Write $B_i = B(x_i, r_i)$. We can assume that r_i are integers. Now set for $1 \le r \le R$

$$I_r = \{i : r_i = r\}.$$

We can find a maximal subset I'_R of I_R such that $\{B_i, i \in I'_R\}$ are disjoint. For a general metric space one may need to use the Hausdorff maximal principle, but here since I_R is countable it can be done by a simple explicit procedure.

Given $I'_R, \dots, I'_{R-m+1}$, let I'_{R-m} be a maximal subset of I_{R-m} such that $\{B_i, i \in I'_{R-m}\}$ are disjoint, and also disjoint from

$$\left\{ B_i : i \in \bigcup_{j=R-m+1}^{R} I'_j \right\}.$$

Now set $J = \cup_{j=1}^{\infty} I'_j$. By construction, the balls B_j, $j \in J$ are disjoint.

We now show $\{B(x_j, 3r_j), j \in J\}$ cover $\cup_I B_i$. Let $i \in I$. If $i \in J$ we are done. If not, then there exists $k \in J$ with $r_k \ge r_i$ such that $B_i \cap B_k \ne \emptyset$. Let $z \in B_i \cap B_k$. Then for any $y \in B_i$

$$d(y, x_k) \le d(y, x_i) + d(x_i, z) + d(z, x_k) \le 2r_i + r_k \le 3r_k,$$

so that $y \in B(x_k, 3r_k)$. □

We now let Γ be a graph satisfying volume doubling. For $f \in C(\mathbb{V})$ set

$$Mf(x) = \sup \left\{ \frac{1}{V(y, r)} \sum_{z \in B(y,r)} |f(z)| \mu_z : x \in B(y, r) \right\}.$$

Theorem A.45 *Let (Γ, μ) satisfy volume doubling. For all $f \in C_0(\mathbb{V})$ we have*

$$||Mf||_2 \le c||f||_2. \tag{A.30}$$

Proof We can assume $f \ge 0$. Let

$$A_t(f) = \{x : Mf(x) > t\}, \qquad \text{for } t > 0.$$

If $x \in A_t(f)$ then there exists a ball B_x containing x such that

$$\sum_{z \in B_x} f(z)\mu_z > t\mu(B_x). \tag{A.31}$$

The balls $\{B_x, x \in A_t(f)\}$ cover $A_t(f)$, and as f has finite support they have a maximal radius. So, by Theorem A.44 there exists $G \subset A_t(f)$ so that $\{B_x,\ x \in G\}$ are disjoint, and

$$A_t(f) \subset \bigcup_{x \in G} 3B_x.$$

Then, using volume doubling and (A.31),

$$\begin{aligned} \mu(A_t(f)) &\le \sum_{x \in G} \mu(3B_x) \le c_1 \sum_{x \in G} \mu(B_x) \\ &\le \frac{c_1}{t} \sum_{x \in G} \sum_{z \in B_x} f(z)\mu_z \le \frac{c_1}{t} ||f||_1. \end{aligned} \tag{A.32}$$

Thus we have proved that M is of weak L^1 type. Since also $||Mf||_\infty \le ||f||_\infty$ we can apply the Marcinkiewicz interpolation theorem to deduce (A.30). However, a direct proof is easy, so we give it here.

Fix (for the moment) $t > 0$ and let $g = t \wedge f$, $h = f - g$. Since $Mf = M(g + h) \le Mg + Mh$ and $g \le t$ we have

$$A_{2t}(f) \subset \{Mg > t\} \cup \{Mh > t\} = \{Mh > t\} = A_t(h).$$

Applying (A.32) to $A_t(h)$, and using the facts that $h = 0$ on $\{f \le t\}$ and $h \le f$, we obtain

$$\mu(A_{2t}(f)) \le \mu(A_t(h)) \le \frac{c_1}{t} \sum_{x:f(x)>t} h(x)\mu_x \le \frac{c_1}{t} \sum_{x:f(x)>t} f(x)\mu_x.$$

Hence

$$\begin{aligned} ||Mf||_2^2 &= \int_0^\infty 8t\mu(Mf > 2t)dt \le 8c_1 \int_0^\infty \sum_{x:f(x)>t} f(x)\mu_x dt \\ &= 8c_1 \sum_{x \in \mathbb{V}} f(x)\mu_x \int_0^\infty \mathbf{1}_{(t<f(x))}dt = 8c_1||f||_2^2. \end{aligned}$$

□

Lemma A.46 *Let (Γ, μ) satisfy volume doubling. Let $B_i = B(x_i, r_i)$, $1 \le i \le m$, be balls in Γ. Let $\lambda \ge 1$. There exists a constant $C(\lambda)$ such that, for any $a_i \ge 0$,*

$$\Big\| \sum_i a_i \mathbf{1}_{\lambda B_i} \Big\|_2 \le C(\lambda) \Big\| \sum_i a_i \mathbf{1}_{B_i} \Big\|_2.$$

Proof Let

$$f = \sum_i a_i \mathbf{1}_{\lambda B_i}, \quad g = \sum_i a_i \mathbf{1}_{B_i}.$$

It is sufficient to show that for any $h \in C_0(\mathbb{V})$

$$||hf||_1 \le c||g||_2||h||_2. \tag{A.33}$$

Given h let

$$H_i = \frac{1}{\mu(\lambda B_i)} \sum_{x \in \lambda B_i} h(x)\mu_x.$$

Then

$$||hf||_1 = \sum_x \sum_i a_i \mathbf{1}_{\lambda B_i}(x) h(x)\mu_x = \sum_i a_i \mu(\lambda B_i) H_i \le c \sum_i a_i H_i \mu(B_i).$$

Now $\mathbf{1}_{B_i}(x) H_i \le \mathbf{1}_{B_i}(x) M h(x)$. So

$$||hf||_1 \le c||\sum_i a_i H_i \mathbf{1}_{B_i}||_1 \le c||\sum_i a_i \mathbf{1}_{B_i} Mh||_1 \le c||g||_2||Mh||_2 \le c||g||_2||h||_2,$$

by Theorem A.45. This proves (A.33) and completes the proof. □

A.9 Poincaré Inequalities

The main result in this section is the proof of the weighted PI Proposition 6.22, which plays an essential role in the proof of the near-diagonal lower bound. We remark that there is a quick proof of a weighted PI from the strong PI in [DK], but this argument does not work if one only has the weak PI. Since we wish to prove stability for the Gaussian bounds HK$(\alpha, 2)$, we need stable hypotheses, and it is not clear that the strong PI, on its own, is stable. We therefore need to derive the weighted PI from the weak PI. This was done by Jerison [Je], and in fact the same technique can be used to prove both the strong and the weighted PI from the weak PI – see Corollary A.51 later in this section. The approach used here is a simplification of that in [Je] – see [SC2] for its presentation in the metric space context, and also for historical notes.

Let (Γ, μ) be a weighted graph and $B = B(o, R)$ be a ball. Recall the definition of φ_B:

$$\varphi_B(y) = \left(1 \wedge \frac{(R+1-d(o,y))^2}{R^2}\right)\mathbf{1}_B(y), \tag{A.34}$$

and set for $A \subset B$

$$\widetilde{\mathscr{E}}_A^{(B)}(f, f) = \tfrac{1}{2} \sum_{x,y \in A} (f(x) - f(y))^2 \mu_{xy} (\varphi_B(x) \wedge \varphi_B(y)).$$

Recall from (1.16) the notation $\mathscr{E}_B(f, f)$, and that the (weak) PI states that for any function f we have

$$\inf_a \sum_{X \in B(x,r)} (f(x) - a)^2 \mu_x \le C_P r^2 \mathscr{E}_{B(x,\lambda r)}(f, f). \tag{A.35}$$

The main result of this section is the following.

Theorem A.47 *Let (Γ, μ) satisfy (H5), volume doubling, and the PI with expansion factor λ. Then for any ball $B = B(x_0, r)$ the following weighted PI holds:*

$$\inf_a \sum_{x \in B} (f(x) - a)^2 \varphi_B(x) \mu_x \le C_1(\lambda) R^2 \widetilde{\mathscr{E}}_B^{(B)}(f, f).$$

If $f : A \to \mathbb{R}$ then we write

$$\overline{f}_A = \mu(A)^{-1} \sum_{x \in A} f(x) \mu_x$$

for the mean value of f on A. For finite sets $A \subset A'$ we write $P(A, A')$ for the smallest real number C such that the following PI holds:

$$\sum_{x \in A} (f(x) - \overline{f}_A)^2 \mu_x \le C \mathscr{E}_{A'}(f, f). \tag{A.36}$$

The next result enables us to use the PI to chain the means of a function f along a sequence of sets.

Lemma A.48 *Let $A_i \subset A_i^*$, $i = 1, 2$ be finite sets in $\mathbb{V}$ and $f \in C(\mathbb{V})$. Then*

$$|\overline{f}_{A_1} - \overline{f}_{A_2}|^2 \le \frac{2P(A_1, A_1^*)\mathscr{E}_{A_1^*}(f, f) + 2P(A_2, A_2^*)\mathscr{E}_{A_2^*}(f, f)}{\mu(A_1 \cap A_2)}.$$

Proof We have

$$\begin{aligned}
\mu(A_1 \cap A_2)|\overline{f}_{A_1} - \overline{f}_{A_2}|^2 \\
&= \sum_{x \in A_1 \cap A_2} |(f(x) - \overline{f}_{A_1}) - (f(x) - \overline{f}_{A_2})|^2 \mu_x \\
&\le 2 \sum_{x \in A_1 \cap A_2} |f(x) - \overline{f}_{A_1}|^2 \mu_x + 2 \sum_{x \in A_1 \cap A_2} |f(x) - \overline{f}_{A_2}|^2 \mu_x \\
&\le 2 \sum_{x \in A_1} |f(x) - \overline{f}_{A_1}|^2 \mu_x + 2 \sum_{x \in A_2} |f(x) - \overline{f}_{A_2}|^2 \mu_x \\
&\le 2P(A_1, A_1^*)\mathscr{E}_{A_1^*}(f, f) + 2P(A_2, A_2^*)\mathscr{E}_{A_2^*}(f, f). \qquad \square
\end{aligned}$$

For the remainder of this section we fix a ball $B = B(o, R)$ in a graph (Γ, μ) which satisfies the hypotheses of Theorem A.47. We can assume that $\lambda \ge 10$, and set $\lambda_1 = \lceil 4\lambda \rceil$, $R_0 = 10$. If $\lambda_1 R_0 \le R$ then let $B_i' \subset B_i^* \subset D_i \subset D_i^*$, $1 \le i \le N$ be the sets given in Definition A.42. If $R < \lambda_1 R_0$ then we just take $M = 0$, $N = 1$, and $B_1^* = D_1 = B(o, R)$. We take $\theta = \log C_V / \log 2$, as in (A.26). Since B is fixed we will write $\widetilde{\mathscr{E}}_A(f, f)$ for $\widetilde{\mathscr{E}}_A^{(B)}(f, f)$, and φ for φ_B.

Lemma A.49 *We have*

$$P(B_i^*, D_i) \le C_P r_i^2, \quad 1 \le i \le M,$$
$$P(B_i^*, D_i) \le C(\lambda R_0)^{2\theta}, \quad M + 1 \le i \le N.$$

Proof If $i \le M$ then $B_i^* = B(x_i, 2r_i)$ and $B(x_i, 2\lambda r_i) \subset D_i$, so this is immediate from the weak PI (A.35).

Now let $i \ge M + 1$, and recall that $B_i^* = D_i = \widetilde{B}_i$. Using (H5) and Lemma 1.3 we have for any $x \sim y$ in D_i that $\mu_{xy} \le cV(x, 1)$. Then by Lemmas A.37 and A.40(c), and the condition (H5),

$$\mu(D_i) \le V(x, 2(2 + \lambda_1)R_0) \le cR_0^\theta \lambda^\theta V(x, 1) \le c_1 R_0^\theta \lambda^\theta \mu_{xy}.$$

Let $\mathscr{M}$ be any family of paths which covers D_i and let $\kappa(\mathscr{M})$ be as in Definition 3.23. Then

$$\kappa(\mathscr{M}) = \max_{e \in E(D_i)} \mu_e^{-1} \sum_{\{(x,y): e \in \gamma(x,y)\}} \mu_x \mu_y \le \max_{e \in E(D_I)} \mu_e^{-1} \mu(D_i)^2.$$

Hence $R_I(D_i)^{-1} \le \kappa(\mathscr{M})/\mu(D_i) \le c(\lambda R_0)^\theta$, and so by Proposition 3.27 we have $P(B_i^*, D_i) = P(D_i, D_i) \le c(\lambda R_0)^{2\theta}$. $\square$

Set

$$P_i = P(B_i^*, D_i), \quad P_* = \max_i P_i;$$

by the preceding lemma we have $P_* \le CR^2$.

Let

$$\varphi_i = \max_{D_i} \varphi(x).$$

Lemma A.50 *There exists c_1, depending only on λ_1 and R_0, such that*

$$\varphi_i \le c_1 \varphi(x), \quad x \in D_i.$$

Hence for any i we have

$$\varphi_i \mathscr{E}_{D_i}(f, f) \le c_1 \widetilde{\mathscr{E}}_{D_i}(f, f).$$

Proof If $i \le M$ this follows easily from Lemma A.39(d) with $\theta = \frac{1}{2}\lambda_1$. If $i > M$ then $\min_{D_i} \varphi(x) \ge R^{-2}$. By Lemma A.40(c) $\rho(x) \le 2(1+\lambda_1)R_0$ for all $x \in D_i$, so $\max_{D_i} \varphi(x)/\min_{D_i} \varphi(x) \le c\lambda_1^2 R_0^2$. □

Proof of Theorem A.47 If $R \le \lambda_1 R_0$ then the weighted PI holds with a constant $c = c(\lambda)$ using Lemmas A.49 and A.50. So now assume that $R > \lambda_1 R_0$.

Let $f : B \to \mathbb{R}$ and write $\overline{f}_i = \overline{f}_{B_i^*}$. Then

$$\begin{aligned}
\sum_{x \in B} |f(x) - \overline{f}_1|^2 \varphi(x)\mu_x
&\le \sum_i \sum_{x \in B_i^*} |f(x) - \overline{f}_1|^2 \varphi(x)\mu_x \\
&\le 2\sum_i \sum_{x \in B_i^*} |f(x) - \overline{f}_i|^2 \varphi(x)\mu_x + 2\sum_i \sum_{x \in B_i^*} |\overline{f}_i - \overline{f}_1|^2 \varphi(x)\mu_x \\
&= S_1 + S_2.
\end{aligned}$$

Using the weak PI for $B_i^* \subset D_i$, and Lemmas A.50 and A.43,

$$\begin{aligned}
S_1 &\le 2\sum_{i=1}^N \sum_{x \in B_i^*} \varphi_i |f(x) - \overline{f}_i|^2 \mu_x \\
&\le 2\sum_{i=1}^N \varphi_i P_i \mathscr{E}_{D_i}(f, f) \le 2C_1 \sum_{i=1}^N P_i \widetilde{\mathscr{E}}_{D_i}(f, f) \le 2C_1 K_2 P_* \widetilde{\mathscr{E}}_B(f, f).
\end{aligned}$$

The sum S_2 is harder. Set

$$g(x) = \sum_i \varphi_i^{1/2} |\overline{f}_1 - \overline{f}_i| \mathbf{1}_{B_i'}(x).$$

As the balls B_i' are disjoint,

$$g(x)^2 = \sum_i \varphi_i |\overline{f}_1 - \overline{f}_i|^2 \mathbf{1}_{B_i'}(x),$$

and

$$S_2 \le 2 \sum_i \varphi_i |\overline{f}_1 - \overline{f}_i|^2 \mu(B_i^*) \le c \sum_i \varphi_i |\overline{f}_1 - \overline{f}_i|^2 \mu(B_i') = c \sum_x g(x)^2 \mu_x.$$

To bound g we first fix $j \in \{1, \dots, N\}$. Let $m = m_j$ be such that $\mathfrak{a}_m(j) = 1$, and write $j_k = \mathfrak{a}_k(j)$ for $k = 0, \dots, m$. Then by Lemma A.48

$$\begin{aligned} |\overline{f}_1 - \overline{f}_j| \varphi_j^{1/2} &\le \sum_{k=0}^{m-1} \varphi_j^{1/2} |\overline{f}_{j_k} - \overline{f}_{j_{k+1}}| \\ &\le c P_*^{1/2} \sum_{k=0}^{m-1} \varphi_j^{1/2} \left(\frac{\mathscr{E}_{D_{j_k}}(f, f) + \mathscr{E}_{D_{j_{k+1}}}(f, f)}{\mu(B_{j_k}^* \cap B_{j_{k+1}}^*)} \right)^{1/2}. \end{aligned}$$

Let

$$b_i^2 = \frac{\widetilde{\mathscr{E}}_{D_i}(f, f)}{\mu(B_i^*)}.$$

Then by Lemma A.41

$$\begin{aligned} \varphi_j \frac{\mathscr{E}_{D_{j_k}}(f, f) + \mathscr{E}_{D_{j_{k+1}}}(f, f)}{\mu(B_{j_k}^* \cap B_{j_{k+1}}^*)} &\le \varphi_j \left(\frac{\mathscr{E}_{D_{j_k}}(f, f)}{\mu(B_{j_k}^*)} + \frac{\mathscr{E}_{D_{j_{k+1}}}(f, f)}{\mu(B_{j_{k+1}}^*)} \right) \\ &\le c(b_{j_k}^2 + b_{j_{k+1}}^2) \le c(b_{j_k} + b_{j_{k+1}})^2. \end{aligned}$$

In the final line we used Lemma A.50 and the fact that $\varphi_j \le c\varphi_{j_k}$ for $k \ge 1$.

By Lemma A.41 $B_j' \subset D_{j_k}^*$, so

$$\begin{aligned} \varphi_j^{1/2} |\overline{f}_1 - \overline{f}_j| \mathbf{1}_{B_j'}(x) &\le c P_*^{1/2} \sum_{k=0}^{m} b_{j_k} \mathbf{1}_{B_j'}(x) \\ &\le c \mathbf{1}_{B_j'}(x) P_*^{1/2} \sum_k b_{j_k} \mathbf{1}_{D_{j_k}^*}(x) \\ &\le c \mathbf{1}_{B_j'}(x) P_*^{1/2} \sum_{i=1}^{N} b_i \mathbf{1}_{D_i^*}(x). \end{aligned}$$

Surprisingly, the constant in the final bound does not depend on j. Using again the fact that B_i' are disjoint, so $\sum \mathbf{1}_{B_j'} \le 1$,

$$\begin{aligned} g(x) &= \sum_j \varphi_j^{1/2} |\overline{f}_1 - \overline{f}_j| \mathbf{1}_{B_j'}(x) \\ &\le c \sum_j \mathbf{1}_{B_j'}(x) \sum_{i=1}^{N} P_*^{1/2} b_i \mathbf{1}_{D_i^*}(x) \le c P_*^{1/2} \sum_{i=1}^{N} b_i \mathbf{1}_{D_i^*}(x). \end{aligned}$$

Hence, using Lemma A.46,

$$S_2 \le c||g||_2^2 \le cP_*||\sum_i b_i \mathbf{1}_{D_i^*}(x)||_2^2 \le c'P_*||\sum_i b_i \mathbf{1}_{B_i'}(x)||_2^2$$
$$= c'P_* \sum_i b_i^2 \mu(B_i') \le c''P_* \sum_i \widetilde{\mathscr{E}}_{D_i}(f, f) \le c'''K_2 P_* \widetilde{\mathscr{E}}_B(f, f).$$

Since $P_* \le cR^2$ this completes the proof. □

Corollary A.51 *Let* (Γ, μ) *satisfy (H5), volume doubling, and the weak PI with scale constant* λ. *Then* (Γ, μ) *satisfies the strong PI.*

Proof The proof is the same as that of Theorem A.47, except that we replace φ_B by 1. □

Remark A.52 Here is an example which shows that we need the condition (H5) to obtain either the strong or weighted PI from the weak PI. Let $\Gamma = \mathbb{Z}^2$, and let ε_n be any sequence of positive reals with $\lim_n \varepsilon_n = 0$. Let $e_1 = (0, 1)$ and $e_2 = (1, 0)$. Choose weights so that $\mu_{xy} = 1$, except that, writing $x_n = (10n, 0)$, $y_n = x_n + e_2$, we set

$$\mu_{x_n+z, y_n+z} = \varepsilon_n \qquad \text{for } z = -e_1, 0, e_1.$$

Then one can verify that weak Poincaré inequality holds with $\lambda = 5$ and a constant C_P of order 1. For example, if $B_n = B(x_n, 1)$ then $B(x_n, 5)$ contains a path of length 5 consisting of edges of weight 1 which connect x_n and y_n.

However, if $A_n = \{(x, y) \in B_n : y \ge 1\}$ then $\mu_E(A_n; B_n - A_n) = 3\varepsilon_n$. It follows that the PI in B_n holds with a constant C_P' of order ε_n, and consequently that the strong PI does not hold for this weighted graph.

The discrete time heat kernel bounds fail, since $\inf_n p_1(x_n, y_n) = 0$. However, one can prove that the continuous time bounds HK(2, 2) do hold – see [BC].

References

[AF] D.J. Aldous and J. Fill. *Reversible Markov Chains and Random Walks on Graphs.* See `http://www.stat.berkeley.edu/~aldous/RWG/book.html`

[Ah] L.V. Ahlfors. *Conformal Invariants: Topics in Geometric Function Theory.* New York, NY: McGraw-Hill Book Co. (1973).

[Ar] D.G. Aronson. Bounds on the fundamental solution of a parabolic equation. *Bull. Amer. Math. Soc.* **73** (1967) 890–896.

[Ba1] M.T. Barlow Which values of the volume growth and escape time exponent are possible for a graph? *Rev. Mat. Iberoamer.* **20** (2004), 1–31.

[Ba2] M.T. Barlow. Some remarks on the elliptic Harnack inequality. *Bull. London Math. Soc.* **37** (2005), no. 2, 200–208.

[BB1] M.T. Barlow and R.F. Bass. Random walks on graphical Sierpinski carpets. In: *Random Walks and Discrete Potential Theory*, eds. M. Piccardello and W. Woess. Symposia Mathematica XXXIX. Cambridge: Cambridge University Press (1999).

[BB2] M.T. Barlow and R.F. Bass. Stability of parabolic Harnack inequalities. *Trans. Amer. Math. Soc.* **356** (2003), no. 4, 1501–1533.

[BC] M.T. Barlow and X. Chen. Gaussian bounds and parabolic Harnack inequality on locally irregular graphs. *Math. Annalen* **366** (2016), no. 3, 1677–1720.

[BCK] M.T. Barlow, T. Coulhon, and T. Kumagai. Characterization of sub-Gaussian heat kernel estimates on strongly recurrent graphs. *Comm. Pure. Appl. Math.* **LVIII** (2005), 1642–1677.

[BH] M.T. Barlow and B.M. Hambly. Parabolic Harnack inequality and local limit theorem for percolation clusters. *Electron. J. Prob.* **14** (2009), no. 1, 1–27.

[BK] M.T. Barlow and T. Kumagai. Random walk on the incipient infinite cluster on trees. *Illinois J. Math.* **50** (Doob volume) (2006), 33–65.

[BJKS] M.T. Barlow, A.A. Járai, T. Kumagai and G. Slade. Random walk on the incipient infinite cluster for oriented percolation in high dimensions. *Commun. Math. Phys.* **278** (2008), no. 2, 385–431.

[BM] M.T. Barlow and R. Masson. Spectral dimension and random walks on the two dimensional uniform spanning tree. *Commun. Math. Phys.* **305** (2011), no. 1, 23–57.

[BMu] M.T. Barlow and M. Murugan. Stability of elliptic Harnack inequality. In prep.

[BT] Zs. Bartha and A. Telcs. Quenched invariance principle for the random walk on the Penrose tiling. *Markov Proc. Related Fields* **20** (2014), no. 4, 751–767.

[Bas] R.F. Bass. On Aronsen's upper bounds for heat kernels. *Bull. London Math. Soc.* **34** (2002), 415–419.

[BPP] I. Benjamini, R. Pemantle, and Y. Peres. Unpredictable paths and percolation. *Ann. Prob.* **26** (1998), 1198–1211.

[BD] A. Beurling and J. Deny. Espaces de Dirichlet. I. Le cas élémentaire. *Acta Math.* **99** (1958), 203–224.

[BL] K. Burdzy and G.F. Lawler. Rigorous exponent inequalities for random walks. *J. Phys. A* **23** (1990), L23–L28.

[Bov] A. Bovier. Metastability: a potential theoretic approach. *Proc. ICM Madrid 2006*. Zurich: European Mathematical Society.

[CKS] E.A. Carlen, S. Kusuoka, and D.W. Stroock. Upper bounds for symmetric Markov transition functions. *Ann. Inst. H. Poincaré Suppl. no. 2* (1987), 245–287.

[Ca] T.K. Carne. A transmutation formula for Markov chains. *Bull. Sci. Math.* **109** (1985), 399–405.

[CRR] A.K. Chandra, P. Raghavan, W.L. Ruzzo, R. Smolensky, and P. Tiwari. The electrical resistance of a graph captures its commute and cover times. In: *Proc. 21st Annual ACM Symposium on Theory of Computing*, Seattle, WA, ed. D.S. Johnson. New York, NY: ACM (1989).

[Ch] X. Chen. Pointwise upper estimates for transition probability of continuous time random walks on graphs. In prep.

[Co1] T. Coulhon. Espaces de Lipschitz et inégalités de Poincaré. *J. Funct. Anal.* **136** (1996), no. 1, 81–113.

[Co2] T. Coulhon. Heat kernel and isoperimetry on non-compact Riemannian manifolds. In: *Heat Kernels and Analysis on Manifolds, Graphs, and Metric Spaces (Paris, 2002)*, eds. P. Auscher, T. Coulhon, and A. Grigor'yan. *Contemporary Mathematics* vol. 338. Providence, RI: AMS (2003), pp. 65–99.

[CG] T. Coulhon and A. Grigor'yan. Random walks on graphs with regular volume growth. *GAFA* **8** (1998), 656–701.

[CGZ] T. Coulhon, A. Grigoryan, and F. Zucca. The discrete integral maximum principle and its applications. *Tohoku Math. J. (2)* **57** (2005), no. 4, 559–587.

[CS] T. Coulhon and A. Sikora. Gaussian heat kernel upper bounds via the Phragmén-Lindelöf theorem. *Proc. Lond. Math. Soc. (3)* **96** (2008), no. 2, 507–544.

[Da] E.B. Davies. Large deviations for heat kernels on graphs. *J. London Math. Soc.(2)* **47** (1993), 65–72.

[D1] T. Delmotte. Parabolic Harnack inequality and estimates of Markov chains on graphs. *Rev. Math. Iberoamer.* **15** (1999), 181–232.

[D2] T. Delmotte. Graphs between the elliptic and parabolic Harnack inequalities. *Potential Anal.* **16** (2002), no. 2, 151–168.

[De] Y. Derriennic. Lois 'zero ou deux' pour les processus de Markov. *Ann. Inst. Henri Poincaré Sec. B* **12** (1976), 111–129.

[DiS] P. Diaconis and D. Stroock. Geometric bounds for eigenvalues of Markov chains. *Ann. Appl. Prob.* **1** (1991), 36–61.

[DS] P. Doyle and J.L. Snell. *Random Walks and Electrical Networks.* Washington D.C.: Mathematical Association of America (1984). Arxiv: .PR/0001057.

[Duf] R.J. Duffin. The extremal length of a network. *J. Math. Anal. Appl.* **5** (1962), 200–215.

[Dur] R. Durrett. *Probability: Theory and Examples, 4th edn.* Cambridge: Cambridge University Press (2010).

[DK] B. Dyda and M. Kassmann. On weighted Poincaré inequalities. *Ann. Acad. Sci. Fenn. Math.* **38** (2013), 721–726.

[DM] E.B. Dynkin and M.B. Maljutov. Random walk on groups with a finite number of generators. (In Russian.) *Dokl. Akad. Nauk SSSR* **137** (1961), 1042–1045.

[FS] E.B. Fabes and D.W. Stroock. A new proof of Moser's parabolic Harnack inequality via the old ideas of Nash. *Arch. Mech. Rat. Anal.* **96** (1986), 327–338.

[Fa] K. Falconer. *Fractal Geometry.* Chichester: Wiley (1990).

[Fe] W. Feller. *An Introduction to Probability Theory and its Applications. Vol. I,* 3rd edn. New York, NY: Wiley (1968).

[Fo1] M. Folz. Gaussian upper bounds for heat kernels of continuous time simple random walks. *Elec. J. Prob.* **16** (2011), 1693–1722, paper 62.

[Fo2] M. Folz. Volume growth and stochastic completeness of graphs. *Trans. Amer. Math. Soc.* **366** (2014), 2089–2119.

[Fos] F.G. Foster. On the stochastic matrices associated with certain queuing processes. *Ann. Math. Statist.* **24** (1953), 355–360.

[FOT] M. Fukushima, Y. Oshima, and M. Takeda, *Dirichlet Forms and Symmetric Markov Processes.* Berlin: de Gruyter (1994).

[Ga] R.J. Gardner. The Brunn-Minkowski inequality. *Bull. AMS* **39** (2002), 355–405.

[Gg1] A.A. Grigor'yan. The heat equation on noncompact Riemannian manifolds. *Math. USSR Sbornik* **72** (1992), 47–77.

[Gg2] A.A. Grigor'yan. Heat kernel upper bounds on a complete non-compact manifold. *Revista Math. Iberoamer.* **10** (1994), 395–452.

[GT1] A. Grigor'yan and A. Telcs. Sub-Gaussian estimates of heat kernels on infinite graphs. *Duke Math. J.* **109** (2001), 452–510.

[GT2] A. Grigor'yan and A. Telcs. Harnack inequalities and sub-Gaussian estimates for random walks. *Math. Annal.* **324** (2002), no. 3, 521–556.

[GHM] A. Grigor'yan, X.-P. Huang, and J. Masamune. On stochastic completeness of jump processes. *Math. Z.* **271** (2012), no. 3, 1211–1239.

[Grom1] M. Gromov. Groups of polynomial growth and expanding maps. *Publ. Math. IHES* **53** (1981), 53–73.

[Grom2] M. Gromov. Hyperbolic groups. In: *Essays in Group Theory*, ed. S.M. Gersten. New York, NY: Springer (1987), pp. 75–263.

[HSC] W. Hebisch and L. Saloff-Coste. Gaussian estimates for Markov chains and random walks on groups. *Ann. Prob.* **21** (1993), 673–709.

[Je] D. Jerison. The weighted Poincaré inequality for vector fields satisfying Hörmander's condition. *Duke Math. J.* **53** (1986), 503–523.

[Kai] V.A. Kaimanovich. Measure-theoretic boundaries of 0-2 laws and entropy. In: *Harmonic Analysis and Discrete Potential Theory (Frascati, 1991)*. New York, NY: Plenum (1992), pp. 145–180.

[Kan1] M. Kanai. Rough isometries and combinatorial approximations of geometries of non-compact riemannian manifolds. *J. Math. Soc. Japan* **37** (1985), 391–413.

[Kan2] M. Kanai. Analytic inequalities, and rough isometries between non-compact reimannian manifolds. In: *Curvature and Topology of Riemannian Manifolds. Proc. 17th Intl. Taniguchi Symp.*, Katata, Japan, eds. K. Shiohama, T. Sakai, Takashi, and T. Sunada. Lecture Notes in Mathematics **1201**. New York, NY: Springer (1986), pp. 122–137.

[KSK] J.G. Kemeny, J.L. Snell, and A.W. Knapp. *Denumerable Markov Chains.* New York, NY: Springer (1976).

[Ki] J. Kigami. Harmonic calculus on limits of networks and its application to dendrites. *J. Funct. Anal.* **128** (1995), no. 1, 48–86.

[Kir1] G. Kirchhoff. Über die Auflösung der Gleichungen, auf welche man bei der Untersuchungen der linearen Vertheilung galvanischer Ströme geführt wird. *Ann. Phys. Chem.* **72** (1847), 497–508.

[Kir2] G. Kirchhoff. On the solution of the equations obtained from the investigation of the linear distribution of galvanic currents. (Translated by J.B. O'Toole.) *IRE Trans. Circuit Theory* **5** (1958), 4–7.

[KN] G. Kozma and A. Nachmias. The Alexander-Orbach conjecture holds in high dimensions. *Invent. Math.* **178** (2009), no. 3, 635–654.

[Kum] T. Kumagai. *Random Walks on Disordered Media and their Scaling Limits.* École d'Été de Probabilités de Saint-Flour XL – 2010. Lecture Notes in Mathematics **2101**. Chan: Springer International (2014).

[LL] G.F. Lawler and V. Limic. *Random Walk: A Modern Introduction.* Cambridge: Cambridge University Press (2010).

[LP] R. Lyons and Y. Peres. *Probability on Trees and Networks.* Cambridge: Cambridge University Press, 2016.

[Ly1] T. Lyons. A simple criterion for transience of a reversible Markov chain. *Ann. Prob.* **11** (1983), 393–402.

[Ly2] T. Lyons. Instability of the Liouville property for quasi-isometric Riemannian manifolds and reversible Markov chains. *J. Diff. Geom.* **26** (1987), 33–66.

[Mer] J.-F. Mertens, E. Samuel-Cahn, and S. Zamir. Necessary and sufficient conditions for recurrence and transience of Markov chains in terms of inequalities. *J. Appl. Prob.* **15** (1978), 848–851.

[Mo1] J. Moser. On Harnack's inequality for elliptic differential equations. *Commun. Pure Appl. Math.* **14** (1961), 577–591.

[Mo2] J. Moser. On Harnack's inequality for parabolic differential equations. *Commun. Pure Appl. Math.* **17** (1964), 101–134.

[Mo3] J. Moser. On a pointwise estimate for parabolic differential equations. *Commun. Pure Appl. Math.* **24** (1971), 727–740.

[N] J. Nash. Continuity of solutions of parabolic and elliptic equations. *Amer. J. Math.* **80** (1958), 931–954.

[Nor] J. Norris. *Markov Chains.* Cambridge: Cambridge University Press (1998).

[NW] C. St J.A. Nash-Williams. Random walks and electric currents in networks. *Proc. Camb. Phil. Soc.* **55** (1959), 181–194.

[Os] H. Osada. Isoperimetric dimension and estimates of heat kernels of pre-Sierpinski carpets. *Prob. Theor Related Fields* **86** (1990), 469–490.

[Pol] G. Polya. Über eine Ausgabe der Wahrscheinlichkeitsrechnung betreffend die Irrfahrt im Strassennetz. *Math. Ann.* **84** (1921), 149–160.

[Rev] D. Revelle. Heat kernel asymptotics on the lamplighter group. *Elec. Commun. Prob.* **8** (2003), 142–154.

[RW] L.C.G. Rogers and D.W. Williams. *Diffusions, Markov Processes, and Martingales. Vol. 1. Foundations.* 2nd edn. Chichester: Wiley (1994).

[Ro] O.S. Rothaus. Analytic inequalities, isoperimetric inequalities and logarithmic Sobolev inequalities. *J. Funct. Anal.* **64** (1985), 296–313.

[SC1] L. Saloff-Coste. A note on Poincaré, Sobolev, and Harnack inequalities. *Inter. Math. Res. Not* **2** (1992), 27–38.

[SC2] L. Saloff-Coste. *Aspects of Sobolev-type Inequalities.* LMS Lecture Notes **289**. Cambridge: Cambridge University Press (2002).

[SJ] A. Sinclair and M. Jerrum. Approximate counting, uniform generation and rapidly mixing Markov chains. *Inform. Comput.* **82** (1989), 93–133.

[So1] P.M. Soardi. Rough isometries and Dirichlet finite harmonic functions on graphs. *Proc. AMS* **119** (1993), 1239–1248.

[So2] P.M. Soardi. *Potential Theory on Infinite Networks.* Berlin: Springer (1994).

[Spi] F. Spitzer. *Principles of Random Walk.* New York, NY: Springer (1976).

[SZ] D.W. Stroock and W. Zheng. Markov chain approximations to symmetric diffusions. *Ann. Inst. H. Poincaré Prob. Stat.* **33** (1997), 619–649.

[T1] A. Telcs. Random walks on graphs, electric networks and fractals. *Prob. Theor. Related Fields* **82** (1989), 435–451.

[T2] A. Telcs. Local sub-Gaussian estimates on graphs: the strongly recurrent case. *Electron. J. Prob.* **6** (2001), no. 22, 1–33.

[T3] A. Telcs. Diffusive limits on the Penrose tiling. *J. Stat. Phys.* **141** (2010), 661–668.

[Tet] P. Tetali. Random walks and the effective resistance of networks. *J. Theor. Prob.* **4** (1991), 101–109.

[Tru] K. Truemper. On the delta-wye reduction for planar graphs. *J. Graph Theory* **13** (1989), 141–148.

[V1] N.Th. Varopoulos. Isoperimetric inequalities and Markov chains. *J. Funct. Anal.* **63** (1985), 215–239.

[V2] N.Th. Varopoulos. Long range estimates for Markov chains. *Bull. Sci. Math., 2^{e} serie* **109** (1985), 225–252.

[W] D. Williams. *Probability with Martingales.* Cambridge: Cambridge University Press (1991).

[Wo] W. Woess. *Random Walks on Infinite Graphs and Groups.* Cambridge: Cambridge University Press (2000).

[Z] A.H. Zemanian. *Infinite Electrical Metworks.* Cambridge: Cambridge University Press (1991).

[Zie] W.P. Ziemer. *Weakly Differentiable Functions.* New York, NY: Springer (1989).

Index